T0212048

Undergraduate Texts in Mathematics

Undergraduate Texts in Mathematics

Undergraduate Texts in Mathematics are generally aimed at third- and fourth-year undergraduate mathematics students at North American universities. These texts strive to provide students and teachers with new perspectives and novel approaches. The books include motivation that guides the reader to an appreciation of interrelations among different aspects of the subject. They feature examples that illustrate key concepts as well as exercises that strengthen understanding.

More information about this series at http://www.springer.com/series/666

Charles H.C. Little • Kee L. Teo • Bruce van Brunt

Real Analysis via Sequences and Series

 Springer

Charles H.C. Little
Research Fellow and former
    Professor of Mathematics
Institute of Fundamental Sciences
Massey University
Palmerston North, New Zealand

Kee L. Teo
Research Fellow and former
    Associate Professor of Mathenatics
Institute of Fundamental Sciences
Massey University
Palmerston North, New Zealand

Bruce van Brunt
Associate Professor of Mathematics
Institute of Fundamental Sciences
Massey University
Palmerston North, New Zealand

ISSN 0172-6056   ISSN 2197-5604 (electronic)
Undergraduate Texts in Mathematics
ISBN 978-1-4939-4181-0   ISBN 978-1-4939-2651-0 (eBook)
DOI 10.1007/978-1-4939-2651-0

Mathematics Subject Classification (2010): 26-01, 03E15, 40A05, 26A03, 26A15, 26A24, 26A42, 40A30, 26A18

Springer New York Heidelberg Dordrecht London
© Springer Science+Business Media New York 2015
Softcover reprint of the hardcover 1st edition 2015

Printed on acid-free paper

Springer Science+Business Media LLC New York is part of Springer Science+Business Media (www.springer.com)

Preface

This book is a text on real analysis for students with a basic knowledge of calculus of a single variable. There are many fine works on analysis, and one must ask what advantages a new book brings. This one contains the standard material for a first course in analysis, but our treatment differs from many other accounts in that concepts such as continuity, differentiation, and integration are approached via sequences. The main analytical concept is thus the convergence of a sequence, and this idea is extended to define infinite series and limits of functions. This approach not only has the merit of simplicity but also places the student in a position to appreciate and understand more sophisticated concepts such as completeness that play a central part in more advanced fields such as functional analysis.

The theory of sequences and series forms the backbone of this book. Much of the material in the book is devoted to this theory and, in contrast to many other texts, infinite series are treated early. The appearance of series in Chap. 3 has the advantages that it provides many straightforward applications of the results for sequences given in Chap. 2 and permits the introduction of the elementary transcendental functions as infinite series. The disadvantage is that certain convergence tests such as the integral test must be postponed until the improper integral is defined in Chap. 7. The Cauchy condensation test is used in Chap. 3 to tackle convergence problems where the integral test is normally applied. Although much of the material in Chap. 2 is standard, there are some unusual features such as the treatment of harmonic, geometric, and arithmetic means and the sequential definition of the exponential function. In Chap. 3 we present results, such as the Kummer–Jensen test, Dirichlet's test, and Riemann's theorem on the rearrangement of series, that are often postponed or not treated in a first course in analysis.

Limits of functions are introduced in Chap. 4 through the use of convergent sequences, and this concept is then used in Chaps. 5 and 6 to introduce continuity and differentiation. As with any analysis book, results such as the intermediate-value theorem and the mean-value theorem can be found in these chapters, but there are also some other features. For instance, the logarithm is introduced in Chap. 5 and then used to prove Gauss's test for infinite series. In Chap. 6 we present a discrete version of l'Hôpital's rule that is seldom found in analysis texts. We conclude this chapter with a short account of the differentiation of power series using differential equations to motivate the discussion.

The Riemann integral is presented in Chap. 7. In the framework of elementary analysis this integral is perhaps more accessible than, say, the Lebesgue integral, and it is still an important concept. This chapter features a number of items beyond the normal fare. In particular, a proof of Wallis's formula followed by Stirling's formula, and a proof that π and e are irrational, appear here. In addition, there is also a short section on numerical integration that further illustrates the definition of the Riemann integral and applications of results such as the mean-value theorem.

Chapter 8 consists of a short account of Taylor series. Much of the analytical apparatus for this topic is established earlier in the book so that, aside from Taylor's theorem, the chapter really covers mostly the mechanics of determining Taylor series. There is an extensive theory on this topic, and it is difficult to limit oneself so severely to these basic ideas. Here, despite the book's emphasis on series, the authors eschew topics such as the theorems of Abel and Tauber and, more importantly, the question of which functions have a Taylor series. A full appreciation of this theory requires complex analysis, which takes us too far afield.

The student encounters Newton's method in a first calculus course as an application of differentiation. This method is based on constructing a sequence, motivated geometrically, that converges (hopefully) to the solution of a given equation. The emphasis in this first encounter is on the mechanics of the method and choosing a sensible "initial guess." In Chap. 9 we look at this method in the wider context of the fixed-point problem. Fixed-point problems provide a practical application of the theory of sequences. The sequence produced by Newton's method is already familiar to the student, and the theory shows how problems such as error estimates and convergence can be resolved.

The final chapter deals with sequences of functions and uniform convergence. By this stage the reader is familiar with the example of power series, but those series have particularly nice properties not shared generally by other series of functions. The material is motivated by a problem in differential equations followed by various examples that illustrate the need for more structure. This chapter forms a short introduction to the field and is meant to prime the reader for more advanced topics in analysis.

An introductory course in analysis is often the first time a student is exposed to the rigor of mathematics. Upon reflection, some students might even view such a course as a rite of passage into mathematics, for it is here that they are taught the need for proofs, careful language, and precise arguments. There are few shortcuts to mastering the subject, but there are certain things a book can do to mitigate difficulties and keep the student interested in the material. In this book we strive to motivate definitions, results, and proofs and present examples that illustrate the new material. These examples are generally the simplest available that fully illuminate the material. Where possible, we also provide examples that show why certain conditions are needed. A simple counterexample is an exceedingly valuable tool for understanding and remembering a result that is laden with technical conditions. There are exercises at the end of most sections. Needless to say, it is here that the student begins to fully understand the material.

The authors appreciate the encouragement and support of their wives. They also thank Fiona Richmond for her help in preparing the figures. The work has also benefited from the thoughtful comments and suggestions of the reviewers.

Palmerston North, New Zealand   Charles H.C. Little
                                                                Kee L. Teo
                                                                Bruce van Brunt

Contents

Chapter 1
Introduction

1.1 Sets

To study analysis successfully, the reader must be conversant with some of the basic concepts of mathematics. Foremost among these is the idea of a set. For our purposes a set may be thought of as a collection of objects. This statement is too imprecise to be regarded as a definition, and in fact it leads to logical difficulties, but it does convey a mental image of a set that is satisfactory for our purposes. The reader who wishes to delve into the nature of this concept more deeply is referred to [10].

Sets are important in that they can be used to construct a host of mathematical concepts. In fact, every mathematical object studied in this book can be constructed from sets. We therefore begin this introductory chapter with some basic properties of sets. Proofs are omitted because most of the properties are evident and their proofs are straightforward.

First, the objects in a set X are called its **elements** or **members**. They are said to be **contained** in X and to **belong** to X. If an object x is contained in X, then we write $x \in X$; otherwise we write $x \notin X$.

We shall assume the existence of a set with no elements. This set is denoted by $\emptyset$, and it is unique. It is said to be **empty**.

If X and Y are sets such that every element of X is also an element of Y, then we say that X is a **subset** of Y and that it is **included** in Y. In this case we write $X \subseteq Y$; otherwise we write $X \nsubseteq Y$. Note that $\emptyset \subseteq X$ for every set X. The reasoning is that since $\emptyset$ has no elements at all, it certainly has no elements that are not in X. Therefore we can safely say, without fear of contradiction, that each of its elements does belong to X. In particular, $\emptyset$ is a subset of itself, and in fact it is its only subset. Observe also that every set includes itself. Moreover, if X, Y, Z are sets such that $X \subseteq Y$ and $Y \subseteq Z$, then $X \subseteq Z$. In this case we write $X \subseteq Y \subseteq Z$.

© Springer Science+Business Media New York 2015
C.H.C. Little et al., *Real Analysis via Sequences and Series*, Undergraduate
Texts in Mathematics, DOI 10.1007/978-1-4939-2651-0_1

If X and Y are sets such that $X \subseteq Y \subseteq X$, then X and Y contain exactly the same elements. In this case we say that these sets are **equal**, and we write $X = Y$. Otherwise X and Y are **distinct**, and we write $X \neq Y$. Any set is equal to itself, and if $X = Y$, then $Y = X$. Furthermore, if X, Y, Z are sets such that $X = Y$ and $Y = Z$, then $X = Z$. In this case we write $X = Y = Z$. Equal sets are treated as identical since they contain the same elements.

The collection of all subsets of a given set X is another set, called the **power set** of X. It is denoted by $\mathcal{P}(X)$. For example, $\mathcal{P}(\emptyset)$ is a set having $\emptyset$ as its only element. This set is denoted by $\{\emptyset\}$. Moreover, $\mathcal{P}(\{\emptyset\})$ is a set containing only the elements $\emptyset$ and $\{\emptyset\}$ and is denoted by $\{\emptyset, \{\emptyset\}\}$.

If X is any set, we may replace the elements of X by other objects and thereby construct a new set. For example, we may replace the unique element $\emptyset$ of the set $\{\emptyset\}$ by any object Y. We then have a new set whose only element is Y. This set is denoted by $\{Y\}$. Similarly, if we replace the elements $\emptyset$ and $\{\emptyset\}$ of the set $\{\emptyset, \{\emptyset\}\}$ by objects Y and Z, respectively, then we obtain a new set whose only elements are Y and Z. This set is denoted by $\{Y, Z\}$. The notation may be extended to an arbitrary number of objects.

If X is a set and P is a property that may be satisfied by some elements of X, then we can construct a subset of X whose elements are precisely the members of X that do satisfy P. This set is denoted by $\{x \in X \mid P\}$. For example, let X and Y be sets, and let P be the property that $x \in Y$, where $x \in X$. Then $\{x \in X \mid P\}$ is the set whose elements are the objects that are in both X and Y. This set is called the **intersection** of X and Y and is denoted by $X \cap Y$. If this intersection happens to be empty, then the sets X and Y are **disjoint**. If $\mathcal{S}$ is a collection of sets (in other words, a set whose elements are themselves sets), then the sets in $\mathcal{S}$ are said to be **mutually disjoint** if the sets A and B are disjoint whenever $A \in \mathcal{S}$ and $B \in \mathcal{S}$.

On the other hand, if P is the property that $x \notin Y$, then $\{x \in X \mid P\}$ is the set whose elements are the members of X that are not in Y. This set is denoted by $X - Y$ and is called the **complement** of Y with respect to X.

Let $\mathcal{S}$ be a collection of sets. Then we may construct another set whose elements are the objects that belong to at least one member of $\mathcal{S}$. This set is called the **union** of $\mathcal{S}$. For example, if $\mathcal{S} = \{X, Y\}$ for some sets X and Y, then the union of $\mathcal{S}$ is the set of all objects that are in X or Y. In particular, it contains all the objects that are in both of those sets. It is denoted by $X \cup Y$.

We may define the **intersection** of $\mathcal{S}$ as the set of all objects in the union of $\mathcal{S}$ that belong to every set in $\mathcal{S}$. If $\mathcal{S} = \{X, Y\}$ for some sets X and Y, then the intersection of $\mathcal{S}$ is the set of all objects that are in both X and Y. Thus it is equal to $X \cap Y$.

1.2 Ordered Pairs, Relations, and Functions

As we said before, sets can be used to construct a large number of mathematical concepts. In this section we show how to construct ordered pairs, relations, and functions from sets, but once again the reader is referred to [10] for the details.

If x and y are any objects, then the **ordered pair** (x, y) is defined as the set $\{\{x\}, \{x, y\}\}$. We refer to x and y as its **components**, x being the **first** component and y the **second**. The important observation to be made here is that the definition does not treat x and y similarly (if in fact they are distinct objects). Instead, we are given a way of distinguishing them: y is a member of just one of the two sets in (x, y), but x belongs to both sets. From this observation it is easy to deduce that the ordered pairs (x, y) and (a, b) are equal if and only if $x = a$ and $y = b$. In other words, for equality to hold it is not sufficient for the sets $\{x, y\}$ and $\{a, b\}$ to be equal. Their elements must also be listed in the same order. This is the only property of ordered pairs that is important to remember. Once it is grasped, the definition may be forgotten. The definition can be extended to ordered triples by defining

$$(x, y, z) = ((x, y), z)$$

for all objects x, y, z. Thus (x, y, z) technically is an ordered pair whose first component is itself an ordered pair. This notation may be extended to an arbitrary number of objects, as we shall see later.

If X and Y are sets, we may construct a set $X \times Y$ whose elements are the ordered pairs (x, y) such that $x \in X$ and $y \in Y$. This set is called the **Cartesian product** of X and Y. A subset of $X \times Y$ is called a **relation** from X to Y. Thus a relation is just a set of ordered pairs. A relation from X to X is sometimes called a relation **on** X. An example is the relation R on X such that $(x, y) \in R$ if and only if $x = y$. This relation is called the **equality** relation. Similarly, we may define the **inclusion** relation by specifying that it contains the ordered pair (x, y) if and only if x and y are sets such that $x \subseteq y$.

If R is a relation and x and y are objects, we often write xRy to indicate that $(x, y) \in R$. For instance, we are already accustomed to using $=$ to denote the equality relation and writing $x = y$ instead of $(x, y) \in =$. We also write $x \not R y$ instead of $(x, y) \notin R$.

If R and S are relations and x, y, z are objects, then we write $xRySz$ if xRy and ySz. This notation may be extended to arbitrarily long chains of relations.

Some kinds of relations are of particular importance, and we discuss them now. If R is a relation on a set X, then R is **reflexive** if xRx for each $x \in X$. Examples include the equality and inclusion relations. The relation R is **symmetric** if yRx whenever x and y are members of X satisfying xRy. Equality has this property, but inclusion does not. We also say that R is **transitive** if xRy whenever there is a $z \in X$ for which $xRzRy$. Equality and inclusion both exhibit this property. A relation with all three of these properties is an **equivalence relation**. Equality is such a relation, but inclusion is not.

Equivalence relations are closely linked to partitions. A **partition** of a set X is a set of nonempty, mutually disjoint subsets of X whose union is X. The elements of a partition are sometimes called its **cells**. It can be shown that with any equivalence relation R on a set X there is an associated unique partition P of X with the property that elements x and y of X belong to the same cell of P if and only if xRy. Conversely, any partition P of X has associated with it a unique equivalence relation

R on X constructed in the same way: xRy if and only if x and y belong to the same cell of P. The cells of P are called the **equivalence classes** of R. For each x in X, we denote by $[x]$ the unique equivalence class to which it belongs. Thus $[x] = [y]$ if and only if y is an element of X that belongs to the equivalence class $[x]$.

There is yet another kind of relation that is of paramount importance. A relation f from a set X to a set Y is called a **function** from X into Y if for each $x \in X$ there is a unique $y \in Y$ for which $(x, y) \in f$. We usually write y as $f(x)$. We say that f **maps** x to y and that y is the **image** of x under f and **corresponds** to x under f. We think of the function f as providing a rule for associating with each x in X a unique corresponding element y of Y. It is this aspect of the concept that is usually important for our purposes, but it is of interest to see how to construct the idea of a function from sets, as we have done.

Given sets X and Y, we sometimes write $f: X \to Y$ to indicate that f is a function from X into Y. The set X is called the **domain** of the function f. The subset of Y consisting of the images of the elements of X is the **range** of f. In this book the range of f will usually be a set of real or complex numbers, in which case we describe the function as **real-valued** or **complex-valued**, respectively. The range of f is not necessarily the whole of Y: Some elements of Y might not correspond to any element of X. The domain of f is denoted by $\mathcal{D}_f$ and the range of f by $\mathcal{R}_f$. If $\mathcal{R}_f = Y$, then f is described as **surjective** and called a **surjection** from X **onto** Y. In the case of a surjection, each member of Y does correspond to an element of X, but this element need not be unique.

Suppose on the other hand that each element of $\mathcal{R}_f$ does correspond to a unique element of X. Then f is **injective** and an **injection** from X into Y. Thus f is injective if and only if $w = x$ whenever $f(w) = f(x)$: Distinct elements of X must be mapped to distinct elements of Y. An injective function is also described as **one-to-one**.

Perhaps f is both injective and a surjection from X onto Y. Then f is a **bijection** from X onto Y and described as **bijective**. In this case each member of Y corresponds to a unique element of X. Thus there exists a function g from Y into X such that $g(f(x)) = x$ for all $x \in X$. This function is called the **inverse** of f and is denoted by f^{-1}. It is a bijection from Y onto X. Note that $f^{-1}(f(x)) = x$ for all $x \in X$. Moreover $f(f^{-1}(y)) = y$ for each $y \in Y$, and $(f^{-1})^{-1} = f$.

A function f with domain X is said to be **constant** (on X) if $f(w) = f(x)$ for each w and x in X. The range of a constant function with nonempty domain therefore consists of a single element.

Now let f be a function from a set X into a set Y and g a function from Y into a set Z. Then $g(f(x))$ is defined for each $x \in X$. Letting $h(x) = g(f(x))$ for each such x, we see that h is a function from X into Z. We call it the **composition** of g and f and denote it by $g \circ f$. Thus

$$(g \circ f)(x) = g(f(x))$$

for each $x \in X$. For instance, if f is a bijection from X onto Y, it follows that $(f^{-1} \circ f)(x) = x$ for each $x \in X$ and $(f \circ f^{-1})(y) = y$ for each $y \in Y$. It is easily checked that a composition of injections is injective and that a composition of surjections is surjective. It follows that a composition of bijections is bijective.

If f is a function and $X \subseteq \mathcal{D}_f$, then we write

$$f(X) = \{f(x) \in \mathcal{R}_f \mid x \in X\}.$$

Thus $f(\mathcal{D}_f) = \mathcal{R}_f$.

If f is a real-valued function whose domain is a set of real numbers, then the **graph** of f is the set of all points $(x, f(x))$ in the Cartesian plane, where $x \in \mathcal{D}_f$. It is also the graph of the equation $f(x) = y$.

1.3   Induction and Inequalities

Natural numbers are those used to count. The set $\{1, 2, \ldots\}$ of natural numbers is denoted by $\mathbb{N}$. We assume familiarity with the basic properties of these numbers and simply highlight those that are of particular importance for our development of analysis. The details of the development of natural numbers, integers, rational numbers, and real numbers in terms of sets are given in [10] and will not be repeated here.

The main properties of $\mathbb{N}$ that we require are that it contains the number 1 and that every natural number has a successor in $\mathbb{N}$. The successor of a natural number n is denoted by $n + 1$. For example, the successor of 1 is $2 = 1 + 1$ and that of 2 is $3 = 2 + 1$. This notion of a successor for each natural number enables us to list the natural numbers in order, beginning with 1. In fact, it can be shown that each natural number $n \neq 1$ is the successor of a unique natural number $n - 1$. If Y is a set of natural numbers such that $1 \in Y$ and $n + 1 \in Y$ for each $n \in Y$, then $Y = \mathbb{N}$.

This observation can be applied to yield an important technique, called **induction**, for proving theorems about natural numbers. In this application, Y is the set of all natural numbers for which the desired result is true. First, prove the desired theorem for the natural number 1. (In other words, prove that $1 \in Y$.) Then assume that it is true for a particular natural number n (so that $n \in Y$) and prove it for $n + 1$ under this assumption. The conclusion that $Y = \mathbb{N}$ then shows that the theorem is indeed true for all natural numbers. This idea is perhaps most easily visualized as follows. Imagine a line of dominoes standing on end and close together so that the line begins with a particular domino and extends indefinitely to the right of that first domino. Knock the first domino onto the second. Then it is easy to see that all the dominoes will fall over, because the first domino will fall and every other domino has one before it that will eventually knock it over. This picture captures the essence of induction.

Before giving examples of induction at work, let us make some observations about the pattern of a proof by induction. One approach to the use of induction to prove a theorem about a natural number n, as we have seen, is to prove the theorem for the natural number 1, then assume it for a particular natural number n (that is, for a particular integer $n \geq 1$), and finally prove it for $n + 1$. The assumption that the theorem holds for a particular n is commonly called the **inductive hypothesis**. We can in fact strengthen it by assuming that the theorem holds for all natural numbers less than n as well. Equivalently, one could prove the theorem first for 1, then assume as an inductive hypothesis that it holds for a particular integer $n - 1$ (or for all natural numbers less than n) where $n \geq 2$, and finally prove it for n under this assumption. Another observation is that the process of induction need not begin with the natural number 1. In fact, it should begin with the smallest integer for which the theorem is to be proved. In other words, suppose we wish to prove a theorem for all integers $n \geq a$ for some fixed integer a. We start an inductive proof in this case by proving the theorem for a. Then there are two equivalent ways to continue. One is to assume as an inductive hypothesis that the theorem holds for some particular integer $n \geq a$ (or for all integers k such that $a \leq k \leq n$) and then prove it for $n + 1$. The alternative is to assume that it holds for $n - 1$ for a particular integer $n > a$ (or for all integers k such that $a \leq k < n$) and then prove it for n under this assumption.

Because of its importance, we now state the principle of induction formally as a theorem. We denote the set of all integers by $\mathbb{Z}$, so that

$$\mathbb{Z} = \{\ldots, -2, -1, 0, 1, 2, \ldots\}.$$

Theorem 1.3.1. *Let $Y \subseteq \mathbb{Z}$ and $a \in Y$. Suppose also that $y + 1 \in Y$ whenever $y \in Y$. Then Y contains every integer greater than a.*

In applications of Theorem 1.3.1 to prove a given assertion about integers, Y is the set of all integers for which the assertion in question is true.

We now offer an example of an inductive proof.

Example 1.3.1. We shall prove by induction that if x is a nonnegative real number, then

$$(1 + x)^n \geq 1 + nx + \frac{n(n-1)}{2}x^2 \tag{1.1}$$

for all integers $n \geq 0$. It is easy to see that equality holds for $n = 0$. Assume that inequality (1.1) holds for a particular integer $n \geq 0$. Then

$$
\begin{aligned}
(1 + x)^{n+1} &= (1 + x)^n (1 + x) \\
&\geq \left(1 + nx + \frac{n(n-1)}{2}x^2\right)(1 + x) \\
&= 1 + (n+1)x + \left(n + \frac{n(n-1)}{2}\right)x^2 + \frac{n(n-1)}{2}x^3
\end{aligned}
$$

$$\geq 1 + (n+1)x + \frac{2n + n^2 - n}{2}x^2$$

$$= 1 + (n+1)x + \frac{n(n+1)}{2}x^2,$$

since $n(n-1)x^3/2 \geq 0$. The proof by induction is now complete.

It follows that

$$(1+x)^n \geq 1 + nx \tag{1.2}$$

and

$$(1+x)^n > \frac{n(n-1)}{2}x^2 \tag{1.3}$$

hold for every nonnegative real number x and nonnegative integer n. Inequality (1.2) is known as **Bernoulli's inequality**. Both of these inequalities will be found to be useful later.   △

Example 1.3.1 involved an inequality. It is assumed that the reader is conversant with inequalities and can work with them comfortably. One of the salient points to remember is that multiplication of both sides of an inequality by a negative number causes a change in the direction of the inequality. For instance, if $x < y$, then $-x > -y$. Similarly, if both sides of an inequality have the same sign, then taking the reciprocals of both sides again induces a change in the direction of the inequality. Thus $1/x > 1/y$ if either $x < y < 0$ or $0 < x < y$.

Induction can be used to make definitions as well as to prove theorems. Let us illustrate this point by defining finite sums inductively. Let m and n be integers with $n > m$. If $a_m, a_{m+1}, \ldots, a_n$ are numbers, we define

$$\sum_{j=m}^{m} a_j = a_m$$

and

$$\sum_{j=m}^{n} a_j = \sum_{j=m}^{n-1} a_j + a_n.$$

Often we write

$$\sum_{j=m}^{n} a_j = a_m + a_{m+1} + \cdots + a_n.$$

This quantity is called the **sum** of $a_m, a_{m+1}, \ldots, a_n$, and those numbers are its **terms**. We also define

$$\sum_{j=m}^{n} a_j = 0$$

if $m > n$, and we regard this expression as a sum with no terms.

We may define the product of the same numbers in a similar way. Thus

$$\prod_{j=m}^{m} a_j = a_m$$

and

$$\prod_{j=m}^{n} a_j = a_n \prod_{j=m}^{n-1} a_j$$

for all $n > m$. We often write

$$\prod_{j=m}^{n} a_j = a_m a_{m+1} \ldots a_n.$$

This number is the **product** of $a_m, a_{m+1}, \ldots, a_n$, and those numbers are its **factors**. If $m > n$, then we define

$$\prod_{j=m}^{n} a_j = 1$$

and regard this expression as a product with no factors.

Similarly, we obtain analogous expressions for unions and intersections of sets by replacing $\sum$ with $\bigcup$ and $\bigcap$, respectively.

Induction may also be used to define exponentiation for powers that are natural numbers: Given a real number a, set $a^1 = a$ and if a^n has been defined for a specific natural number n, let

$$a^{n+1} = a^n \cdot a.$$

We may also define $a^0 = 1$, though this definition is normally made only if $a \neq 0$. Furthermore we define

$$a^{-n} = \frac{1}{a^n}$$

if $n \in \mathbb{N}$ and $a \neq 0$.

One can easily prove by induction that

$$(ab)^n = a^n b^n \tag{1.4}$$

for any real numbers a and b and natural number n, and similarly that

$$\left(\frac{a}{b}\right)^n = \frac{a^n}{b^n} \tag{1.5}$$

if $b \neq 0$. If $ab \neq 0$, then these two equations extend to the case where n is any integer. For each fixed $m \in \mathbb{N}$, it is also easy to prove by induction on n that

$$a^m a^n = a^{m+n} \tag{1.6}$$

for all $n \in \mathbb{N}$ and then that

$$(a^m)^n = a^{mn} \tag{1.7}$$

for all $n \in \mathbb{N}$. If $a \neq 0$, then these two rules may also be extended to the case where m and n are any integers.

In order to extend these ideas, we need the following definition.

Definition 1.3.1. If n is a nonnegative integer and $a_0, a_1, \ldots, a_n$ are constants, then the function given by

$$\sum_{j=0}^{n} a_j x^j \tag{1.8}$$

for all numbers x is called a **polynomial**. Its **degree** is n if $a_n \neq 0$. The numbers $a_0, a_1, \ldots, a_n$ are its **coefficients**. If

$$\sum_{j=0}^{n} a_j c^j = 0,$$

then c is called a **root** of the polynomial (1.8).

It is shown in [10] that every nonnegative number a has a unique nonnegative square root. This square root is denoted by $a^{1/2}$ or $\sqrt{a}$. Thus we have $(a^{1/2})^2 = a$. It is also shown in [10] that every polynomial of odd degree whose coefficients are real has a real root. For example, if m is an odd positive integer and a is a real number, then there is a real number c satisfying the equation $c^m = a$. The uniqueness of c follows from the fact that if x and y are real numbers such that $x < y$, then $x^m < y^m$ since m is odd. (This fact is obvious if $x \leq 0$ and $y \geq 0$, for then $x^m \leq 0$, $y^m \geq 0$, and at least one of x^m and y^m is nonzero. If $0 < x < y$, then use induction to prove it for all positive integers m. If $x < y < 0$, then note that $0 < -y < -x$ and apply the previous case with m odd.) We write $a^{1/m} = c$. In the case where $m = 1$, this definition is consistent with the equation $a^1 = a$. In any case, we have $(a^{1/m})^m = a$, and so we refer to $a^{1/m}$ as the mth root of a. Note that $0^{1/m} = 0$ and that $a^{1/m}$ has the same sign as a if $a \neq 0$.

An arbitrary positive integer can be written in the form $2^k m$, where k and m are nonnegative integers and m is odd. If $n = 2^k m$, then we can show by induction on k that each $a \geq 0$ has a unique nonnegative nth root. Indeed, we have already observed this fact if $k = 0$. Suppose that $k > 0$ and that a has a unique nonnegative rth root c, where $r = 2^{k-1} m = n/2$. Then

$$a = c^r = \left(\left(c^{\frac{1}{2}} \right)^2 \right)^r = \left(c^{\frac{1}{2}} \right)^{2r} = \left(c^{\frac{1}{2}} \right)^n.$$

On the other hand, if $a = b^n = b^{2r} = (b^2)^r$, where $b \geq 0$, then $b^2 = c$ by the uniqueness of c, and so $b = c^{1/2}$. We conclude that $\sqrt{c}$ is the unique nonnegative nth root of a. We write it as $a^{1/n}$. Hence

$$\left(a^{\frac{1}{n}} \right)^n = a. \tag{1.9}$$

Moreover $0^{1/n} = 0$ and $1^{1/n} = 1$.

An alternative proof of the existence of the nonnegative nth root of a will be given in Example 5.3.5.

Let $0 \leq a < b$, and for some positive integer n let $c = a^{1/n}$ and $d = b^{1/n}$. If $c \geq d$, then we should have the contradiction that $a = c^n \geq d^n = b$. We conclude that $c < d$. For instance, if $a > 1$, then $a^{1/n} > 1$.

Let $a \geq 0$, let r and s be positive integers, and let $c = a^{1/r}$. Then

$$\left(c^{\frac{1}{s}} \right)^{rs} = \left(\left(c^{\frac{1}{s}} \right)^s \right)^r = c^r = a,$$

so that

$$a^{\frac{1}{rs}} = c^{\frac{1}{s}} = \left(a^{\frac{1}{r}} \right)^{\frac{1}{s}}.$$

Next, let $a \geq 0$, let $n \in \mathbb{N}$, and let m be a nonnegative integer. We then define

$$a^{\frac{m}{n}} = \left(a^{\frac{1}{n}} \right)^m.$$

This definition generalizes Eq. (1.9) and is consistent with the equation $a^1 = a$ in the case where $m = 1$ or $n = 1$. It is also consistent with the equation $a^0 = 1$ in the case where $m = 0$. If $m = jk$ and $n = jl$ for some positive integers j, k, l, then

$$a^{\frac{jk}{jl}} = \left(a^{\frac{1}{jl}} \right)^{jk} = \left(\left(\left(a^{\frac{1}{l}} \right)^{\frac{1}{j}} \right)^j \right)^k = \left(a^{\frac{1}{l}} \right)^k = a^{\frac{k}{l}},$$

a result that is consistent with the equation $jk/(jl) = k/l$. If $a > 0$, then we define

$$a^{-\frac{m}{n}} = \frac{1}{a^{\frac{m}{n}}} = \frac{1}{\left(a^{\frac{1}{n}}\right)^m} = \left(a^{\frac{1}{n}}\right)^{-m}.$$

We proceed to the generalization of Eqs. (1.4)–(1.7). Let $a \geq 0$, let m be an integer, and let $n \in \mathbb{N}$. Since

$$\left(a^{\frac{m}{n}}\right)^n = \left(\left(a^{\frac{1}{n}}\right)^m\right)^n = \left(a^{\frac{1}{n}}\right)^{nm} = \left(\left(a^{\frac{1}{n}}\right)^n\right)^m = a^m,$$

it follows that

$$(a^m)^{\frac{1}{n}} = a^{\frac{m}{n}} = \left(a^{\frac{1}{n}}\right)^m.$$

Thus if p and q are also integers and $q > 0$, then

$$\left(a^{\frac{m}{n}}\right)^{\frac{p}{q}} = \left(\left(\left(a^{\frac{1}{n}}\right)^m\right)^{\frac{1}{q}}\right)^p = \left(\left(\left(a^{\frac{1}{n}}\right)^{\frac{1}{q}}\right)^m\right)^p = \left(a^{\frac{1}{nq}}\right)^{mp} = a^{\frac{mp}{nq}}.$$

This result generalizes Eq. (1.7). Equation (1.6) also generalizes, since

$$a^{\frac{m}{n}} a^{\frac{p}{q}} = a^{\frac{qm}{qn}} a^{\frac{pn}{qn}} = \left(a^{\frac{1}{qn}}\right)^{qm} \left(a^{\frac{1}{qn}}\right)^{pn} = \left(a^{\frac{1}{qn}}\right)^{qm+pn} = a^{\frac{qm+pn}{qn}} = a^{\frac{m}{n}+\frac{p}{q}}.$$

Now suppose that b is also a nonnegative real number. As

$$ab = \left(a^{\frac{1}{n}}\right)^n \left(b^{\frac{1}{n}}\right)^n = \left(a^{\frac{1}{n}} b^{\frac{1}{n}}\right)^n,$$

it then follows that

$$(ab)^{\frac{1}{n}} = a^{\frac{1}{n}} b^{\frac{1}{n}}.$$

Therefore

$$(ab)^{\frac{m}{n}} = ((ab)^m)^{\frac{1}{n}} = (a^m b^m)^{\frac{1}{n}} = (a^m)^{\frac{1}{n}} (b^m)^{\frac{1}{n}} = a^{\frac{m}{n}} b^{\frac{m}{n}},$$

a result that generalizes Eq. (1.4). Thus if $b > 0$, then

$$\left(\frac{a}{b}\right)^{\frac{m}{n}} = (ab^{-1})^{\frac{m}{n}} = a^{\frac{m}{n}} (b^{-1})^{\frac{m}{n}} = a^{\frac{m}{n}} b^{-\frac{m}{n}} = \frac{a^{\frac{m}{n}}}{b^{\frac{m}{n}}},$$

and so Eq. (1.5) generalizes as well.

We have now defined the real number a^x for every number $a > 0$ and every rational power x. Later we shall extend this definition to the case where x is any real number. In fact, similar ideas may be used to define w^z for any complex numbers w and z provided that if w is real, then it is positive.

As another example of an inductive definition, we may define the **factorial** $n!$ of a nonnegative integer n by writing $0! = 1$ and

$$n! = n(n-1)!$$

for all $n \in \mathbb{N}$.

Given objects $x_1, x_2, \ldots, x_n$, where $n > 1$, we may define the **ordered set**

$$(x_1, x_2, \ldots, x_n)$$

by induction: The ordered pair (x_1, x_2) has already been defined, and for each $n > 2$ we let

$$(x_1, x_2, \ldots, x_n) = ((x_1, x_2, \ldots, x_{n-1}), x_n).$$

Let $X_1, X_2, \ldots, X_n$ be sets, where $n > 1$. We can also define the **Cartesian product** $X_1 \times X_2 \times \cdots \times X_n$ of $X_1, X_2, \ldots, X_n$ by induction: The definition has already been made for $n = 2$, and for each $n > 2$ define

$$X_1 \times X_2 \times \cdots \times X_n = (X_1 \times X_2 \times \cdots \times X_{n-1}) \times X_n.$$

The result is the collection of all ordered sets $(x_1, x_2, \ldots, x_n)$ such that $x_j \in X_j$ for each j. If $X_j = X$ for each j, then we write

$$X^n = X_1 \times X_2 \times \cdots \times X_n.$$

The absolute value of a real number x plays a prominent role in analysis. Denoted by $|x|$, it is defined as x if $x \geq 0$ and $-x$ otherwise. Thus it is always the case that $|x| \geq 0$ and that

$$|x| = \sqrt{x^2} = |-x|.$$

For every $a > 0$ it is also easy to see that $|x| < a$ if and only if $-a < x < a$. From these inequalities it follows that $|x| > a$ if and only if $x > a$ or $x < -a$, and that $|x - y| < a$ if and only if $y - a < x < y + a$, where y is another real number. In addition, observe that $|x| \geq x$ and $|x| \geq -x$ for all real x. Thus $|x| + |y| \geq x + y$ and $|x| + |y| \geq -(x + y)$, so that

$$|x + y| \leq |x| + |y|.$$

This result is known as the **triangle inequality**.

Inequalities can be used to define intervals on the real line. If a and b are real numbers with $a < b$, then we write

$$(a, b) = \{x \mid a < x < b\},$$
$$[a, b] = \{x \mid a \le x \le b\},$$
$$(a, b] = \{x \mid a < x \le b\},$$
$$[a, b) = \{x \mid a \le x < b\}.$$

These sets are called **intervals**. The first is **open** and the second **closed**; the others are **half-open**. It should be clear from the context whether the notation (a, b) specifies an open interval or an ordered pair. The numbers a and b are called the **ends** of each of these intervals. In addition, we write

$$(a, \infty) = \{x \mid a < x\},$$
$$[a, \infty) = \{x \mid a \le x\},$$
$$(-\infty, a) = \{x \mid x < a\},$$
$$(-\infty, a] = \{x \mid x \le a\}.$$

These sets are also reckoned as intervals, and a is regarded as an end of each.

Let f be a real-valued function whose domain includes an interval I. Then f is said to be **increasing** on I if $f(x_1) < f(x_2)$ whenever x_1 and x_2 are numbers in I such that $x_1 < x_2$. If, on the other hand, $f(x_1) \le f(x_2)$ for each such x_1 and x_2, then f is **nondecreasing** on I. We define functions that are **decreasing** or **nonincreasing** on I similarly. All these functions are said to be **monotonic** on I, and those that are increasing on I or decreasing on I are **strictly monotonic** on I.

Exercises 1.1.

1. Show that if $a < b$, then

$$a \le \alpha a + (1 - \alpha)b \le b$$

   for all α such that $0 \le \alpha \le 1$.

2. Prove that

$$|xy| = |x||y|$$

   for all real numbers x and y. If $y \ne 0$, prove also that

$$\left| \frac{x}{y} \right| = \frac{|x|}{|y|}.$$

3. Use the triangle inequality to prove that

$$||x| - |y|| \le |x - y|,$$

where x and y are real numbers.
4. Solve the following equations and inequalities, where x is real:
    (a) $|3x - 2| = |5x + 4|$;   (d) $|3x + 4| \ge 2$;
    (b) $|x + 4| = -4x$;   (e) $|3x + 2| < 3|x|$.
    (c) $|2x - 3| < 4$;
5. Prove the following by induction for all positive integers n and real numbers $a, a_1, a_2, \ldots, a_n$:

    (a) $|a^n| = |a|^n$.
    (b) $\left| \prod_{j=1}^{n} a_j \right| = \prod_{j=1}^{n} |a_j|$.
    (c) $\left| \sum_{j=1}^{n} a_j \right| \le \sum_{j=1}^{n} |a_j|$.

6. Prove that $\sqrt{2}$ is irrational by writing $\sqrt{2} = a/b$, where a and b are positive integers with no common factor, and obtaining a contradiction by showing that a and b must both be even.

1.4   Complex Numbers

Throughout this book we will assume that any numbers we are working with are complex unless an indication to the contrary is given by either the context or an explicit statement. For example, the numbers $a_0, a_1, \ldots, a_n, x$ in the definition of a polynomial (Definition 1.3.1) need not be real. As complex numbers might not be as familiar to the reader as real numbers, we define them here and prove some basic properties.

The equation

$$x^2 = -1 \tag{1.10}$$

has no real solution. We seek to extend the real number system to include a number i such that $i^2 = -1$ while preserving familiar operations such as addition and multiplication. If we put $x + iy = 0$, where x and y are real numbers, then $x = -iy$, so that $x^2 = i^2 y^2 = -y^2$ and hence $x = y = 0$. Now if

$$x + iy = a + ib, \tag{1.11}$$

where a and b are also real, then

$$x - a + i(y - b) = 0.$$

The argument above shows that $x - a = y - b = 0$ and we conclude that Eq. (1.11) holds if and only if $x = a$ and $y = b$. Consequently $x + iy$ may be identified with the ordered pair (x, y) of real numbers.

We therefore define a **complex number** as an ordered pair of real numbers. Hence every complex number can be visualized as a point in the Cartesian plane. If $z = (x, y)$, where x and y are real, then we define Re $(z) = x$, and Im $(z) = y$, and we refer to x and y as the **real** and **imaginary** parts, respectively, of z. We define addition and multiplication by the rules

$$(u, v) + (x, y) = (u + x, v + y)$$

and

$$(u, v)(x, y) = (ux - vy, uy + vx),$$

respectively, where u, v, x, y are all real. These rules are motivated by the calculations

$$(u + iv) + (x + iy) = u + x + i(v + y)$$

and

$$(u + iv)(x + iy) = ux - vy + i(uy + vx).$$

Thus

$$(u, 0) + (x, 0) = (u + x, 0)$$

and

$$(u, 0)(x, 0) = (ux, 0).$$

Therefore we may identify $(x, 0)$ with the real number x. In particular, it follows that $(0, 0) = 0$ and $(1, 0) = 1$. Note also that

$$\text{Re} (w + z) = \text{Re} (w) + \text{Re} (z)$$

and

$$\text{Im} (w + z) = \text{Im} (w) + \text{Im} (z)$$

for any two complex numbers w and z. It is also worth observing that the addition of complex numbers may be interpreted geometrically as vector addition. In Chap. 6 we will give a geometric interpretation of the multiplication of complex numbers.

The required solution i to Eq. (1.10) is identified with the ordered pair $(0, 1)$. According to our definition of multiplication, it has the desired property:

$$i^2 = (0, 1)(0, 1) = (-1, 0) = -1.$$

Note also that if x and y are real, then the definitions of addition and multiplication do indeed give

$$x + iy = (x, 0) + (0, 1)(y, 0) = (x, 0) + (0, y) = (x, y).$$

These results show that the algebraic manipulation of complex numbers is identical to that of real numbers with the additional rule that $i^2 = -1$. In particular, we have the laws that

$$a + b = b + a, \tag{1.12}$$

$$a + (b + c) = (a + b) + c, \tag{1.13}$$

and

$$a(b + c) = ab + ac \tag{1.14}$$

for all complex numbers a, b, c. We summarise Eqs. (1.12) and (1.13) by asserting that the addition of complex numbers is **commutative** and **associative**, respectively. Similarly, multiplication of complex numbers is commutative and associative. Equation (1.14) asserts that multiplication is **distributive** over addition.

We also define

$$-(x + iy) = -x - iy.$$

Then subtraction is defined by the equation

$$w - z = w + (-z)$$

for all complex numbers w and z.

We have now extended the real number system to include a number i that gives a solution to the equation $x^2 = -1$. More generally, $i\sqrt{c}$ gives a solution to the equation $x^2 = -c$, where $c \geq 0$. In fact, it can be shown (see [10]) that every nonconstant polynomial with complex coefficients has at least one complex root. This result is known as the fundamental theorem of algebra. For example, suppose that

$$az^2 + bz + c = 0,$$

where a, b, c, z are all complex numbers and $a \neq 0$. As the left-hand side of this equation is a polynomial of degree 2, the equation is said to be **quadratic**. It can be solved for z by the following procedure. Since $a \neq 0$, we have

$$0 = z^2 + \frac{bz}{a} + \frac{c}{a}$$

$$= \left(z + \frac{b}{2a}\right)^2 - \frac{b^2}{4a^2} + \frac{c}{a}$$

$$= \left(z + \frac{b}{2a}\right)^2 - \frac{b^2 - 4ac}{4a^2},$$

so that

$$\left(z + \frac{b}{2a}\right)^2 = \frac{b^2 - 4ac}{4a^2}.$$

Thus

$$z + \frac{b}{2a} = \pm \frac{1}{2a}\sqrt{b^2 - 4ac},$$

and we conclude that

$$z = \frac{-b \pm \sqrt{b^2 - 4ac}}{2a}.$$

The expression $b^2 - 4ac$ is called the **discriminant** of the polynomial $az^2 + bz + c$. If a, b, c are real, then the polynomial has just two real roots if its discriminant is positive, just one if its discriminant is 0, and none otherwise.

We proceed to the exponentiation of complex numbers. As in the case of real numbers, we define $z^1 = z$ for every complex number z, and if z^n has been defined for a specific natural number n, we write $z^{n+1} = z^n \cdot z$. We also define $z^0 = 1$ if $z \neq 0$.

Next, let

$$z = x + iy \neq 0,$$

where x and y are real. Then

$$(x + iy)(x - iy) = x^2 + y^2 \neq 0,$$

and so

$$\frac{1}{z} = \frac{1}{x + iy} = \frac{x - iy}{(x + iy)(x - iy)} = \frac{x - iy}{x^2 + y^2}.$$

Thus if we define

$$z^{-1} = \frac{x - iy}{x^2 + y^2},$$

then we have $zz^{-1} = 1$. For every positive integer n we now define $z^{-n} = (z^{-1})^n$ if $z \neq 0$. This definition agrees with the equation $z^1 = z$ in the case where $n = 1$. We also define $w/z = wz^{-1}$ for every complex number w.

We turn our attention now to two numbers that are associated with a given complex number. The first is the conjugate of a complex number $z = x + iy$, where x and y are real. The **conjugate** of z is defined as $x - iy$ and is denoted by $\bar{z}$. Thus $\bar{\bar{z}} = z$, and $z = \bar{z}$ if and only if z is real. Geometrically, the function that maps z to $\bar{z}$ is a reflection about the x-axis.

If we also have $w = u + iv$, where u and v are real, then

$$\overline{w + z} = \overline{u + x + i(v + y)} = u + x - i(v + y) = u - iv + x - iy = \bar{w} + \bar{z}.$$

Since

$$\overline{-z} = \overline{-x - iy} = -x + iy = -(x - iy) = -\bar{z},$$

it also follows that

$$\overline{w - z} = \overline{w + (-z)} = \bar{w} + \overline{-z} = \bar{w} + (-\bar{z}) = \bar{w} - \bar{z}.$$

Moreover

$$\overline{wz} = \overline{ux - vy + i(uy + vx)} = ux - vy - i(uy + vx),$$

and so

$$\bar{w} \cdot \bar{z} = (u - iv)(x - iy) = ux - vy + i(-uy - vx) = \overline{wz}.$$

If $z \neq 0$, then

$$\overline{1/z} = \frac{x + iy}{x^2 + y^2} = 1/\bar{z},$$

and it follows that

$$\overline{w/z} = \overline{wz^{-1}} = \bar{w}\overline{z^{-1}} = \bar{w} \cdot \overline{1/z} = \frac{\bar{w}}{\bar{z}}.$$

Note also that

$$z + \bar{z} = 2x = 2\mathrm{Re}\,(z)$$

and, similarly,

$$z - \bar{z} = 2i \operatorname{Im}(z);$$

hence

$$\operatorname{Re}(z) = \frac{z + \bar{z}}{2}$$

and

$$\operatorname{Im}(z) = \frac{z - \bar{z}}{2i}.$$

A further observation is that

$$z\bar{z} = (x + iy)(x - iy) = x^2 + y^2.$$

We now define

$$|z| = \sqrt{x^2 + y^2}.$$

This number is called the **modulus** of z. For example, $|i| = 1$. We note that

$$z\bar{z} = |z|^2,$$

$$|z| \geq \sqrt{x^2} = |x| = |\operatorname{Re}(z)|,$$

and, similarly,

$$|z| \geq \sqrt{y^2} = |\operatorname{Im}(z)|.$$

Moreover if $z = x$, then

$$|z| = \sqrt{x^2} = |x|.$$

This observation shows that $|z|$ is a generalization of the absolute value of a real number. It follows that

$$z + \bar{z} = 2\operatorname{Re}(z) \leq 2|\operatorname{Re}(z)| \leq 2|z|.$$

If w is as defined in the previous paragraph, then

$$|z - w| = |x + iy - u - iv| = |x - u + i(y - v)| = \sqrt{(x - u)^2 + (y - v)^2};$$

hence

$$|z - w| = |w - z|$$

and we perceive $|z - w|$ geometrically as the distance between z and w. In particular, $|z|$ is the distance between z and the origin. Note also that

$$|z - c| = r,$$

where c is a complex number and $r \geq 0$ is the equation of a circle with center c and radius r.

We also have $|z| \geq 0$ for all z, and $|z| = 0$ if and only if $z = 0$. In addition,

$$|z| = |-z| = |\bar{z}|.$$

Since

$$|wz|^2 = wz \cdot \overline{wz} = wz\bar{w}\bar{z} = w\bar{w}z\bar{z} = |w|^2|z|^2 = (|w||z|)^2,$$

we deduce that

$$|wz| = |w||z|.$$

If $z \neq 0$, it follows that

$$|w| = \left|\frac{wz}{z}\right| = \left|\frac{w}{z}\right||z|,$$

and so

$$\left|\frac{w}{z}\right| = \frac{|w|}{|z|}.$$

The triangle inequality

$$|w + z| \leq |w| + |z|$$

also holds, as can be inferred from the calculation

$$
\begin{aligned}
|w + z|^2 &= (w + z)\overline{w + z} \\
&= (w + z)(\bar{w} + \bar{z}) \\
&= w\bar{w} + w\bar{z} + z\bar{w} + z\bar{z} \\
&= |w|^2 + w\bar{z} + \overline{w\bar{z}} + |z|^2 \\
&\leq |w|^2 + 2|w\bar{z}| + |z|^2
\end{aligned}
$$

$$= |w|^2 + 2|w||\bar{z}| + |z|^2$$
$$= |w|^2 + 2|w||z| + |z|^2$$
$$= (|w| + |z|)^2.$$

Thus

$$|w| = |w - z + z| \le |w - z| + |z|,$$

so that

$$|w - z| \ge |w| - |z|.$$

But we also have

$$|w - z| = |z - w| \ge |z| - |w| = -(|w| - |z|),$$

and so we conclude that

$$|w - z| \ge ||w| - |z||.$$

Finally, the triangle inequality also shows that

$$|z| \le |x| + |iy| = |x| + |y| = |\text{Re}\,(z)| + |\text{Im}\,(z)|.$$

The notion of an inequality for real numbers does not extend to complex numbers. We have $x^2 \ge 0$ for all real x, but $i^2 < 0$. It is meaningless to write $w < z$ if either w or z is not real. The inequality $|z| < a$ is equivalent to $-a < z < a$ if and only if a and z are both real.

Throughout the book we will denote by $\mathbb{Z}$, $\mathbb{Q}$, $\mathbb{R}$, and $\mathbb{C}$ the sets of integers, rational numbers, real numbers, and complex numbers, respectively.

Let f and g be functions whose domains are subsets of $\mathbb{C}$. For all $z \in \mathcal{D}_f \cap \mathcal{D}_g$, we define

$$(f + g)(z) = f(z) + g(z),$$

$$(f - g)(z) = f(z) - g(z),$$

and

$$(fg)(z) = f(z)g(z).$$

We also define

$$\left(\frac{f}{g}\right)(z) = \frac{f(z)}{g(z)}$$

for all $z \in \mathcal{D}_f \cap \mathcal{D}_g$ such that $g(z) \neq 0$. Thus $f + g$, $f - g$, fg, and f/g are all functions. In addition, if $c \in \mathbb{C}$, then we define cf to be the function such that

$$(cf)(z) = cf(z)$$

for all $z \in \mathcal{D}_f$.

Exercises 1.2.

1. Express in the form $x + iy$, where x and y are real,

   (a) $3 + 4i + (1 - i)(1 + i)$;
   (b) $(2 - 5i)^2$;
   (c) $\frac{1-2i}{3+4i}$.

2. Compute i^n for every integer n.
3. For every complex number z, find the real and imaginary parts of the following expressions:

   (a) $\frac{z+2}{z-2i}$;
   (b) iz.

4. Let $z = x + iy$ and $w = a + ib$, where x, y, a, b are real. Suppose that $z^2 = w$. Show that

$$x^2 = \frac{a + |w|}{2}$$

and

$$y^2 = \frac{-a + |w|}{2}.$$

Deduce that

$$z = \pm(\alpha + \lambda\beta i),$$

where

$$\alpha = \sqrt{\frac{a + |w|}{2}},$$

$$\beta = \sqrt{\frac{-a + |w|}{2}},$$

$\lambda = 1$ if $b \geq 0$ and $\lambda = -1$ if $b < 0$. Hence, conclude that the square root of a complex number is real if and only if the complex number is real and positive.

5. Solve the following equations:

   (a) $z^2 = 1 - i$;
   (b) $z^4 = i$.

6. Show that

$$\frac{|z|}{|u+v|} \leq \frac{|z|}{||u| - |v||},$$

   where u, v, z are complex numbers and $|u| \neq |v|$.

7. If z and w are complex numbers, show that

$$|z + w| = |z| + |w|$$

   if and only if $z = \alpha w$ for some real number α.

8. Recall that the function $d(z, w) = |z - w|$ measures the distance between the points representing the complex numbers z and w. Prove the inequality

$$d(z, w) \leq d(z, v) + d(v, w),$$

   where w, v, z are complex numbers. More generally, let $z_1, z_2, \ldots, z_n$ be complex numbers. Show that

$$d(z_1, z_n) \leq d(z_1, z_2) + d(z_2, z_3) + \ldots + d(z_{n-1}, z_n).$$

9. Show that

$$|d(z, v) - d(v, w)| \leq d(z, w)$$

   for complex numbers v, w, z.

10. Give a condition for

$$d(z, w) = d(z, v) + d(v, w),$$

   where v, w, z are complex numbers.

1.5   Finite Sums

Our development of analysis is based on the concepts of sequences and series. In order to be able to deal with series, we need to be familiar with the properties of finite sums. This section is therefore devoted to the development of their basic properties.

Let m and n be integers with $n \geq m$. Recall that

$$\sum_{j=m}^{n} a_j = a_m + a_{m+1} + \cdots + a_n,$$

where $a_m, a_{m+1}, \ldots, a_n$ are numbers. Observe that j is a dummy variable. In other words, the sum is independent of j, so that if k is another letter, then

$$\sum_{k=m}^{n} a_k = \sum_{j=m}^{n} a_j. \tag{1.15}$$

(Sometimes j is referred to as an index.) More generally, let r be any integer and put $k = j + r$. Then $j = k - r$. Moreover $k = m + r$ when $j = m$, and $k = n + r$ when $j = n$, and so we may write

$$\sum_{j=m}^{n} a_j = \sum_{k=m+r}^{n+r} a_{k-r}.$$

Using Eq. (1.15), we therefore obtain the following result.

Proposition 1.5.1. *If m, n, r are integers with $m \leq n$ and a_j is a number for each integer j such that $m \leq j \leq n$, then*

$$\sum_{j=m+r}^{n+r} a_{j-r} = \sum_{j=m}^{n} a_j.$$

Moreover the associativity of addition shows that if m, n and r are integers such that $m \leq r < n$, then

$$\sum_{j=m}^{n} a_j = \sum_{j=m}^{r} a_j + \sum_{j=r+1}^{n} a_j.$$

The next theorem is a basic property of finite sums.

Theorem 1.5.2. *Let m, n be integers with $m \leq n$, and let a_j and b_j be numbers for each integer j for which $m \leq j \leq n$. For all numbers s, t, we have*

$$\sum_{j=m}^{n} (sa_j + tb_j) = s \sum_{j=m}^{n} a_j + t \sum_{j=m}^{n} b_j. \tag{1.16}$$

Proof. We fix m and use induction on n. For $n = m$ we have

$$\sum_{j=m}^{m}(sa_j + tb_j) = sa_m + tb_m$$

$$= s\sum_{j=m}^{m}a_j + t\sum_{j=m}^{m}b_j.$$

Now suppose that Eq. (1.16) holds for some integer $n \geq m$. Then

$$\sum_{j=m}^{n+1}(sa_j + tb_j) = \sum_{j=m}^{n}(sa_j + tb_j) + sa_{n+1} + tb_{n+1}$$

$$= s\sum_{j=m}^{n}a_j + t\sum_{j=m}^{n}b_j + sa_{n+1} + tb_{n+1}$$

$$= s\left(\sum_{j=m}^{n}a_j + a_{n+1}\right) + t\left(\sum_{j=m}^{n}b_j + b_{n+1}\right)$$

$$= s\sum_{j=m}^{n+1}a_j + t\sum_{j=m}^{n+1}b_j,$$

and the proof by induction is complete. □

For example, by taking $t = 0$, we obtain the distributive law:

$$\sum_{j=m}^{n}sa_j = s\sum_{j=m}^{n}a_j. \tag{1.17}$$

Similarly, putting $s = t = 1$, we find that

$$\sum_{j=m}^{n}(a_j + b_j) = \sum_{j=m}^{n}a_j + \sum_{j=m}^{n}b_j,$$

and by setting $s = 1$ and $t = -1$, we have

$$\sum_{j=m}^{n}(a_j - b_j) = \sum_{j=m}^{n}a_j - \sum_{j=m}^{n}b_j.$$

Thus if $m \le n$ and $a_m, a_{m+1}, \ldots, a_{n+1}$ are numbers, then

$$
\begin{aligned}
\sum_{j=m}^{n} (a_{j+1} - a_j) &= \sum_{j=m}^{n} a_{j+1} - \sum_{j=m}^{n} a_j \\
&= \sum_{j=m+1}^{n+1} a_j - \sum_{j=m}^{n} a_j \\
&= \sum_{j=m+1}^{n} a_j + a_{n+1} - \left(a_m + \sum_{j=m+1}^{n} a_j \right) \\
&= a_{n+1} - a_m,
\end{aligned}
$$

a result known as the **telescoping property**. We state it as a theorem.

Theorem 1.5.3. *Let m, n be integers such that $m \le n$, and let a_j be a number for each j such that $m \le j \le n + 1$. Then*

$$
\sum_{j=m}^{n} (a_{j+1} - a_j) = a_{n+1} - a_m.
$$

This theorem in fact is intuitively clear, since the sum can be written as

$$
(a_{n+1} - a_n) + (a_n - a_{n-1}) + \cdots + (a_{m+1} - a_m)
$$

and cancellation yields $a_{n+1} - a_m$. Note also that

$$
\sum_{j=m}^{n} (a_j - a_{j+1}) = -\sum_{j=m}^{n} (a_{j+1} - a_j) = -(a_{n+1} - a_m) = a_m - a_{n+1}.
$$

As an example, if $a_j = j$ for each j, we obtain

$$
\sum_{j=m}^{n} 1 = \sum_{j=m}^{n} (j + 1 - j) = n + 1 - m
$$

by the telescoping property. In particular, if $m = 1$ and n is a positive integer, then

$$
\sum_{j=1}^{n} 1 = n + 1 - 1 = n,
$$

as expected, because we are simply adding n copies of the number 1.

For a less trivial application of the telescoping property, let us evaluate

$$\sum_{j=1}^{n}(2j + 1).$$

Note first that

$$(j + 1)^2 - j^2 = j^2 + 2j + 1 - j^2$$
$$= 2j + 1.$$

In the telescoping property we therefore take $a_j = j^2$ for each j and thereby obtain

$$\sum_{j=1}^{n}(2j + 1) = \sum_{j=1}^{n}((j + 1)^2 - j^2)$$
$$= (n + 1)^2 - 1$$
$$= n^2 + 2n.$$

But we also have

$$\sum_{j=1}^{n}(2j + 1) = 2\sum_{j=1}^{n}j + \sum_{j=1}^{n}1$$
$$= 2\sum_{j=1}^{n}j + n,$$

and so

$$2\sum_{j=1}^{n}j = \sum_{j=1}^{n}(2j + 1) - n$$
$$= n^2 + 2n - n$$
$$= n^2 + n.$$

Hence we obtain the following theorem.

Theorem 1.5.4. *For every positive integer n*

$$\sum_{j=1}^{n}j = \frac{n(n + 1)}{2}.$$

A similar argument can be used to show that

$$\sum_{j=1}^{n} j^2 = \frac{n(n+1)(2n+1)}{6}.$$

The next theorem also illustrates the telescoping property in action.

Theorem 1.5.5. *If n is a nonnegative integer and $a \neq b$, then*

$$\sum_{j=0}^{n} a^j b^{n-j} = \frac{a^{n+1} - b^{n+1}}{a - b} = \frac{b^{n+1} - a^{n+1}}{b - a},$$

where the convention that $0^0 = 1$ is used.

Proof. By distributivity and the telescoping property, we have

$$(a - b) \sum_{j=0}^{n} a^j b^{n-j} = \sum_{j=0}^{n} (a^{j+1} b^{n-j} - a^j b^{n-j+1})$$

$$= a^{n+1} - b^{n+1},$$

and the result follows.   □

Putting $b = 1$, we draw the following conclusion.

Corollary 1.5.6. *If $a \neq 1$, then*

$$\sum_{j=0}^{n} a^j = \frac{a^{n+1} - 1}{a - 1} = \frac{1 - a^{n+1}}{1 - a},$$

where $0^0 = 1$.

There is one further theorem concerning finite sums that we include in this section. It is called the binomial theorem and gives a formula for $(a + b)^n$ for every nonnegative integer n. It furnishes another example of a proof by induction.

First we introduce some new notation. If n and r are integers such that $0 \leq r \leq n$, then we define

$$\binom{n}{r} = \frac{n!}{r!(n-r)!}$$

$$= \frac{n(n-1)\cdots(n-r+1)}{r!}.$$

For example,

$$\binom{n}{0} = \frac{n!}{0!n!} = 1.$$

Similarly,

$$\binom{n}{n} = 1.$$

Because of their role in the binomial theorem, these numbers are often called the **binomial coefficients**. They satisfy the following lemma.

Lemma 1.5.7. *If n and r are positive integers and $r \leq n$, then*

$$\binom{n+1}{r} = \binom{n}{r} + \binom{n}{r-1}.$$

Proof. By direct calculation, we have

$$\begin{aligned}
\binom{n}{r} + \binom{n}{r-1} &= \frac{n!}{r!(n-r)!} + \frac{n!}{(r-1)!(n-r+1)!} \\
&= \frac{n!(n-r+1+r)}{r!(n-r+1)!} \\
&= \frac{(n+1)!}{r!(n-r+1)!} \\
&= \binom{n+1}{r}.
\end{aligned}$$

$\square$

Theorem 1.5.8 (Binomial Theorem). *Let a and b be numbers and n a nonnegative integer. Then*

$$(a+b)^n = \sum_{j=0}^{n} \binom{n}{j} a^j b^{n-j}, \tag{1.18}$$

where the convention that $0^0 = 1$ is used.

Proof. Both sides are equal to 1 if $n = 0$. Assume that the theorem holds for some integer $n \geq 0$. Then

$$(a + b)^{n+1} = (a + b)(a + b)^n$$

$$= (a + b) \sum_{j=0}^{n} \binom{n}{j} a^j b^{n-j}$$

$$= \sum_{j=0}^{n} \binom{n}{j} a^{j+1} b^{n-j} + \sum_{j=0}^{n} \binom{n}{j} a^j b^{n-j+1}$$

$$= \sum_{j=1}^{n+1} \binom{n}{j-1} a^j b^{n-j+1} + \sum_{j=0}^{n} \binom{n}{j} a^j b^{n-j+1}$$

$$= \sum_{j=1}^{n} \binom{n}{j-1} a^j b^{n+1-j} + a^{n+1} + b^{n+1} + \sum_{j=1}^{n} \binom{n}{j} a^j b^{n+1-j}$$

$$= b^{n+1} + \sum_{j=1}^{n} \left(\binom{n}{j-1} + \binom{n}{j} \right) a^j b^{n+1-j} + a^{n+1}$$

$$= b^{n+1} + \sum_{j=1}^{n} \binom{n+1}{j} a^j b^{n+1-j} + a^{n+1}$$

$$= \sum_{j=0}^{n+1} \binom{n+1}{j} a^j b^{n+1-j}.$$

The result follows by induction. $\square$

The sum on the right-hand side of Eq. (1.18) is often referred to as the **binomial expansion** of $(a + b)^n$. Note that the result of Example 1.3.1 can easily be deduced by considering three terms of the binomial expansion with $a = 1$ and $b = x$.

Exercises 1.3.

1. Use induction to prove the following formulas for all positive integers n:
   (a) $\sum_{j=1}^{n} j = \frac{n(n+1)}{2}$;
   (b) $\sum_{j=1}^{n} j^2 = \frac{n(n+1)(2n+1)}{6}$;
   (c) $\sum_{j=1}^{n} \frac{1}{j(j+1)} = \frac{n}{n+1}$;
   (d) $\sum_{j=1}^{n} \frac{j}{(j+1)!} = 1 - \frac{1}{(n+1)!}$;
   (e) $\sum_{j=1}^{n} j^3 = \left(\sum_{j=1}^{n} j \right)^2$;

(f) $\sum_{j=1}^{n} j^4 = \frac{n(n+1)(2n+1)(3n^2+3n-1)}{30}$;

(g) $\sum_{j=1}^{n-1} jx^j = \frac{x-nx^n+(n-1)x^{n+1}}{1-x^2}$ for every real $x \neq \pm 1$.

2. Let $n \in \mathbb{N}$ and let $a_1, a_2, \ldots, a_n$ be real numbers such that $0 \leq a_j \leq 1$ for all j. Prove that

$$\prod_{j=1}^{n} (1 - a_j) \geq 1 - \sum_{j=1}^{n} a_j.$$

3. Discover and prove a theorem about the relative sizes of 3^n and $n!$, where n is a positive integer.

4. A function $f : \mathbb{R} \to \mathbb{R}$ is said to be **convex** if for all nonnegative real numbers α_1 and α_2 with sum equal to 1 we have

$$f(\alpha_1 x_1 + \alpha_2 x_2) \leq \alpha_1 f(x_1) + \alpha_2 f(x_2)$$

for each $x_1 \in \mathbb{R}$ and $x_2 \in \mathbb{R}$. Prove that

$$f\left(\sum_{j=1}^{n} \alpha_j x_j\right) \leq \sum_{j=1}^{n} \alpha_j f(x_j)$$

whenever n is an integer greater than 1 and $\alpha_1, \alpha_2, \ldots, \alpha_n$ are nonnegative real numbers with sum 1. This result is known as Jensen's inequality. More generally, show that

$$f\left(\sum_{j=1}^{n} (\alpha_j x_j) \Big/ \sum_{k=1}^{n} \alpha_k\right) \leq \sum_{j=1}^{n} (\alpha_j f(x_j)) \Big/ \sum_{k=1}^{n} \alpha_k.$$

5. Use the telescoping property to prove the following identities for all $n \in \mathbb{N}$:

(a) $\sum_{j=1}^{n} \frac{1}{j(j+2)} = \frac{3}{4} - \frac{1}{2}\left(\frac{1}{n+1} + \frac{1}{n+2}\right)$;

(b) $\sum_{j=1}^{n} \frac{1}{\sqrt{j}+\sqrt{j+1}} = -1 + \sqrt{n+1}$;

(c) $\sum_{j=1}^{n} j \cdot j! = (n+1)! - 1$;

(d) $\sum_{j=1}^{n} \frac{6^j}{(3^{j+1}-2^{j+1})(3^j-2^j)} = 3 - \frac{3^{n+1}}{3^{n+1}-2^{n+1}}$;

(e) $\sum_{j=1}^{n} \frac{1}{j^2+4j+3} = \frac{5}{12} - \frac{1}{2}\left(\frac{1}{n+2} - \frac{1}{n+3}\right)$;

(f) $\sum_{j=1}^{n} \frac{1}{(j+1)\sqrt{j}+j\sqrt{j+1}} = 1 - \frac{1}{\sqrt{n+1}}$;

(g) $\sum_{j=1}^{n} \frac{j}{j^4+j^2+1} = \frac{1}{2} - \frac{1}{2n(n+1)+2}$;

(h) $\sum_{j=1}^{n} \frac{j}{4j^4+1} = \frac{1}{4} - \frac{1}{8n^2+8n+4}$.

6. Let d and a_1 be numbers. Define a_j for each integer $j > 1$ by the equation $a_{j+1} = a_j + d$, and suppose that $a_j \neq 0$ for all $j \in \mathbb{N}$. Find the sum

$$\sum_{j=1}^{n} \frac{1}{a_j a_{j+1}}$$

for all $n \in \mathbb{N}$.

7. Prove that

$$\sum_{j=1}^{n} j!(j^2 + j + 1) = n!(n+1)^2 - 1$$

for all $n \in \mathbb{N}$.

8. Evaluate the sum

$$\sum_{j=3}^{n} \frac{1}{j^2 - 4}$$

for all integers $n \geq 3$.

9. Suppose that $|na_n| \leq M$ for all $n \geq 0$. Show that

$$S_n - \sigma_n = \frac{1}{n+1} \sum_{j=1}^{n} j a_j \leq M$$

for all n, where $S_n = \sum_{j=0}^{n} a_j$ and $\sigma_n = \sum_{j=0}^{n} S_j$ for all n.

Chapter 2
Sequences

Analysis is based on the notion of a limit, a concept that can be defined in terms of sequences. Moreover, elementary functions, such as trigonometric, exponential, and logarithm functions and many algebraic functions, can be approximated by using sequences. With modern computers, such approximations can be made accurate enough for most practical purposes.

2.1 Definitions and Examples

A sequence is a function whose domain is the set of all integers greater than or equal to some fixed integer. More formally, we have the following definition.

Definition 2.1.1. Let a be a fixed integer and A the set of all integers greater than or equal to a. A function s from A into a set $\mathbb{F}$ is called a **sequence** in $\mathbb{F}$. The images of the members of A are called the **terms** of the sequence. They are ordered by the ordering of A itself so that, for example, $s(a)$ is the **first** term of the sequence, $s(a + 1)$ is the **second**, and so forth.

Throughout this book $\mathbb{F}$ will denote one of the three fields $\mathbb{Q}, \mathbb{R}$, or $\mathbb{C}$. The sequence will then be said to be **rational**, **real**, or **complex**, respectively. Results established for complex sequences will therefore be valid for rational and real sequences as they are special cases.

Given a sequence $s : A \to \mathbb{F}$, we usually write s_n instead of $s(n)$, where $n \in A$. Moreover we denote the sequence by $\{s_n\}_{n \in A}$ or simply $\{s_n\}$ if A is either clear from the context or immaterial. If A is the set of all integers $n \geq a$, then we also write the sequence as $\{s_n\}_{n \geq a}$. Occasionally, the first few terms of the sequence may be listed in order, so that we write $s_a, s_{a+1}, \ldots$, for instance.

© Springer Science+Business Media New York 2015
C.H.C. Little et al., *Real Analysis via Sequences and Series*, Undergraduate
Texts in Mathematics, DOI 10.1007/978-1-4939-2651-0_2

We now present some examples of sequences.

Example 2.1.1 (Sign Sequence). The sequence given by $\{(-1)^{n+1}\}_{n\geq 1}$ is

$$1, -1, 1, -1, \ldots.$$

$\triangle$

Example 2.1.2 (Harmonic Sequence). The sequence given by $\{1/n\}_{n\geq 1}$ is

$$1, \frac{1}{2}, \frac{1}{3}, \frac{1}{4}, \ldots.$$

$\triangle$

Example 2.1.3 (Complex Harmonic Sequence). The sequence given by $\{i^n/n\}_{n\geq 1}$ is

$$i, -\frac{1}{2}, -\frac{i}{3}, \frac{1}{4}, \frac{i}{5}, -\frac{1}{6}, -\frac{i}{7}, \frac{1}{8}, \ldots.$$

$\triangle$

Example 2.1.4 (Geometric Sequence). If we temporarily adopt the convention that $0^0 = 1$, then for all complex numbers a and r, the sequence given by $\{ar^n\}_{n\geq 0}$ is

$$a, ar, ar^2, ar^3, \ldots.$$

$\triangle$

Example 2.1.5 (Arithmetic Sequence). For each complex a and d the sequence given by $\{a + nd\}_{n\geq 0}$ is

$$a, a + d, a + 2d, a + 3d, \ldots.$$

$\triangle$

Example 2.1.6. The sequence given by $\{(1 + 1/n)^n\}_{n\geq 1}$ is

$$2, \frac{9}{4}, \frac{64}{27}, \frac{625}{256}, \ldots.$$

We shall return to this sequence later. $\triangle$

Often, particularly in computer applications, a sequence is defined inductively by specifying the first few terms and then defining the remaining terms by means of the preceding ones. For example, the geometric sequence $\{ar^n\}$ can be defined inductively by writing $s_0 = a$ and $s_k = rs_{k-1}$ for all $k > 0$. Likewise, the arithmetic sequence $\{a + nd\}$ can be defined by setting $s_0 = a$ and $s_k = s_{k-1} + d$ for all $k > 0$.

Example 2.1.7. A sequence that is often used in botany is the **Fibonacci** sequence (see [6]). It is specified by putting $F_0 = F_1 = 1$ and

$$F_k = F_{k-1} + F_{k-2}$$

for all $k > 1$. The sequence so defined is

$$1, 1, 2, 3, 5, 8, 13, 21, \ldots.$$

We show by induction that

$$F_n = \frac{1}{\sqrt{5}} \left(\left(\frac{1 + \sqrt{5}}{2} \right)^{n+1} - \left(\frac{1 - \sqrt{5}}{2} \right)^{n+1} \right)$$

for all $n \geq 0$. First, for convenience let us define $\alpha = (1 + \sqrt{5})/2$ and $\beta = (1 - \sqrt{5})/2$. These are the solutions of the equation

$$x^2 - x - 1 = 0, \tag{2.1}$$

so that $\alpha^2 = \alpha + 1$ and $\beta^2 = \beta + 1$. Since $\alpha - \beta = \sqrt{5}$, the required formula holds for $n = 0$. It also holds for $n = 1$, since $\alpha^2 - \beta^2 = \alpha + 1 - (\beta + 1) = \alpha - \beta = \sqrt{5}$. Assuming that $n > 1$ and that the result holds for all positive integers less than n, it follows that

$$\begin{aligned}
F_n &= F_{n-1} + F_{n-2} \\
&= \frac{1}{\sqrt{5}} (\alpha^n - \beta^n + \alpha^{n-1} - \beta^{n-1}) \\
&= \frac{1}{\sqrt{5}} (\alpha^{n-1}(\alpha + 1) - \beta^{n-1}(\beta + 1)) \\
&= \frac{1}{\sqrt{5}} (\alpha^{n+1} - \beta^{n+1}),
\end{aligned}$$

as required. It is somewhat surprising that the formula yields an integer. The number α is known as the **golden ratio**.   △

Example 2.1.8. Fix a positive integer k. Define $s_1 = 1$ and

$$s_n = \frac{s_{n-1} + k}{s_{n-1} + 1}$$

for all $n > 1$. The resulting sequence is

$$1, \frac{k+1}{2}, \frac{3k+1}{k+3}, \frac{k^2+6k+1}{4k+4}, \ldots$$

We will see later (Example 2.3.2) that its terms give a good approximation for $\sqrt{k}$ when n is large. For $k = 2$, the sequence was familiar to the ancient Greeks, having appeared in Chapter 31 of the first volume of the manuscript *Expositio rerum mathematicarum ad legendum Platonem utilium* by Theon of Smyrna in 130 AD. The reader is invited to show by induction that

$$s_n = \sqrt{k} \cdot \frac{(1+\sqrt{k})^n + (1-\sqrt{k})^n}{(1+\sqrt{k})^n - (1-\sqrt{k})^n} \tag{2.2}$$

for all n.   △

Remark. The reader may wonder how the expressions for the general term were obtained in the two preceding examples. One method using matrices is given in [15]. Here is another way of obtaining the formula for the Fibonacci sequence. Let us write

$$F_n = A\alpha^n + B\beta^n$$

for some constants A, B, α, β such that $\alpha \neq 0$ and $\beta \neq 0$. For the equation

$$F_n = F_{n-1} + F_{n-2}$$

to hold, we need to ensure that

$$A\alpha^n + B\beta^n = A\alpha^{n-1} + B\beta^{n-1} + A\alpha^{n-2} + B\beta^{n-2},$$

that is,

$$A\alpha^{n-2}(\alpha^2 - \alpha - 1) + B\beta^{n-2}(\beta^2 - \beta - 1) = 0.$$

For this purpose it is sufficient to take α and β to be the solutions of Eq. (2.1). This observation gives the values for α and β in Example 2.1.7. Next, the initial condition that $F_0 = F_1 = 1$ yields the equations

$$A + B = 1$$

and

$$A\alpha + B\beta = 1.$$

The solution gives

$$B = \frac{\alpha - 1}{\alpha - \beta} = \frac{\sqrt{5} - 1}{2\sqrt{5}} = -\frac{\beta}{\sqrt{5}},$$

so that

$$A = 1 - B = \frac{\alpha}{\sqrt{5}}.$$

The desired expression for F_n is now obtained by substitution.

Exercises 2.1.

1. Prove Eq. (2.2).

2.2 Convergence of Sequences

We are often concerned with the limiting behavior of a sequence $\{s_n\}$ as n becomes large. The process of enlarging n indefinitely is indicated by the notation $n \to \infty$.

In what follows the sequences will be assumed to be complex unless an indication to the contrary is given.

Definition 2.2.1. Let $\{s_n\}$ be a sequence and L a number. We say that $\{s_n\}$ **converges** to L, and that L is the **limit** of $\{s_n\}$, if for each $\varepsilon > 0$ there exists an integer N such that

$$|s_n - L| < \varepsilon$$

whenever $n \geq N$. If $\{s_n\}$ converges to L, then we also say that s_n **approaches L as n approaches infinity**, and we write

$$\lim_{n \to \infty} s_n = L$$

or $s_n \to L$ as $n \to \infty$.

A sequence is said to be **convergent** if there exists a number to which it converges. A sequence is said to **diverge**, and to be **divergent**, if it does not converge.

In Definition 2.2.1 the integer N may of course depend on ε. It is also clear that the inequality $n \geq N$ may be replaced by the corresponding strict inequality, for $n \geq N$ if and only if $n > N - 1$.

Given L and $\varepsilon > 0$, we may define the set of all z for which $|z - L| < \varepsilon$ as the ε-**neighborhood** of L. This set is denoted by $N_\varepsilon(L)$. Roughly speaking, the definition of the convergence of a sequence to L says that for each prescribed ε-neighborhood of L, the terms of the sequence will eventually enter the neighborhood and remain there. In other words, the terms of the sequence become arbitrarily close to L as n increases.

Note that the inequality $|s_n - L| < \varepsilon$ is equivalent in $\mathbb{R}$ to

$$L - \varepsilon < s_n < L + \varepsilon.$$

A sequence that converges to 0 is said to be **null**.

Example 2.2.1. For each number a the constant sequence $\{a\}$ converges to a.

Proof. Define $s_n = a$ for each integer $n \in \mathbb{N}$. For every $\varepsilon > 0$ and every $n > 0$ we have $|s_n - a| = |a - a| = 0 < \varepsilon$, as desired.

$\triangle$

Example 2.2.2. Let us show that

$$\lim_{n \to \infty} \frac{i^n}{n} = 0.$$

Given any $\varepsilon > 0$, we need to find an integer N such that

$$\left| \frac{i^n}{n} - 0 \right| < \varepsilon$$

for all $n \geq N$. Since $|i| = 1$, we need $1/n < \varepsilon$, and so we choose any $N > 1/\varepsilon > 0$. For all $n \geq N$ we deduce that

$$\frac{1}{n} \leq \frac{1}{N} < \varepsilon,$$

and the desired result follows.

$\triangle$

Remark. The only property of i used in this proof is that $|i| = 1$. Hence the argument can also be used to show that

$$\lim_{n \to \infty} \frac{1}{n} = 0.$$

Example 2.2.3. We show that the sequence $\{(-1)^n\}$ diverges. This result is intuitively clear, for the values of $(-1)^n$ oscillate between 1 and -1 as n increases. The distance between these numbers is 2, and so every number L must be at a distance of at least 1 from one or the other of them. But if L were the limit of our sequence, then the terms would become arbitrarily close to L as n increases. We infer that no number can be the limit of the sequence, and the sequence therefore diverges.

We can also cast this intuitive argument in terms of neighborhoods. For every number L we can choose ε so small that the ε-neighborhood of L does not contain both 1 and -1. Therefore there is no term in the sequence beyond which the values in the sequence remain in the ε-neighborhood.

We now transform this intuition into a formal argument. Suppose the sequence were to have a limit L, and choose any ε such that $0 < \varepsilon < 1$. There would be an integer N such that $|(-1)^n - L| < \varepsilon$ whenever $n \geq N$, no matter whether n is even or odd. It would follow that

$$|1 - L| < \varepsilon$$

and

$$|1 + L| = |-1 - L| < \varepsilon.$$

Using the triangle inequality, we would therefore reach the contradiction that

$$2 = |1 + L + 1 - L| \leq |1 + L| + |1 - L| < 2\varepsilon < 2.$$

$\triangle$

Example 2.2.4. For every complex number z such that $|z| < 1$, we will show that

$$\lim_{n \to \infty} z^n = 0.$$

If $z = 0$, then the result follows from Example 2.2.1 with $a = 0$, and so we assume that $z \neq 0$. Choose $\varepsilon > 0$. We must find an integer N such that $|z^n| < \varepsilon$ for all $n \geq N$. Since $0 < |z| < 1$, we have $1/|z| > 1$. Let us write

$$\frac{1}{|z|} = 1 + p$$

for some $p > 0$. By Example 1.3.1,

$$(1 + p)^n \geq 1 + np$$

for all positive integers n. Choose N large enough so that

$$1 + Np > \frac{1}{\varepsilon}.$$

Then, for all $n \geq N$, we have

$$\frac{1}{|z^n|} = \left(\frac{1}{|z|}\right)^n = (1 + p)^n \geq 1 + np \geq 1 + Np > \frac{1}{\varepsilon} > 0,$$

so that $|z^n| < \varepsilon$ for each such n, as required.

$\triangle$

Remark. Since $1 < \sqrt{5} < 3$, we have $0 > 1 - \sqrt{5} > -2$. Thus $|1 - \sqrt{5}| < 2$, and so

$$\left| \frac{1 - \sqrt{5}}{2} \right| < 1.$$

Applying the result of the last example, we conclude that

$$\lim_{n \to \infty} \left(\frac{1 - \sqrt{5}}{2} \right)^n = 0.$$

This observation motivates us to approximate F_n in Example 2.1.7 by

$$\frac{1}{\sqrt{5}} \left(\frac{1 + \sqrt{5}}{2} \right)^{n+1}.$$

In fact, this approximation turns out to be accurate to within one decimal place for all $n \geq 4$.

Since the convergence of a sequence $\{s_n\}$ depends only on the behavior of the terms s_n where n is large, we would expect intuitively that the first few terms are immaterial. This expectation is encapsulated in the following lemma.

Lemma 2.2.1. *Let k be a positive integer. Then a sequence $\{s_n\}$ approaches a number L as n approaches infinity if and only if $\{s_{n+k}\}$ also approaches L.*

Proof. Suppose first that $s_n \to L$ as $n \to \infty$ and choose $\varepsilon > 0$. Since

$$\lim_{n \to \infty} s_n = L,$$

there exists N such that $|s_n - L| < \varepsilon$ whenever $n \geq N$. But for each $n \geq N$ we have $n + k > n \geq N$, and so $|s_{n+k} - L| < \varepsilon$. Therefore

$$\lim_{n \to \infty} s_{n+k} = L,$$

as required.

Conversely, suppose $s_{n+k} \to L$ as $n \to \infty$. For every $\varepsilon > 0$ there exists N such that $|s_{n+k} - L| < \varepsilon$ for all $n \geq N$. Choose $n \geq N + k$. Then $|s_n - L| < \varepsilon$, and we conclude that $s_n \to L$ as $n \to \infty$.

In view of this lemma, every theorem or definition concerning limits that postulates s_n to satisfy a specified property for all n will remain valid if in fact s_n satisfies the specified property only for all n greater than some fixed integer.

We show next that if a sequence converges, then the limit is unique. Certainly, our intuition leads us to expect this result. The reasoning is akin to that used in

Example 2.2.3. If the sequence $\{s_n\}$ were to converge to two distinct numbers L_1 and L_2, which are a distance $|L_1-L_2|$ apart, then each term s_n would be at a distance of at least $|L_1-L_2|/2$ from one or the other of them. Therefore s_n could not become arbitrarily close to both L_1 and L_2 as n increases. In terms of neighborhoods, we choose ε so small that the ε-neighborhoods of L_1 and L_2 are disjoint. It is therefore impossible that as n increases the terms of the sequence enter both neighborhoods and remain there. The proof of the uniqueness of the limit just formalizes this intuitive argument.

Theorem 2.2.2. *A sequence has at most one limit.*

Proof. Let $\{s_n\}$ be a sequence with limits L_1 and L_2 and choose $\varepsilon > 0$. Since s_n approaches L_1, there is an integer N_1 such that

$$|s_n - L_1| < \varepsilon$$

for all $n \geq N_1$. Likewise, there is an integer N_2 such that

$$|s_n - L_2| < \varepsilon$$

for all $n \geq N_2$. For every $n \geq \max\{N_1, N_2\}$, both of the preceding inequalities hold. For such n we therefore have

$$|L_1 - L_2| = |L_1 - s_n + s_n - L_2| \leq |s_n - L_1| + |s_n - L_2| < 2\varepsilon.$$

But ε is an arbitrary positive number. If $L_1 \neq L_2$, then we could choose

$$\varepsilon = \frac{|L_1 - L_2|}{2} > 0,$$

thereby obtaining the contradiction that $|L_1 - L_2| < |L_1 - L_2|$. We deduce that $L_1 = L_2$, as required.

The next proposition gives another way of expressing the definition of a limit of a sequence. It is often more convenient to use

Proposition 2.2.3. *Let $\{s_n\}$ be a sequence, L a number, and c a positive number. Then $\lim_{n \to \infty} s_n = L$ if and only if for each $\varepsilon > 0$ there exists an integer N such that*

$$|s_n - L| < c\varepsilon$$

for all $n \geq N$.

Proof. Since $c\varepsilon > 0$, it is immediate from the definition of the limit that the stated condition holds if $\lim_{n \to \infty} s_n = L$. Let us suppose therefore that for each $\varepsilon > 0$

there exists an integer N such that $|s_n - L| < c\varepsilon$ for all $n \geq N$. Choose $\varepsilon > 0$.
Since $\varepsilon/c > 0$, there exists N such that

$$|s_n - L| < c \cdot \frac{\varepsilon}{c} = \varepsilon$$

for all $n \geq N$. Hence $\lim_{n \to \infty} s_n = L$.   □

Sometimes we wish to consider more than one sequence, in which case the
following proposition is sometimes useful.

Proposition 2.2.4. *Let $\{s_n\}$ and $\{t_n\}$ be sequences with respective limits K and
L. Then for each $\varepsilon > 0$ there is an integer N such that both $|s_n - K| < \varepsilon$ and
$|t_n - L| < \varepsilon$ for all $n \geq N$.*

Proof. Given $\varepsilon > 0$, there exists N_1 such that $|s_n - K| < \varepsilon$ for all $n \geq N_1$.
Similarly, there exists N_2 such that $|t_n - L| < \varepsilon$ for all $n \geq N_2$. If we now take
$N = \max\{N_1, N_2\}$ and choose $n \geq N$, then $n \geq N_1$ and $n \geq N_2$, so that both
required inequalities follow.   □

Remark. This result can clearly be generalized to handle situations where more than
two sequences are under consideration.

Exercises 2.2.

1. Show from the definition that the following sequences are divergent:
   (a) $\{(-1)^n + \frac{1}{n}\}$;
   (b) $\left\{ \frac{1+(-1)^n}{2} \right\}$;
   (c) $\{i^n + \frac{1}{2^n}\}$.

2. Show that

$$\lim_{n \to \infty} \left(\frac{1+i}{2} \right)^n = 0.$$

3. Suppose that $x_n > 0$ for all $n \in \mathbb{N}$ and that $\lim_{n \to \infty} x_n = 0$. Show that the set

$$S = \{x_n \mid n \in \mathbb{N}\}$$

   contains a maximum member. Must S always contain a minimum member?
4. Let

$$s_n(x) = \frac{x + x^n}{1 + x^n}$$

   for all $n \in \mathbb{N}$ and $x \in \mathbb{R} - \{-1\}$. Find each real number x for which the sequence
   $\{s_n\}$ is convergent and find the limit of the sequence.

5. Test the convergence of the real sequence

$$\left\{ \frac{a^n - b^n}{a^n + b^n} \right\},$$

where a and b are numbers such that $a^n + b^n \neq 0$ for all n.

6. Suppose that

$$\lim_{n \to \infty} \frac{z_n - L}{z_n + L} = 0,$$

where $L \in \mathbb{C}$ and $z_n \in \mathbb{C} - \{-L\}$ for all n. Show that

$$\lim_{n \to \infty} \frac{|z_n| - |L|}{|z_n| + |L|} = 0$$

and $\lim_{n \to \infty} z_n = L$.

7. Let w be a complex number such that $|w| \neq 1$ and let

$$z_n = \frac{w^n}{1 + w^{2n}}$$

for all n. Show that the sequence $\{z_n\}$ converges to 0. Is this statement still true if $|w| = 1$?

2.3 Algebra of Limits

Using the definition of convergence to determine whether a sequence converges, and if so to what limit, can be a tedious process. However, if a given sequence is a sum, difference, product, or quotient of other sequences, then the behavior of the given sequence can be investigated by studying simpler sequences. Our next objective is to see how this simplification is effected. We begin with some results concerning the size of terms in a convergent sequence.

A sequence $\{s_n\}$ is said to be **bounded** if there exists M such that $|s_n| < M$ for all n.

Theorem 2.3.1. *Every convergent sequence is bounded.*

Proof. Let $\{s_n\}$ be a convergent sequence with limit L. For each $\varepsilon > 0$ there is an integer N such that $|s_n - L| < \varepsilon$ for all $n \geq N$. For every such n we have

$$||s_n| - |L|| \leq |s_n - L| < \varepsilon,$$

so that

$$|L| - \varepsilon < |s_n| < |L| + \varepsilon.$$

Setting

$$M = \max\{|s_1|, |s_2|, \ldots, |s_{N-1}|, |L| + \varepsilon\},$$

we find that $|s_n| \leq M$ for all n. Hence $\{s_n\}$ is bounded. □

The preceding proof shows that if

$$\lim_{n \to \infty} s_n = L,$$

then

$$\lim_{n \to \infty} |s_n| = |L|.$$

On the other hand, we observe from Example 2.2.3 that the converse does not necessarily hold. Nevertheless, if $\lim_{n \to \infty} |s_n| = 0$, then for each $\varepsilon > 0$ there exists an integer N such that $|s_n| < \varepsilon$ for all $n \geq N$, and we conclude that $\lim_{n \to \infty} s_n = 0$. We summarize these observations in the following proposition.

Proposition 2.3.2. *1. If* $\lim_{n \to \infty} s_n = L$, *then* $\lim_{n \to \infty} |s_n| = |L|$.
2. If $\lim_{n \to \infty} |s_n| = 0$, *then* $\lim_{n \to \infty} s_n = 0$.

Proposition 2.3.3. *Suppose* $\{s_n\}$ *is a real sequence such that*

$$\lim_{n \to \infty} s_n = L > c$$

for some number c. Then there exist numbers N and k > c such that $s_n > k$ for all $n \geq N$.

Proof. Choose ε such that $0 < \varepsilon < L - c$. There exists N such that $|s_n - L| < \varepsilon$ for all $n \geq N$. For each such n we have

$$L - \varepsilon < s_n < L + \varepsilon.$$

The conclusion of the theorem is therefore satisfied by $k = L - \varepsilon > c$. □

Similarly, one can establish the following result.

Proposition 2.3.4. *Suppose* $\{s_n\}$ *is a real sequence such that*

$$\lim_{n \to \infty} s_n = L < c.$$

Then there exist numbers N and k < c such that $s_n < k$ for all $n \geq N$.

Also, if $\{s_n\}$ is a real sequence such that

$$\lim_{n \to \infty} s_n = L > 0,$$

then we may take $c = L/2$ in Proposition 2.3.3 and infer the existence of a number N such that $s_n > L/2$ for all $n \geq N$. A corresponding statement holds if $L < 0$. Thus we have the following result.

Proposition 2.3.5. *Let $\{s_n\}$ be a real sequence such that*

$$\lim_{n\to\infty} s_n = L \neq 0.$$

1. If $L > 0$, then there is an integer N such that $s_n > L/2$ for all $n \geq N$.
2. If $L < 0$, then there is an integer N such that $s_n < L/2$ for all $n \geq N$.

Corollary 2.3.6. *Let $\{s_n\}$ be a (real or complex) sequence such that*

$$\lim_{n\to\infty} s_n = L \neq 0.$$

Then there is an integer N such that $|s_n| > |L|/2$ for all $n \geq N$.

Proof. From the hypothesis it follows that

$$\lim_{n\to\infty} |s_n| = |L| > 0.$$

An appeal to Proposition 2.3.5(1) completes the proof. □

We are now ready to establish the sum, product, and quotient rules for limits.

Theorem 2.3.7. *Let $\{s_n\}$ and $\{t_n\}$ be sequences, and suppose that $\lim_{n\to\infty} s_n = K$ and $\lim_{n\to\infty} t_n = L$. Then*

1.

$$\lim_{n\to\infty} (s_n + t_n) = K + L;$$

2.

$$\lim_{n\to\infty} s_n t_n = KL;$$

3.

$$\lim_{n\to\infty} \frac{s_n}{t_n} = \frac{K}{L}$$

if $L \neq 0$ and $t_n \neq 0$ for all n.

Proof. 1. Choose $\varepsilon > 0$. Using Proposition 2.2.4, we find an integer N such that both

$$|s_n - K| < \varepsilon$$

and

$$|t_n - L| < \varepsilon$$

for all $n \geq N$. For every such n we deduce that

$$
\begin{aligned}
|(s_n + t_n) - (K + L)| &= |(s_n - K) + (t_n - L)| \\
&\leq |s_n - K| + |t_n - L| \\
&< 2\varepsilon,
\end{aligned}
$$

and the result follows from Proposition 2.2.3.

2. Choosing $\varepsilon > 0$, we find an integer N with the property stated in the proof of part (1). Moreover, the convergent sequence $\{s_n\}$ is bounded, and so there exists M such that $|s_n| < M$ for all n. For all $n \geq N$ we therefore have

$$
\begin{aligned}
|s_n t_n - KL| &= |s_n t_n - Ls_n + Ls_n - KL| \\
&\leq |s_n||t_n - L| + |L||s_n - K| \\
&< M\varepsilon + |L|\varepsilon \\
&= (M + |L|)\varepsilon,
\end{aligned}
$$

and again the result follows from Proposition 2.2.3.

3. By Corollary 2.3.6 there exists an integer N_1 such that $|t_n| > |L|/2 > 0$ for all $n \geq N_1$. Thus

$$\frac{1}{|t_n|} < \frac{2}{|L|}.$$

We prove next that

$$\lim_{n \to \infty} \frac{1}{t_n} = \frac{1}{L}.$$

To this end, choose $\varepsilon > 0$. There is an integer N_2 such that $|t_n - L| < \varepsilon$ for all $n \geq N_2$. For every $n \geq \max\{N_1, N_2\}$ we therefore have

$$
\begin{aligned}
\left|\frac{1}{t_n} - \frac{1}{L}\right| &= \left|\frac{L - t_n}{Lt_n}\right| \\
&= \frac{|t_n - L|}{|L||t_n|} \\
&< \varepsilon \cdot \frac{1}{|L|} \cdot \frac{2}{|L|} \\
&= \frac{2}{|L|^2}\varepsilon,
\end{aligned}
$$

as required.

We conclude the proof by using part (2) to deduce that

$$\lim_{n \to \infty} \frac{s_n}{t_n} = \lim_{n \to \infty} s_n \cdot \lim_{n \to \infty} \frac{1}{t_n} = \frac{K}{L}.$$

$\square$

Note that the sequences $\{s_n + t_n\}$, $\{s_n t_n\}$, and $\{s_n/t_n\}$ may converge even when neither $\{s_n\}$ nor $\{t_n\}$ does so. For instance, $\{(-1)^n\}$ and $\{(-1)^{n-1}\}$ both diverge, but $\{(-1)^n + (-1)^{n-1}\}$, $\{(-1)^n(-1)^{n-1}\}$, and $\{(-1)^n/(-1)^{n-1}\}$ are all constant sequences and therefore converge.

Before presenting some examples, we note three corollaries.

Corollary 2.3.8. *Using the notation of the theorem, we have*

$$\lim_{n \to \infty} (s_n - t_n) = K - L.$$

Proof. From parts (1) and (2) of the theorem, we infer that

$$\lim_{n \to \infty} (s_n - t_n) = \lim_{n \to \infty} (s_n + (-1)t_n)$$
$$= K + (-1)L$$
$$= K - L.$$

$\square$

Corollary 2.3.9. *Let s_n, t_n, K, L be as in the statement of the theorem.*

1. *If $s_n \geq t_n$ for all n, then $K \geq L$.*
2. *If $s_n \leq t_n$ for all n, then $K \leq L$.*

Proof. 1. By Corollary 2.3.8 we have

$$\lim_{n \to \infty} (s_n - t_n) = K - L.$$

If $K - L < 0$, then, by Proposition 2.3.5(2), there would be an integer N such that

$$s_n - t_n < \frac{K - L}{2} < 0$$

for all $n \geq N$. This contradiction shows that $K - L \geq 0$.

2. The proof of part (2) is similar.

$\square$

Putting $\{t_n\}$ equal to a constant sequence $\{c\}$ yields the following corollary.

Corollary 2.3.10. *Let $\{s_n\}$ be a real sequence converging to K, and let c be a real constant.*

1. *If $s_n \geq c$ for all n, then $K \geq c$.*
2. *If $s_n \leq c$ for all n, then $K \leq c$.*

Remark. It should not be thought that if $s_n > c$ for all n then $K > c$. For instance, $1/n \to 0$ as $n \to \infty$, but $1/n > 0$ for all $n > 0$. Similarly, if $s_n < c$ for all n, then we can conclude only that $K \leq c$.

Example 2.3.1. Let

$$s_n = \frac{4n(4n+3)}{(4n+1)(4n+2)}$$

for each $n \in \mathbb{N}$. Dividing both numerator and denominator by n^2, we obtain

$$s_n = \frac{4\left(4 + \frac{3}{n}\right)}{\left(4 + \frac{1}{n}\right)\left(4 + \frac{2}{n}\right)}.$$

As

$$\lim_{n \to \infty} \frac{1}{n} = 0,$$

we may apply Theorem 2.3.7 to find that

$$\lim_{n \to \infty} s_n = \frac{4(4+0)}{(4+0)(4+0)} = 1.$$

$\triangle$

Example 2.3.2. For every positive integer k we have

$$|1 - \sqrt{k}| = \sqrt{k} - 1 < \sqrt{k} + 1 = |1 + \sqrt{k}|,$$

and so

$$\left| \frac{1 - \sqrt{k}}{1 + \sqrt{k}} \right| < 1.$$

Using the result of Example 2.2.4, we therefore find that

$$\lim_{n \to \infty} \left(\frac{1 - \sqrt{k}}{1 + \sqrt{k}} \right)^n = 0.$$

Applying an argument similar to that of the preceding example, we conclude that the sequence in Example 2.1.8 converges to $\sqrt{k}$. $\triangle$

Example 2.3.3. Let a and r be real numbers and suppose that $|r| < 1$. Let

$$s_n = \sum_{j=0}^{n} r^j = \frac{1 - r^{n+1}}{1 - r}$$

for all n. By Example 2.2.4 and Theorem 2.3.7, we deduce that

$$\lim_{n \to \infty} s_n = \frac{1}{1 - r}.$$

$\triangle$

Moreover if p denotes a polynomial and $\{s_n\}$ is a sequence converging to L, then Theorem 2.3.7 shows that

$$\lim_{n \to \infty} p(s_n) = p(L).$$

If p and q are polynomials such that $q(L) \neq 0$ and $q(s_n) \neq 0$ for all n, then it also follows that

$$\lim_{n \to \infty} \frac{p(s_n)}{q(s_n)} = \frac{p(L)}{q(L)}.$$

We show next that the study of a complex sequence can be reduced to the study of two real sequences.

Let $\{s_n\}$ be a complex sequence and suppose that $s_n = a_n + ib_n$ for all n, where each a_n and b_n is real. Then the sequences $\{a_n\}$ and $\{b_n\}$ are called the **real** and **imaginary** parts, respectively, of $\{s_n\}$. They satisfy the following theorem.

Theorem 2.3.11. *A complex sequence converges to a number L if and only if its real and imaginary parts converge, respectively, to the real and imaginary parts of L.*

Proof. Let $s_n = a_n + ib_n$ for all n, where each a_n and b_n is real. If the sequences $\{a_n\}$ and $\{b_n\}$ converge to A and B, respectively, then $\{s_n\}$ converges to $A + iB$, by Theorem 2.3.7.

Conversely, suppose that

$$\lim_{n \to \infty} s_n = L = A + iB,$$

where A and B are real. We must show that $\{a_n\}$ and $\{b_n\}$ converge to A and B, respectively. Choose $\varepsilon > 0$. There exists N such that $|s_n - L| < \varepsilon$ for all $n \geq N$. Recalling that $|\text{Re}(z)| \leq |z|$ for all $z \in \mathbb{C}$, we find that

$$|a_n - A| \leq |s_n - L| < \varepsilon$$

for all $n \geq N$. Similarly, $|b_n - B| < \varepsilon$ for all $n \geq N$, and the theorem follows. $\square$

Corollary 2.3.12. *If a complex sequence $\{s_n\}$ converges to a number L, then $\{\overline{s_n}\}$ converges to $\overline{L}$.*

The following algebraic result is also frequently used in calculations.

Theorem 2.3.13. *Let $\{s_n\}$ be a sequence of nonnegative real numbers converging to L. Then*

$$\lim_{n \to \infty} s_n^{1/m} = L^{1/m}$$

for every positive integer m.

Proof. Note that $L \geq 0$ by Corollary 2.3.10.

Suppose first that $L = 0$. For every $\varepsilon > 0$ there exists N such that $s_n = |s_n| < \varepsilon^m$ for all $n \geq N$. Thus $s_n^{1/m} < \varepsilon$ for all such n, and it follows that $s_n^{1/m} \to 0$ as $n \to \infty$.

Suppose therefore that $L > 0$, and choose $\varepsilon > 0$. There is an integer N such that $|s_n - L| < \varepsilon$ for all $n \geq N$. For each such n define $a_n = s_n^{1/m} > 0$, and let $b = L^{1/m} > 0$. Then for all $n \geq N$ it follows from Theorem 1.5.5 that

$$a_n^m - b^m = (a_n - b) \sum_{j=0}^{m-1} a_n^j b^{m-j-1}$$

$$= (a_n - b) K_n$$

even if $a_n = b$, where

$$K_n = \sum_{j=0}^{m-1} a_n^j b^{m-j-1}$$

$$= b^{m-1} + \sum_{j=1}^{m-1} a_n^j b^{m-j-1}.$$

Now $K_n \geq b^{m-1} > 0$ since $a_n^j b^{m-j-1} > 0$ for each $j > 0$. Therefore

$$|s_n^{1/m} - L^{1/m}| = |a_n - b|$$

$$= \frac{|a_n^m - b^m|}{K_n}$$

$$\leq \frac{|s_n - L|}{b^{m-1}}$$

$$< \frac{\varepsilon}{b^{m-1}},$$

as required. $\square$

Remark. If $s_n > 0$ for all n and $L > 0$, then the theorem also holds for each negative integer m, for if $m < 0$, then $-m > 0$ and

$$s_n^{1/m} = \frac{1}{s_n^{-1/m}} \to \frac{1}{L^{-1/m}} = L^{1/m}.$$

Example 2.3.4. Using Theorem 2.3.13, we find that

$$
\begin{aligned}
\lim_{n \to \infty} \left(\sqrt{n+1} - \sqrt{n} \right) &= \lim_{n \to \infty} \frac{n+1-n}{\sqrt{n+1} + \sqrt{n}} \\
&= \lim_{n \to \infty} \frac{1}{\sqrt{n+1} + \sqrt{n}} \\
&= \lim_{n \to \infty} \frac{\sqrt{\frac{1}{n}}}{\sqrt{1 + \frac{1}{n}} + 1} \\
&= \frac{0}{\sqrt{1+0} + 1} \\
&= 0.
\end{aligned}
$$

$\triangle$

Example 2.3.5. For all integers $n > 1$ we have

$$\frac{1}{n-1} = \frac{\frac{1}{n}}{1 - \frac{1}{n}} \to 0$$

as $n \to \infty$. Hence

$$\lim_{n \to \infty} \frac{1}{\sqrt{n-1}} = 0$$

by Theorem 2.3.13.

$\triangle$

We conclude this section by showing that taking an arithmetic mean does not alter the limit. In other words, the limit of a convergent sequence is also the limit of the arithmetic mean of the terms of the sequence. This result is due to Cauchy. We will give a corresponding result for geometric means later.

Theorem 2.3.14. *If* $\lim_{n \to \infty} s_n = L$, *then*

$$\lim_{n \to \infty} \frac{1}{n} \sum_{j=1}^{n} s_j = L.$$

Proof. Putting $t_n = s_n - L$ for all $n \in \mathbb{N}$, we find that

$$\frac{1}{n}\sum_{j=1}^{n} s_j = \frac{1}{n}\sum_{j=1}^{n}(t_j + L) = \frac{1}{n}\left(\sum_{j=1}^{n} t_j + \sum_{j=1}^{n} L\right) = \frac{1}{n}\sum_{j=1}^{n} t_j + L.$$

It therefore suffices to show that

$$\lim_{n\to\infty}\frac{1}{n}\sum_{j=1}^{n} t_j = 0.$$

Since

$$\lim_{n\to\infty} t_n = \lim_{n\to\infty}(s_n - L) = 0,$$

for each $\varepsilon > 0$ there is a positive integer N such that $|t_n| < \varepsilon$ for each $n \geq N$. For all $n \geq N$ we also have

$$\left|\frac{1}{n}\sum_{j=1}^{n} t_j\right| \leq \frac{1}{n}\sum_{j=1}^{n} |t_j|$$

$$= \frac{1}{n}\left(\sum_{j=1}^{N-1} |t_j| + \sum_{j=N}^{n} |t_j|\right).$$

Since $1/n \to 0$, we may choose n so large that

$$\frac{1}{n}\sum_{j=1}^{N-1} |t_j| < \varepsilon.$$

For large enough n it follows that

$$\left|\frac{1}{n}\sum_{j=1}^{n} t_j\right| \leq \frac{1}{n}\sum_{j=1}^{N-1} |t_j| + \frac{1}{n}\sum_{j=N}^{n} |t_j|$$

$$< \varepsilon + \frac{1}{n}\sum_{j=N}^{n} \varepsilon$$

$$= \varepsilon + \frac{n - (N-1)}{n}\varepsilon$$

$$\leq 2\varepsilon,$$

and the proof is complete.   □

Exercises 2.3.

1. Find the limits of the following sequences:

(a) $\left\{\dfrac{\sqrt{n}-1}{\sqrt{n}+1}\right\}$; (g) $\left\{\dfrac{\sqrt{n}+(n+1)i}{n+3}\right\}$;

(b) $\left\{\dfrac{n+3i}{n} + \left(\dfrac{1}{1+i}\right)^n\right\}$; (h) $\left\{\sqrt{\sqrt{n+k}-\sqrt{n}}\right\}$ where $k \geq 0$;

(c) $\left\{\dfrac{in}{n+2i}\right\}$; (i) $\left\{\dfrac{2n}{n+2} + \dfrac{i^n\sqrt{n}}{n+1}\right\}$;

(d) $\left\{\dfrac{n+i^n}{n+1}\right\}$, (j) $\left\{\sqrt{n^2+1}-n\right\}$;

(e) $\left\{\left(\dfrac{1}{4} + \dfrac{i}{2}\right)^n\right\}$; (k) $\left\{(\sqrt{n+1}-\sqrt{n})\sqrt{n}\right\}$;

(f) $\left\{\dfrac{(2n-1)!}{(2n+1)!}\right\}$; (l) $\left\{n - \sqrt{(n+a)(n+b)}\right\}$.

2. Suppose that $\lim_{n\to\infty} x_n = L > 0$. Show that there is a number N such that

$$9L < 10x_n < 11L$$

for all $n \geq N$. Obtain a similar result for $L < 0$.

2.4 Subsequences

Roughly speaking, a subsequence of a sequence is obtained by discarding some terms. Recall from the definition of an increasing function that a real sequence $\{s_n\}$ is increasing if $s_n < s_{n+1}$ for all n. In order to construct a subsequence $\{t_n\}$ of a sequence $\{s_n\}$, we use an increasing sequence of positive integers to pick out the terms of $\{s_n\}$ that are to appear in $\{t_n\}$. Thus we define a sequence $\{t_n\}$ to be a **subsequence** of a sequence $\{s_n\}$ if there is an increasing sequence $\{k_n\}$ of positive integers such that $t_n = s_{k_n}$ for all n.

Example 2.4.1. Let $\{s_n\}$ be the sequence $\{1/n^2\}$. Its terms are

$$1, \frac{1}{4}, \frac{1}{9}, \frac{1}{16}, \frac{1}{25}, \frac{1}{36}, \frac{1}{49}, \frac{1}{64}, \ldots$$

Then $\{s_{2n}\}$ is the subsequence

$$\frac{1}{4}, \frac{1}{16}, \frac{1}{36}, \frac{1}{64}, \ldots,$$

whereas $\{s_{2n-1}\}$ is the subsequence

$$1, \frac{1}{9}, \frac{1}{25}, \frac{1}{49}, \ldots$$

$\triangle$

We show next that subsequences enjoy the same convergence behavior as the parent sequence.

Theorem 2.4.1. *If a sequence converges to a number L, then so does each of its subsequences.*

Proof. Given that $\{s_n\}$ is a sequence that converges to L, for every $\varepsilon > 0$ there exists an integer N such that $|s_n - L| < \varepsilon$ for all $n \geq N$. If $\{s_{k_n}\}$ is a subsequence of $\{s_n\}$, then $|s_{k_n} - L| < \varepsilon$ whenever $k_n \geq N$. Since $\{k_n\}$ is an increasing sequence in $\mathbb{N}$, we have $k_n \geq k_N \geq N$ for all $n \geq N$. For all such n it therefore follows that $|s_{k_n} - L| < \varepsilon$, and the proof is complete.   □

This theorem provides a useful test for divergence of a sequence.

Corollary 2.4.2. *Any sequence possessing subsequences that converge to distinct limits must be divergent.*

Example 2.4.2. Let $s_n = i^n$ for all n. Since $s_{4n} = 1$ and $s_{4n+1} = i$ for all n, the sequence $\{s_n\}$ has subsequences converging to distinct limits and hence diverges.
                                                                                          △

Example 2.4.3. The sequence

$$\left\{\frac{1}{n} + (-1)^n\right\}$$

has subsequences

$$\left\{\frac{1}{2n} + 1\right\}$$

and

$$\left\{\frac{1}{2n+1} - 1\right\}$$

that converge to 1 and -1, respectively. The sequence is therefore divergent.   △

Theorem 2.4.3. *Let $\{s_n\}$ be a sequence. If the subsequences $\{s_{2n}\}$ and $\{s_{2n+1}\}$ converge to the same number L, then so does $\{s_n\}$.*

Proof. Choose $\varepsilon > 0$. There is an integer N_1 such that

$$|s_{2n} - L| < \varepsilon$$

for all $n \geq N_1$. Similarly, there is an N_2 such that

$$|s_{2n+1} - L| < \varepsilon$$

for all $n \geq N_2$. Now let

$$N = \max\{2N_1, 2N_2 + 1\}.$$

For all $n \geq N$ it follows that

$$|s_n - L| < \varepsilon.$$

$\square$

Exercises 2.4.

1. For all $n \in \mathbb{N}$ let

$$x_n = \begin{cases} \frac{n}{n^2+1} & \text{if } n \text{ is divisible by 3,} \\ \left(-\frac{1}{2}\right)^n & \text{otherwise.} \end{cases}$$

Show that $\lim_{n\to\infty} x_n = 0$.

2. Show that the sequence

$$\left\{ \left(\frac{n-1}{n+1}\right) i^n \right\}$$

is divergent.

3. Test the convergence of the following sequences:

(a) $\left\{ \frac{1+(-1)^n}{2+i^n} \right\}$;

(b) $\left\{ (-1)^n \left(1 + \frac{1}{n}\right) \right\}$.

2.5 The Sandwich Theorem

Sometimes a real sequence is flanked by two sequences that are known to converge to the same limit. It is natural to expect that the given sequence also converges to that limit. In this section we confirm that expectation. We begin with the following special case.

Lemma 2.5.1. *Let $\{a_n\}$ and $\{b_n\}$ be real sequences such that $0 \leq a_n \leq b_n$ for all n. If $\lim_{n\to\infty} b_n = 0$, then $\lim_{n\to\infty} a_n = 0$.*

Proof. Choose $\varepsilon > 0$. There is an integer N such that $|b_n| < \varepsilon$ for all $n \geq N$. For each such n it follows that $|a_n| \leq |b_n| < \varepsilon$. $\square$

Theorem 2.5.2 (Sandwich Theorem). *Let $\{a_n\}$, $\{b_n\}$, and $\{c_n\}$ be real sequences such that $a_n \leq b_n \leq c_n$ for all n. If*

$$\lim_{n \to \infty} a_n = \lim_{n \to \infty} c_n = L,$$

then

$$\lim_{n \to \infty} b_n = L.$$

Proof. The hypotheses show that

$$0 \leq b_n - a_n \leq c_n - a_n$$

for all n, and that

$$\lim_{n \to \infty} (c_n - a_n) = L - L = 0.$$

The previous lemma therefore shows that

$$\lim_{n \to \infty} (b_n - a_n) = 0.$$

Thus

$$
\begin{aligned}
\lim_{n \to \infty} b_n &= \lim_{n \to \infty} (b_n - a_n + a_n) \\
&= \lim_{n \to \infty} (b_n - a_n) + \lim_{n \to \infty} a_n \\
&= 0 + L \\
&= L.
\end{aligned}
$$

$\square$

Example 2.5.1. We shall show that

$$\lim_{n \to \infty} n^{1/n} = 1.$$

Let

$$s_n = n^{1/n} - 1$$

for all $n > 0$. It suffices to prove that the sequence $\{s_n\}$ is null. Note that $s_n \geq 0$ for all positive n. Therefore

$$n = (s_n + 1)^n > \frac{n(n-1)}{2} s_n^2$$

for all $n > 0$, by Eq. (1.3). Thus

$$s_n^2 < \frac{2}{n-1}$$

for all $n > 1$, and so

$$0 \leq s_n < \frac{\sqrt{2}}{\sqrt{n-1}}.$$

The required result now follows from the sandwich theorem, since

$$\lim_{n \to \infty} \frac{1}{\sqrt{n-1}} = 0$$

by Example 2.3.5.   △

Example 2.5.2. We show that

$$\lim_{n \to \infty} b^{1/n} = 1$$

for each constant $b > 0$.

Let us begin with the case where $b \geq 1$. Fix an integer $n \geq b$. Then

$$1 \leq b^{1/n} \leq n^{1/n}.$$

The required result follows in this case by using the previous example and the sandwich theorem.

For b such that $0 < b < 1$, we have $1/b > 1$. Hence

$$\lim_{n \to \infty} \frac{1}{b^{1/n}} = \lim_{n \to \infty} \left(\frac{1}{b}\right)^{1/n} = 1.$$

Therefore the desired result follows in this case as well.   △

Example 2.5.3. Let $0 < x < 1$. We know from Example 2.2.4 that

$$\lim_{n \to \infty} x^n = 0.$$

We now prove this result using the sandwich theorem.

Put

$$p = \frac{1-x}{x} > 0.$$

Then $xp = 1 - x$, so that $x(1 + p) = 1$; hence

$$x = \frac{1}{1+p}.$$

Using inequality (1.2), we conclude that

$$0 < x^n = \frac{1}{(1+p)^n} \le \frac{1}{1+np} < \frac{1}{np}.$$

Since

$$\lim_{n \to \infty} \frac{1}{np} = 0,$$

the desired result follows by the sandwich theorem.

If z is a complex number such that $|z| < 1$, it follows that

$$\lim_{n \to \infty} |z^n| = \lim_{n \to \infty} |z|^n = 0.$$

Therefore

$$\lim_{n \to \infty} z^n = 0$$

by Proposition 2.3.2(2).   △

Example 2.5.4. Given real numbers $x_1, x_2, \ldots, x_k$, define

$$s_n = \left(\sum_{j=1}^{k} |x_j|^n \right)^{1/n}$$

for all positive integers n. It is easy to show that the sequence $\{s_n\}$ converges to

$$M = \max\{|x_1|, |x_2|, \ldots, |x_k|\}.$$

Note first that

$$M^n \le \sum_{j=1}^{k} |x_j|^n \le kM^n.$$

The result now follows from the sandwich theorem applied to the inequalities

$$M = (M^n)^{1/n} \leq s_n \leq (kM^n)^{1/n} = k^{1/n}M$$

upon noting that

$$\lim_{n \to \infty} k^{1/n}M = M.$$

$\triangle$

We conclude this section with more applications of the sandwich theorem.

Theorem 2.5.3. *Suppose that $\{s_n\}$ and $\{t_n\}$ are sequences such that $\lim_{n \to \infty} s_n = 0$ and $\{t_n\}$ is bounded. Then*

$$\lim_{n \to \infty} s_n t_n = 0.$$

Proof. Since $\{t_n\}$ is bounded, there exists M such that $|t_n| < M$ for all n. Hence $0 \leq |s_n t_n| \leq M|s_n| \to 0$ as $n \to \infty$. Therefore $|s_n t_n| \to 0$ as $n \to \infty$, by the sandwich theorem, and the result follows. $\square$

Theorem 2.5.4. *Let $\{s_n\}$ be a sequence of nonzero real numbers and suppose that*

$$\lim_{n \to \infty} \left| \frac{s_{n+1}}{s_n} \right| = L < 1.$$

Then

$$\lim_{n \to \infty} s_n = 0.$$

Proof. Taking

$$\varepsilon = \frac{1-L}{2} > 0,$$

we find that there is an integer N such that

$$\left| \left| \frac{s_{n+1}}{s_n} \right| - L \right| < \frac{1-L}{2}$$

for each $n \geq N$. For each such n it follows that

$$\frac{|s_{n+1}|}{|s_n|} < L + \frac{1-L}{2} = \frac{1+L}{2},$$

and so

$$0 < |s_{n+1}| < \frac{1+L}{2}|s_n|.$$

An inductive argument therefore shows that

$$0 < |s_{N+k}| < \left(\frac{1+L}{2}\right)^k |s_N|$$

for every positive integer k. Now

$$\lim_{k\to\infty} \left(\frac{1+L}{2}\right)^k = 0$$

as $(1+L)/2 < 1$. The sandwich theorem therefore implies that

$$\lim_{k\to\infty} |s_{N+k}| = 0,$$

and the result in question follows immediately. □

Example 2.5.5. Let $s_n = n^p/c^n$ for all positive integers n, where p is any rational number and $c > 1$. Then

$$
\begin{aligned}
0 < \frac{s_{n+1}}{s_n} \\
= \frac{(n+1)^p}{c^{n+1}} \cdot \frac{c^n}{n^p} \\
= \frac{1}{c}\left(\frac{n+1}{n}\right)^p \\
= \frac{1}{c}\left(1 + \frac{1}{n}\right)^p \\
\to \frac{1}{c}
\end{aligned}
$$

as $n \to \infty$. As $1/c < 1$, the sequence $\{s_n\}$ converges to 0. △

Example 2.5.6. Let $b > 0$ and $s_n = b^n/n!$ for all nonnegative integers n. Then

$$0 < \frac{s_{n+1}}{s_n} = \frac{b^{n+1}}{(n+1)!} \cdot \frac{n!}{b^n} = \frac{b}{n+1}.$$

As the sequence $\{b/(n+1)\}$ converges to 0, so does $\{s_n\}$. △

Remark. The condition that

$$\left| \frac{s_{n+1}}{s_n} \right| < 1$$

for all n does not suffice to ensure that $\lim_{n\to\infty} s_n = 0$. For example, let

$$s_n = \frac{n}{n-1}$$

for all $n > 1$. Then

$$\lim_{n\to\infty} s_n = \lim_{n\to\infty} \frac{1}{1 - \frac{1}{n}} = 1,$$

but

$$\frac{s_{n+1}}{s_n} = \frac{n+1}{n} \cdot \frac{n-1}{n} = \frac{n^2 - 1}{n^2} < 1.$$

The sandwich theorem can be used to prove that the limit of a convergent sequence is also the limit of the geometric mean of the terms of the sequence. First we establish the following lemma.

Lemma 2.5.5. *Let $\{s_n\}$ be a sequence of positive numbers. For all integers $n > 0$ define*

$$H_n = n \left/ \sum_{j=1}^{n} \frac{1}{s_j} \right.,$$

$$G_n = \left(\prod_{j=1}^{n} s_j \right)^{1/n},$$

and

$$A_n = \frac{1}{n} \sum_{j=1}^{n} s_j.$$

Then $H_n \leq G_n \leq A_n$ for all n, with equality holding for a given n if and only if $s_1 = s_2 = \ldots = s_n$.

Proof. Certainly, $G_1 = A_1$. In order to prove that $G_n \leq A_n$ for a given $n > 1$, we use induction on the number k of subscripts $j \leq n$ for which $s_j \neq A_n$. If $k = 0$, then $s_j = A_n$ for all $j \leq n$. In this case $G_n = (A_n^n)^{1/n} = A_n$, as desired. Assume therefore that $k > 0$ and that the result holds whenever fewer than k of the first n terms of a sequence are different from the arithmetic mean of those terms. Since $k > 0$, we may assume without loss of generality that $s_1 < A_n$, the argument being similar if $s_1 > A_n$. As A_n is the average of $s_1, s_2, \ldots, s_n$, we may therefore assume, again without losing generality, that $s_2 > A_n$. The n numbers $A_n, s_1 + s_2 - A_n, s_3, s_4, \ldots, s_n$ have the same average A_n as $s_1, s_2, \ldots, s_n$ since they have the same sum, but fewer than k of them are different from A_n. It therefore follows from the inductive hypothesis that

$$(A_n(s_1 + s_2 - A_n)s_3 s_4 \ldots s_n)^{1/n} \le A_n.$$

But

$$A_n(s_1 + s_2 - A_n) - s_1 s_2 = A_n s_1 + A_n s_2 - A_n^2 - s_1 s_2$$
$$= (A_n - s_1)(s_2 - A_n)$$
$$> 0,$$

and so

$$s_1 s_2 < A_n(s_1 + s_2 - A_n).$$

Therefore

$$G_n = (s_1 s_2 \ldots s_n)^{1/n}$$
$$< (A_n(s_1 + s_2 - A_n)s_3 s_4 \ldots s_n)^{1/n}$$
$$\le A_n,$$

as required.

It follows that

$$0 < \left(\prod_{j=1}^{n} \frac{1}{s_j} \right)^{1/n} \le \frac{1}{n} \sum_{j=1}^{n} \frac{1}{s_j},$$

equality holding if and only if $s_1, s_2, \ldots, s_n$ are equal. Hence, taking reciprocals,

$$n \Big/ \sum_{j=1}^{n} \frac{1}{s_j} \le 1 \Big/ \left(\prod_{j=1}^{n} \frac{1}{s_j} \right)^{1/n} = \left(1 \Big/ \left(1 \Big/ \prod_{j=1}^{n} s_j \right) \right)^{1/n} = \left(\prod_{j=1}^{n} s_j \right)^{1/n},$$

and we have proved that $H_n \le G_n$. Once again, equality holds for n if and only if $s_1, s_2, \ldots, s_n$ are equal. $\qquad\square$

The proof that $G_n \le A_n$ was suggested to us by J. Hudson and is a modification of an argument due to Ehlers. The numbers H_n, G_n, and A_n are, respectively, the **harmonic**, **geometric**, and **arithmetic** means of the first n terms of the sequence.

Theorem 2.5.6. *Let $\{s_n\}$ be a sequence of positive numbers that converges to some number $L > 0$. Then*

$$\lim_{n \to \infty} \left(\prod_{j=1}^{n} s_j \right)^{1/n} = L.$$

Proof. With the notation of Lemma 2.5.5, Theorem 2.3.14 shows that $A_n \to L$ as $n \to \infty$. Moreover

$$\lim_{n \to \infty} \frac{1}{s_n} = \frac{1}{L},$$

and so the same theorem shows that

$$\frac{1}{H_n} = \frac{1}{n} \sum_{j=1}^{n} \frac{1}{s_j} \to \frac{1}{L}$$

as $n \to \infty$; hence $\lim_{n \to \infty} H_n = L$. By the sandwich theorem it follows that

$$\lim_{n \to \infty} \left(\prod_{j=1}^{n} s_j \right)^{1/n} = L.$$

$\square$

Exercises 2.5.

1. Test the sequence

$$\left\{ \frac{2^n}{n!} + \frac{i^n}{2^n} \right\}$$

for convergence.

2. Show that the sequence

$$\left\{ n^{1/n} + inw^n \right\}$$

is convergent if $|w| < 1$.

3. Suppose that $|a_n| \le b_n$ for all n and that

$$\lim_{n \to \infty} b_n = 0.$$

Show that

$$\lim_{n\to\infty} a_n = 0.$$

4. Show that the sequence

$$\left\{ \left(\frac{n}{2}\right)^{1/n} \right\}$$

converges.
5. Show that the sequence

$$\{(2^n + 3^n)^{1/n}\}$$

converges to 3.
6. Show that

$$\lim_{n\to\infty} \frac{1}{n} \sum_{j=1}^{n} j^{1/j} = 1.$$

7. Suppose that $x_n > 0$ for all $n \geq 0$ and that

$$\lim_{n\to\infty} \frac{x_{n+1}}{x_n} = L > 0.$$

Show that

$$\lim_{n\to\infty} x_n^{1/n} = L.$$

Is the converse true? (Hint: Use Theorem 2.5.6.)
8. For all $k \in \mathbb{N}$ show that

$$\lim_{n\to\infty} \left(\frac{nk}{n}\right)^{1/n} = \frac{k^k}{(k-1)^{k-1}},$$

where $0^0 = 1$. (Use question 7.)
9. Test the convergence of the following sequences:

(a) $\left\{ \sum_{j=0}^{n} \frac{n}{n^2+j} \right\}$;

(b) $\left\{ \sum_{j=1}^{n} \frac{j}{n^2+j} \right\}$;

(c) $\left\{ \sum_{j=0}^{n} \frac{n^2}{n^3+2n+j} \right\}$.

2.6   The Cauchy Principle

Sometimes it is possible to establish that a sequence is convergent without actually finding its limit. We shall show that a sequence is convergent if and only if its terms are ultimately close to one another even when they are not consecutive. This idea leads to the concept of a Cauchy sequence.

A sequence $\{s_n\}$ is called a **Cauchy sequence** if for every $\varepsilon > 0$ there is an integer N such that

$$|s_n - s_m| < \varepsilon$$

whenever $m \geq N$ and $n \geq N$. Of course, ε may be replaced in this definition by $c\varepsilon$ for any $c > 0$.

Cauchy sequences are characterized in the next theorem.

Theorem 2.6.1. *A sequence $\{s_n\}$ is Cauchy if and only if for all $\varepsilon > 0$ there exists an integer N such that*

$$|s_n - s_N| < \varepsilon$$

for all $n \geq N$.

Proof. The necessity follows immediately by taking $m = N$ in the definition. To prove the sufficiency of the stated condition, choose $\varepsilon > 0$. By hypothesis there is an integer N such that

$$|s_n - s_N| < \varepsilon$$

for all $n \geq N$. Choosing $m \geq N$ and $n \geq N$, we have

$$|s_n - s_m| = |s_n - s_N + s_N - s_m|$$
$$\leq |s_n - s_N| + |s_m - s_N|$$
$$< 2\varepsilon.$$

Therefore $\{s_n\}$ is a Cauchy sequence by definition.   □

We now establish some properties of Cauchy sequences.

Theorem 2.6.2. *Every Cauchy sequence is bounded.*

Proof. Let $\{s_n\}$ be a Cauchy sequence and choose $\varepsilon > 0$. By Theorem 2.6.1 there is an integer N such that

$$|s_n - s_N| < \varepsilon$$

for all $n \geq N$. The proof is now completed by the argument of Theorem 2.3.1, replacing L by s_N. $\square$

Since the terms of a Cauchy sequence are ultimately close to one another, it is intuitively clear that if some subsequence of the sequence converges to a number L, then the whole sequence must converge to L.

Theorem 2.6.3. *If $\{s_n\}$ is a Cauchy sequence with a subsequence that converges to L, then $\{s_n\}$ also converges to L.*

Proof. Let $\{s_{k_n}\}$ be a subsequence that converges to L and choose $\varepsilon > 0$. There exist N_1 and N_2 such that

$$|s_{k_n} - L| < \varepsilon$$

for each $n \geq N_1$ and

$$|s_n - s_m| < \varepsilon$$

whenever $m \geq N_2$ and $n \geq N_2$. Take

$$N = \max\{N_1, N_2\}.$$

Then for each $n \geq N$ we have $n \geq N_1$ and $k_n \geq n \geq N_2$ since $\{k_n\}$ is an increasing sequence of positive integers. Hence

$$
\begin{aligned}
|s_n - L| &= |s_n - s_{k_n} + s_{k_n} - L| \\
&\leq |s_n - s_{k_n}| + |s_{k_n} - L| \\
&< 2\varepsilon,
\end{aligned}
$$

and so

$$\lim_{n \to \infty} s_n = L.$$

$\square$

Theorem 2.6.4. *Every convergent sequence is Cauchy.*

Proof. Let $\{s_n\}$ be a sequence that converges to L and replace s_N by L in the proof of Theorem 2.6.1. $\square$

But is it true that every Cauchy sequence converges? It is possible to find a rational Cauchy sequence that converges to an irrational number. We have already seen this phenomenon in Example 2.3.2. (It is shown in Exercises 1.1 that $\sqrt{2}$ is irrational.) However, every real Cauchy sequence does converge to a real number. In fact, one way of defining real numbers is via Cauchy sequences of rational numbers, as is done in [10]. The first step is to define two sequences $\{s_n\}$ and $\{t_n\}$ to be **equivalent** if the sequence $\{s_n - t_n\}$ is null. It can be shown that this notion defines an equivalence relation on the set of all rational Cauchy sequences, and the real numbers are defined as the corresponding equivalence classes. The convergence of real Cauchy sequences and the fact that every real number is the limit of a rational Cauchy sequence can be established as consequences. The property that all real Cauchy sequences converge to real numbers is referred to as the **completeness** of the real number system.

Let x and y be real numbers such that $x < y$. From the definition of the real number field we see that $(x + y)/2$ is a limit of a Cauchy sequence of rational numbers. Hence the interval (x, y) contains at least one rational number. Consequently, the interval $(x/\sqrt{2}, y/\sqrt{2})$ contains a rational number r. We may assume that $r \neq 0$: If $x < 0 < y$, then replace y by 0. Thus the interval (x, y) contains the irrational number $r\sqrt{2}$. We conclude that every open interval (x, y), where $x < y$, contains at least one rational number and at least one irrational number. This condition is described as the **density** property of the real number system.

The completeness property of real numbers is equivalent to another property known as the supremum property. In order to explain it, we need some additional definitions. A set S of real numbers is said to be **bounded above** if there exists a real number b such that $s \leq b$ for all $s \in S$. Any number b with this property is called an **upper bound** of S. For example, 0 is an upper bound for the set of all negative real numbers. Any positive number is also an upper bound for that set. An upper bound b of S is the **least upper bound** or **supremum** of S if $b \leq c$ for every upper bound c of S. The least upper bound of S, if it exists, is unique, for if b and c are both least upper bounds of S, then $b \leq c$ and $c \leq b$. It is denoted by sup S. For example, if S is the set of all negative numbers, then sup $S = 0$. Sets that are **bounded below, lower bounds** of such sets, and **greatest lower bounds** or **infima** are defined analogously. The greatest lower bound of a set S, if it exists, is unique and is denoted by inf S. If S is the range of a bounded sequence $\{s_n\}$, then its least upper bound is also denoted by sup$\{s_n\}$ and its greatest lower bound by inf$\{s_n\}$.

The **supremum property** alluded to above asserts that every nonempty set of real numbers that is bounded above has a least upper bound. The proof of the equivalence of the completeness and supremum properties is our next goal. Before tackling it, however, we insert an example of a Cauchy sequence.

Example 2.6.1. Let a and b be complex numbers. Define $s_0 = a$, $s_1 = b$, and

$$s_{n+1} = \frac{s_n + s_{n-1}}{2}$$

for all $n > 0$. If $a = b$, then $s_n = a$ for all n. Suppose $a \neq b$. It is not hard to show by induction that

$$s_{n+1} - s_n = (-1)^n \frac{b - a}{2^n} \qquad (2.3)$$

for all $n \geq 0$. For every m and n such that $n > m$, the use of the telescoping property combined with the triangle inequality shows that

$$
\begin{aligned}
|s_n - s_m| &= \left| \sum_{j=m}^{n-1} (s_{j+1} - s_j) \right| \\
&\leq \sum_{j=m}^{n-1} |s_{j+1} - s_j| \\
&= \sum_{j=m}^{n-1} \frac{|b - a|}{2^j} \\
&= \frac{|b - a|}{2^m} \sum_{j=m}^{n-1} \frac{1}{2^{j-m}} \\
&< \frac{|b - a|}{2^{m-1}},
\end{aligned}
$$

since

$$
\sum_{j=m}^{n-1} \frac{1}{2^{j-m}} = \sum_{j=0}^{n-m-1} \frac{1}{2^j} = \frac{1 - \frac{1}{2^{n-m}}}{1 - \frac{1}{2}} < \frac{1}{\frac{1}{2}} = 2.
$$

Example 2.2.4 shows that $|b - a|/2^{m-1} \to 0$ as $m \to \infty$, and so for each $\varepsilon > 0$ we may choose N such that

$$
\frac{|b - a|}{2^{N-1}} < \varepsilon.
$$

For all $m \geq N$ and $n > m \geq N$, it follows that

$$
|s_n - s_m| < \frac{|b - a|}{2^{m-1}} \leq \frac{|b - a|}{2^{N-1}} < \varepsilon.
$$

Therefore $\{s_n\}$ is a Cauchy sequence. By the completeness property it converges to some number L.

In order to find L, one might attempt to make use of the recurrence relation that defines the sequence, but such an attempt leads to the equation

$$
L = \frac{L + L}{2} = L,
$$

which does not give any information about L. Instead, we use induction to show from Eq. (2.3) that for all $k \geq 0$,

$$s_{2k} = a + \sum_{j=0}^{k-1} \frac{b-a}{2^{2j+1}}$$

$$= a + \frac{b-a}{2} \sum_{j=0}^{k-1} \frac{1}{4^j}$$

$$= a + \frac{b-a}{2} \left(\frac{1 - \frac{1}{4^k}}{1 - \frac{1}{4}} \right)$$

$$= a + \frac{b-a}{2} \cdot \frac{4}{3} \cdot \left(1 - \frac{1}{4^k} \right)$$

$$\rightarrow a + \frac{2(b-a)}{3}$$

$$= \frac{a + 2b}{3}$$

as $k \rightarrow \infty$. By Theorem 2.4.1 it follows that

$$\lim_{n \to \infty} s_n = \frac{a + 2b}{3}.$$

$\triangle$

Theorem 2.6.5. *The completeness property for real numbers implies the supremum property.*

Proof. Let S be a nonempty set of real numbers that is bounded above. We must show that S has a least upper bound. If an element $s \in S$ happens to be an upper bound of S, then s is in fact the least upper bound of S. We may therefore assume that no member of S is an upper bound of S.

Since S is nonempty and bounded above, we may choose $a_0 \in S$ and an upper bound b_0 of S. By assumption, a_0 is not an upper bound of S. If $(a_0 + b_0)/2$ is an upper bound of S, then let $a_1 = a_0$ and $b_1 = (a_0 + b_0)/2$; otherwise let $a_1 = (a_0 + b_0)/2$ and $b_1 = b_0$. In both cases b_1 is an upper bound of S, but a_1 is not. Proceeding inductively, we obtain a nested sequence

$$[a_0, b_0] \supseteq [a_1, b_1] \supseteq [a_2, b_2] \supseteq \cdots$$

of closed intervals such that no a_n is an upper bound of S, but each b_n is.

We shall establish that the sequence $\{b_n\}$ is Cauchy. First, an inductive argument shows that

$$b_n - a_n = \frac{b_0 - a_0}{2^n}$$

and

$$0 \leq b_n - b_{n+1} \leq \frac{b_0 - a_0}{2^{n+1}}$$

for all $n \geq 0$. For each positive integer p it therefore follows from the telescoping property that

$$|b_{n+p} - b_n| = b_n - b_{n+p}$$

$$= \sum_{j=0}^{p-1}(b_{n+j} - b_{n+j+1})$$

$$\leq \sum_{j=0}^{p-1} \frac{b_0 - a_0}{2^{n+j+1}}$$

$$= \frac{b_0 - a_0}{2^{n+1}} \sum_{j=0}^{p-1} \frac{1}{2^j}$$

$$< \frac{b_0 - a_0}{2^n},$$

since

$$\sum_{j=0}^{p-1} \frac{1}{2^j} = \frac{1 - \frac{1}{2^p}}{1 - \frac{1}{2}} < 2.$$

Example 2.2.4 shows that $1/2^n \to 0$ as $n \to \infty$. Given $\varepsilon > 0$, we may therefore choose N such that

$$\frac{b_0 - a_0}{2^N} < \varepsilon.$$

For all $n \geq N$ and all positive integers p it follows that

$$|b_{n+p} - b_n| < \frac{b_0 - a_0}{2^n} \leq \frac{b_0 - a_0}{2^N} < \varepsilon,$$

and we infer that $\{b_n\}$ is indeed a Cauchy sequence. A similar argument demonstrates that $\{a_n\}$ is also a Cauchy sequence.

By the completeness principle $\{a_n\}$ and $\{b_n\}$ converge to some numbers a and b, respectively. For all n we have $a_n \leq a_{n+1} < b_{n+1} \leq b_n$, and so

$$a_n \leq a \leq b \leq b_n,$$

by Corollary 2.3.10. Therefore $b_n - a_n \geq b - a$. If $b > a$, then there is an n such that

$$b_n - a_n = \frac{b_0 - a_0}{2^n} < b - a.$$

This contradiction shows that $a = b$.

We shall establish that b is the required least upper bound of S. In order to show that it is an upper bound, choose $s \in S$ and suppose that $s > b$. Since $b_n \to b$, we may use Proposition 2.3.4 to choose n such that $b_n < s$, in contradiction to the fact that b_n is an upper bound of S. Thus $s \leq b$ for all $s \in S$, and so b is indeed an upper bound of S.

Next, choose $t < b = a$. The proof will be completed by showing that t is not an upper bound of S. Since $a_n \to a > t$, we may choose n such that $t < a_n$. As a_n is not an upper bound of S, neither is t. Thus b is the least upper bound of S, and the proof is complete. $\square$

Corollary 2.6.6. *Any nonempty set of real numbers that is bounded below has a greatest lower bound.*

Proof. Let S be a nonempty set that is bounded below, and apply the supremum property to the set $\{-s \mid s \in S\}$. $\square$

We now know that the completeness property implies the supremum property. We prove next that the converse is also true. Throughout this discussion we therefore assume that the supremum property holds. First we show that, under this assumption, bounded real sequences always possess convergent subsequences.

Theorem 2.6.7. *Every bounded sequence of real numbers has a convergent subsequence.*

Proof. Let $\{s_n\}$ be a bounded sequence of real numbers. If its range is a finite set, then some term b is repeated infinitely many times. In other words, there are positive integers $k_1, k_2, \ldots$ such that $\{k_n\}$ is an increasing sequence and $s_{k_n} = b$ for all n. Hence $\lim_{n \to \infty} s_{k_n} = b$.

Suppose therefore that the range $\mathcal{R}_s$ of $\{s_n\}$ is infinite. Since the sequence is bounded, we may choose m and M such that $m < s_n < M$ for all n. Let V be the set of all $x \in [m, M)$ such that $\mathcal{R}_s \cap (x, M)$ is infinite. Certainly, $m \in V$, so that $V \neq \emptyset$. But V is also bounded, and so the supremum property implies that V has a least upper bound $b \leq M$. We distinguish two cases.

Case 1: Suppose $b = M$. Since $b - 1$ is not an upper bound of V, there is a $v_1 \in V$ such that $b - 1 < v_1 < M$. Thus $\mathcal{R}_s \cap (v_1, M)$ is infinite, and so we may choose k_1 such that $s_{k_1} \in (v_1, M)$. Similarly, $b - 1/2$ is not an upper bound of V, and so there exists $v_2 \in V$ such that $b - 1/2 < v_2 < M$. Moreover $\mathcal{R}_s \cap (v_2, M)$ is infinite, so that there exists $k_2 > k_1$ such that $s_{k_2} \in (v_2, M)$. Continuing inductively, we construct an increasing sequence $\{k_n\}$ of positive integers such that $s_{k_n} \in (b - 1/n, M)$ for all $n > 0$. Thus

$$b - \frac{1}{n} < s_{k_n} < M = b.$$

By the sandwich theorem it follows that

$$\lim_{n \to \infty} s_{k_n} = b.$$

Case 2: If $b < M$, then we may choose N such that

$$b < b + \frac{1}{N} < M.$$

Thus

$$b < b + \frac{1}{N + n} < M$$

for all $n \geq 0$. Furthermore, as $b - 1/(N + n)$ is not an upper bound of V, there exists $u_n \in V$ such that

$$b - \frac{1}{N + n} < u_n \leq b.$$

Since b is an upper bound of V, we have $b + 1/(N + n) \notin V$, and so

$$\mathcal{R}_s \cap \left(b + \frac{1}{N + n}, M \right)$$

is finite since

$$m \leq b < b + \frac{1}{N + n} < M.$$

But $\mathcal{R}_s \cap (u_n, M)$ is infinite since $u_n \in V$, and as $b - 1/(N + n) < u_n$, we infer that

$$\mathcal{R}_s \cap \left(b - \frac{1}{N + n}, b + \frac{1}{N + n} \right)$$

is infinite. Putting $n = 0$, we may therefore choose k_0 so that

$$b - \frac{1}{N} < s_{k_0} < b + \frac{1}{N},$$

and for each $n > 0$ there exists $k_n > k_{n-1}$ such that

$$b - \frac{1}{N+n} < s_{k_n} < b + \frac{1}{N+n}.$$

It follows from the sandwich theorem that

$$\lim_{n \to \infty} s_{k_n} = b.$$

$\square$

Corollary 2.6.8. *Every bounded sequence of complex numbers has a convergent subsequence.*

Proof. Let $\{z_n\}$ be a bounded sequence of complex numbers. Then, for each n, it follows that $z_n = x_n + iy_n$ for some real numbers x_n and y_n. Using the facts that $|\text{Re}(z)| \leq |z|$ and $|\text{Im}(z)| \leq |z|$ for each complex number z, we deduce that $\{x_n\}$ and $\{y_n\}$ are also bounded. By Theorem 2.6.7 $\{x_n\}$ has a subsequence, $\{x_{k_n}\}$, that converges to some number a. Similarly, $\{y_{k_n}\}$ has a subsequence, $\{y_{k_{l_n}}\}$, that converges to some number b. The subsequence $\{x_{k_{l_n}}\}$ of $\{x_{k_n}\}$ also converges to a, by Theorem 2.4.1. Hence $\{z_{k_{l_n}}\}$ converges to $a + ib$. $\square$

We have proved that if the supremum property holds, then every bounded sequence contains a convergent subsequence. Since every Cauchy sequence is bounded, a Cauchy sequence $\{s_n\}$ therefore has a convergent subsequence. It follows by Theorem 2.6.3 that $\{s_n\}$ converges as well. In other words, the supremum property implies completeness. For real numbers these two properties therefore imply each other. Henceforth we assume they both hold, referring the reader to [10] for a proof.

We now have the following theorem.

Theorem 2.6.9 (Cauchy Principle). *A real sequence is convergent if and only if it is a Cauchy sequence.*

In fact, by considering separately the real and imaginary parts of the terms of a complex sequence, one can show that this principle holds even for complex sequences. The details are left as an exercise.

Roughly speaking, the Cauchy principle tells us that convergent sequences are those whose terms are getting closer and closer. In particular, for a sequence to converge it is necessary that the distance between successive terms diminishes. In fact, if $\{s_n\}$ is convergent, then

$$\lim_{n \to \infty} (s_{n+1} - s_n) = 0.$$

However, the converse is not true, as we see in the next example.

Example 2.6.2. Let

$$s_n = \sum_{j=1}^{n} \frac{1}{j}$$

for all n. We show that $\{s_n\}$ is divergent.

Suppose that $\lim_{n\to\infty} s_n = s$. Then

$$t_n = \sum_{j=1}^{n} \frac{1}{2j} = \frac{1}{2} \sum_{j=1}^{n} \frac{1}{j} \to \frac{s}{2}$$

as $n \to \infty$, and

$$u_n = \sum_{j=1}^{n} \frac{1}{2j-1} = s_{2n} - t_n \to s - \frac{s}{2} = \frac{s}{2}$$

as $n \to \infty$. These results cannot both hold: Since

$$\frac{1}{2j-1} - \frac{1}{2j} > 0$$

for all $j > 0$, we have

$$\lim_{n\to\infty} (u_n - t_n) > u_1 - t_1 = \frac{1}{2} > 0.$$

Therefore $\{s_n\}$ is indeed divergent. However,

$$s_{n+1} - s_n = \frac{1}{n+1} \to 0$$

as $n \to \infty$.   △

Example 2.6.3. The sequence $\{z^n\}$, where z is a complex number, is convergent if and only if $|z| < 1$ or $z = 1$.

Proof. We have shown in Example 2.2.4 that if $|z| < 1$, then the sequence is convergent. It is certainly so if $z = 1$.

Conversely, suppose that $|z| \geq 1$ and $z \neq 1$. Then

$$|z^{n+1} - z^n| = |z|^n |z-1| \geq |z-1| > 0.$$

Therefore $z^{n+1} - z^n$ does not approach 0 as $n \to \infty$. We conclude that the sequence is divergent.   △

A sequence is injective if and only if its terms are distinct. The following theorem concerns the existence of an injective convergent sequence.

Theorem 2.6.10 (Bolzano, Weierstrass). *Every bounded infinite set of numbers contains an injective convergent sequence.*

Proof. Any infinite set S of numbers certainly contains an injective sequence. If S is bounded, then this sequence is also bounded and therefore contains a convergent subsequence that is necessarily injective.   □

A number b is a **limit point** or an **accumulation point** of a set S if there is an injective sequence $\{s_n\}$ in S that converges to b. It follows from the Bolzano–Weierstrass theorem that every bounded infinite set of numbers has a limit point.

The following theorem gives an alternative definition of a limit point. For every $\varepsilon > 0$ we define

$$N_\varepsilon^*(b) = N_\varepsilon(b) - \{b\}.$$

Theorem 2.6.11. *Let S be a set of numbers and b a number. The following two statements are equivalent:*

1. *b is a limit point of S;*
2. *for each $\varepsilon > 0$,*

$$S \cap N_\varepsilon^*(b) \neq \emptyset.$$

Proof. Let b be a limit point of S. Then there exists an injective sequence $\{s_n\}$ in S converging to b. Choose $\varepsilon > 0$. There exists N such that

$$|s_n - b| < \varepsilon$$

for all $n \geq N$. As $\{s_n\}$ is injective, there exists $n \geq N$ such that $s_n \neq b$. Thus $s_n \in N_\varepsilon^*(b)$, as required.

Conversely, suppose (2) holds. Letting $\varepsilon_0 = 1$, by hypothesis we may choose $s_0 \in S \cap N_{\varepsilon_0}^*(b)$. Thus $0 < |s_0 - b| < 1$. Continuing by induction, suppose that k is a positive integer and that $s_0, s_1, \ldots, s_{k-1}$ are distinct real numbers satisfying

$$0 < |s_{k-1} - b| < \frac{1}{2^{k-1}}$$

and

$$|s_{k-1} - b| < |s_j - b|$$

for each nonnegative integer $j < k - 1$. Let

$$\varepsilon_k = \frac{|s_{k-1} - b|}{2} > 0.$$

By hypothesis, there exists $s_k \in S \cap N_{\varepsilon_k}^*(b)$. Thus

$$|s_k - b| < \varepsilon_k < |s_{k-1} - b| < |s_j - b|$$

for each $j < k - 1$. Therefore $|s_k - b| < |s_j - b|$ for each $j < k$, so that $s_k \neq s_j$ for each such j. Moreover

$$0 < |s_k - b| < \frac{|s_{k-1} - b|}{2} < \frac{1}{2^k}.$$

We have now defined an injective sequence $\{s_n\}$ in S such that

$$0 < |s_n - b| < \frac{1}{2^n}$$

for each n. By the sandwich theorem, the sequence $\{s_n - b\}$ converges to 0, and so

$$\lim_{n \to \infty} s_n = b.$$

$\square$

A subset of $\mathbb{C}$ is **closed** if it contains all its limit points.

Example 2.6.4. Let a and b be real numbers with $a < b$. We can confirm that the interval $I = [a, b]$ satisfies this definition, thereby establishing agreement of this terminology with that introduced earlier in connection with intervals. Indeed, choose $z = (x, y) \in \mathbb{C} - I$. If $y \neq 0$, then

$$N_\varepsilon^*(z) \cap I = \emptyset \tag{2.4}$$

for each positive $\varepsilon < |y|$. Suppose therefore that z is real. If $z < a$, then Eq. (2.4) holds for every positive $\varepsilon < a - z$. If $z > b$, then apply the same argument with $0 < \varepsilon < z - b$. We conclude from Theorem 2.6.11 that I is closed.

A similar argument may be applied to show that the rectangle $R = [a, b] \times [c, d]$ is a closed set in $\mathbb{R}^2 = \mathbb{C}$. Here $a < b$ and $c < d$, and R is the set of points (x, y) such that $a \leq x \leq b$ and $c \leq y \leq d$. Choose a point $z = (x, y) \in \mathbb{C} - R$. Then for some $r > 0$ at least one of the following cases arises:

$$x = b + r, x = a - r, y = d + r, y = c - r.$$

Choose such an r as small as possible, and let $\varepsilon = r/2 > 0$. Then $N_\varepsilon(z) \cap R = \emptyset$, and so z is not a limit point of R. $\triangle$

Exercises 2.6.

1. Show from the definition that the sequence $\{1/n^2\}$ is Cauchy.
2. Show that if for each $\varepsilon > 0$ there exists N such that $|x_n - x_N| < \varepsilon$ whenever $n > N$, then $\{x_n\}$ is a Cauchy sequence.
3. (a) Let $\{x_n\}$ be a sequence and suppose there exists a constant $r \in (0, 1)$ such that

$$|x_{n+1} - x_n| \leq r^n$$

for all n. Show that $\{x_n\}$ is Cauchy.

(b) Give an example of a divergent sequence $\{y_n\}$ with the property that

$$\lim_{n \to \infty} (y_{n+1} - y_n) = 0.$$

4. Show that if

$$|x_{n+2} - x_{n+1}| \le r|x_{n+1} - x_n|$$

for some $r \in (0, 1)$ and all n, then $\{x_n\}$ is Cauchy.
5. Show that if

$$\lim_{n \to \infty} \sup\{|x_j - x_k| \mid j > k \ge n\} = 0,$$

then $\{x_n\}$ is Cauchy.
6. Let $x_n \ne 0$ for all n and suppose that $\lim_{n \to \infty} x_n \ne 0$. Prove that

$$\inf\{|x_n| \mid n \in \mathbb{N}\} > 0.$$

7. Show that a sequence $\{x_n\}$ is Cauchy if and only if for each $\varepsilon > 0$ there exists N such that

$$\frac{|x_n - x_m|}{1 + |x_n - x_m|} < \varepsilon$$

whenever $n > m \ge N$.

2.7 Monotonic Sequences

We now study an important family of sequences whose convergence can often be determined without any knowledge of their limits, namely, monotonic sequences. We begin with some examples that illustrate techniques for establishing the monotonicity of sequences. Note that a sequence $\{s_n\}$ is increasing if and only if $\{-s_n\}$ is decreasing. Similarly, $\{s_n\}$ is nondecreasing if and only if $\{-s_n\}$ is nonincreasing.

Example 2.7.1. Let us show that the sequence $\{s_n\}$ is decreasing, where

$$s_n = \frac{3n + 1}{2n - 3}$$

for all $n \ge 2$.

One approach is to show that $s_{n+1} - s_n < 0$ for all $n \geq 2$. This result follows from the calculation

$$
\begin{aligned}
s_{n+1} - s_n &= \frac{3(n+1)+1}{2(n+1)-3} - \frac{3n+1}{2n-3} \\
&= \frac{3n+4}{2n-1} - \frac{3n+1}{2n-3} \\
&= \frac{-11}{(2n-1)(2n-3)} \\
&< 0.
\end{aligned}
$$

An alternative is to demonstrate that $s_{n+1}/s_n < 1$ for all $n \geq 2$:

$$
\begin{aligned}
\frac{s_{n+1}}{s_n} &= \frac{3n+4}{2n-1} \cdot \frac{2n-3}{3n+1} \\
&= \frac{6n^2 - n - 12}{6n^2 - n - 1} \\
&< 1.
\end{aligned}
$$

$\triangle$

Example 2.7.2. We show by induction that the sequence $\{s_n\}$ is increasing, where $s_1 = 1$ and

$$
s_n = \sqrt{s_{n-1} + 1}
$$

for all $n > 1$.

Certainly, $s_2 = \sqrt{2} > 1 = s_1$. Suppose $s_n > s_{n-1}$ for some $n \geq 2$. Then

$$
s_{n+1} = \sqrt{s_n + 1} > \sqrt{s_{n-1} + 1} = s_n,
$$

as required. $\triangle$

Our next goal is to show that a monotonic sequence is convergent if and only if it is bounded. We need the following notions concerning unbounded sequences.

Definition 2.7.1. Let $\{s_n\}$ be a real sequence. We write

$$
\lim_{n \to \infty} s_n = \infty
$$

if for each M there exists N such that $s_n > M$ for all $n \geq N$. We also say that s_n **approaches infinity** as n approaches infinity, and we write $s_n \to \infty$ as $n \to \infty$.

Likewise, we write

$$\lim_{n \to \infty} s_n = -\infty$$

if for each M there exists N such that $s_n < M$ for all $n \geq N$. In this case we say that s_n **approaches minus infinity** as n approaches infinity, and we write $s_n \to -\infty$ as $n \to \infty$.

Remark 1. Clearly, we may assume that $M > 0$ in the former definition and that $M < 0$ in the latter. We may also replace the inequalities $s_n > M$ and $s_n < M$ by $s_n > cM$ and $s_n < cM$, respectively, for all $c \neq 0$.

Remark 2. Let $t_n = -s_n$ for all n. Then $t_n < M$ if and only if $s_n > -M$. Consequently $s_n \to \infty$ as $n \to \infty$ if and only if $t_n \to -\infty$ as $n \to \infty$.

Theorem 2.7.1. *Let $\{s_n\}$ be a nondecreasing sequence. Then*

$$\lim_{n \to \infty} s_n = \sup\{s_n\}$$

if $\{s_n\}$ is bounded, and

$$\lim_{n \to \infty} s_n = \infty$$

otherwise.

Proof. Suppose first that $\{s_n\}$ is bounded. Then its range has a least upper bound L. We must show that $s_n \to L$.

Choose $\varepsilon > 0$. Then $L - \varepsilon$ is not an upper bound of $\{s_n\}$. Therefore there exists N such that $s_N > L - \varepsilon$. As $\{s_n\}$ is nondecreasing, $s_n \geq s_N$ for all $n \geq N$. Thus $s_n > L - \varepsilon$ for all such n. Hence

$$L - \varepsilon < s_n < L + \varepsilon$$

for all $n \geq N$, and we conclude that

$$\lim_{n \to \infty} s_n = L.$$

Finally, suppose the sequence is not bounded. Since s_1 is a lower bound, $\{s_n\}$ must not be bounded above. Thus for each $M > 0$ there exists N such that $s_N > M$. As $\{s_n\}$ is nondecreasing, we have $s_n \geq s_N > M$ for all $n \geq N$, and it follows that

$$\lim_{n \to \infty} s_n = \infty.$$

$\square$

The next result can be proved in a similar way or by applying the previous result to the sequence $\{-s_n\}$.

Theorem 2.7.2. *Let $\{s_n\}$ be a nonincreasing sequence. Then*

$$\lim_{n \to \infty} s_n = \inf\{s_n\}$$

if $\{s_n\}$ is bounded, and

$$\lim_{n \to \infty} s_n = -\infty$$

otherwise.

Example 2.7.3. In Example 2.7.1 we saw that the sequence $\{s_n\}$ is decreasing, where

$$s_n = \frac{3n + 1}{2n - 3}$$

for all $n \geq 2$. This sequence is bounded below by 0 and therefore converges to its greatest lower bound. In fact, its limit is easily calculated:

$$s_n = \frac{3 + \frac{1}{n}}{2 - \frac{3}{n}} \to \frac{3}{2}$$

as $n \to \infty$.   △

Example 2.7.4. In Example 2.7.2 we saw that the sequence $\{s_n\}$ is increasing, where $s_1 = 1$ and

$$s_n = \sqrt{s_{n-1} + 1} \tag{2.5}$$

for all $n > 1$. It is also easy to see by induction that the sequence is bounded above by 2 and therefore converges to some number L. In order to find L, we begin by taking limits of both sides of Eq. (2.5), thereby obtaining

$$L = \sqrt{L + 1}.$$

Hence

$$L^2 - L - 1 = 0,$$

and so

$$L = \frac{1 \pm \sqrt{5}}{2}.$$

But $s_n > 0$ for all n, and so we conclude that

$$\lim_{n \to \infty} s_n = \frac{1 + \sqrt{5}}{2}.$$

△

Example 2.7.5. In Example 2.1.8 we studied the sequence $\{s_n\}$, where $s_1 = 1$ and

$$s_n = \frac{s_{n-1} + k}{s_{n-1} + 1} \tag{2.6}$$

for all $n > 1$. Here k is a positive integer. We saw in Example 2.3.2 that this sequence converges to $\sqrt{k}$. We now present a different proof. Note that the sequence is the constant sequence $\{1\}$ if $k = 1$, and so we assume that $k > 1$. Moreover $s_n > 0$ for all n, and $s_1 < \sqrt{k}$.

Suppose that

$$\sqrt{k} < s_n = \frac{s_{n-1} + k}{s_{n-1} + 1}$$

for some $n > 1$. Then

$$\sqrt{k} \cdot s_{n-1} + \sqrt{k} < s_{n-1} + k,$$

so that

$$(\sqrt{k} - 1)s_{n-1} < k - \sqrt{k} = \sqrt{k}(\sqrt{k} - 1);$$

hence

$$s_{n-1} < \sqrt{k}.$$

Similarly, if $s_n < \sqrt{k}$, then $s_{n-1} > \sqrt{k}$. Since $s_1 < \sqrt{k}$, it follows that $s_n < \sqrt{k}$ if n is odd and $s_n > \sqrt{k}$ if n is even. Thus the subsequence $\{s_{2n}\}$ is bounded below by $\sqrt{k}$, whereas $\{s_{2n+1}\}$ is bounded above by $\sqrt{k}$. Moreover it is easy to see from the recurrence relation (2.6) that

$$s_{n+2} = \frac{s_{n+1} + k}{s_{n+1} + 1} = \frac{\frac{s_n + k}{s_n + 1} + k}{\frac{s_n + k}{s_n + 1} + 1} = \frac{(k + 1)s_n + 2k}{2s_n + k + 1} \tag{2.7}$$

for all n, and so

$$s_{n+2} - s_n = \frac{(k + 1)s_n + 2k}{2s_n + k + 1} - s_n = \frac{2(k - s_n^2)}{2s_n + k + 1}.$$

Thus $s_{n+2} > s_n$ if $s_n < \sqrt{k}$, whereas $s_{n+2} < s_n$ if $s_n > \sqrt{k}$. Hence $\{s_{2n}\}$ is a decreasing subsequence and $\{s_{2n+1}\}$ is increasing. Both of these subsequences therefore converge.

Let L be the limit of the subsequence $\{s_{2n}\}$. From Eq. (2.7) we see that

$$L = \frac{(k+1)L + 2k}{2L + k + 1},$$

whence

$$2L^2 + (k+1)L = (k+1)L + 2k,$$

so that $L = \pm\sqrt{k}$. But $L \geq 0$, and so $L = \sqrt{k}$. The same argument shows that the subsequence $\{s_{2n+1}\}$ also converges to $\sqrt{k}$. Since both subsequences converge to $\sqrt{k}$, so does the sequence $\{s_n\}$, by Theorem 2.4.3.   △

In the next example we use a sequence to obtain an algorithm that was used in Babylon around 1700 BC for finding the square root of a positive number. It is sometimes known as the divide-and-average method and can also be derived from Newton's method, which we shall study in Sect. 9.3.

Example 2.7.6. Let k and a_1 be positive numbers, and for each $n \in \mathbb{N}$ define

$$a_{n+1} = \frac{1}{2}\left(a_n + \frac{k}{a_n}\right) = \frac{a_n^2 + k}{2a_n}.$$

We shall show that $a_n \to \sqrt{k}$ as $n \to \infty$.

Certainly, $a_n > 0$ for all n. Moreover

$$a_n^2 - 2a_{n+1}a_n + k = 0. \tag{2.8}$$

This equation must have a real solution for a_n, and so the discriminant

$$4a_{n+1}^2 - 4k$$

must be nonnegative. Hence $a_{n+1}^2 \geq k$, and since $a_{n+1} > 0$, it follows that

$$a_{n+1} \geq \sqrt{k} \tag{2.9}$$

for all n.

The sequence (excluding the first term) is nonincreasing: For all $n > 1$, we have

$$a_{n+1} - a_n = \frac{a_n^2 + k}{2a_n} - a_n = \frac{k - a_n^2}{2a_n} \leq 0.$$

This nonincreasing sequence, which is bounded below, therefore converges to some number $L \geq 0$. From Eq. (2.8) we see that $L^2 = k$. Since $L \geq 0$, it follows that $L = \sqrt{k}$. $\triangle$

Remark. Define $h_n = k/a_n > 0$ for all $n \in \mathbb{N}$. Since $a_n \to \sqrt{k}$ as $n \to \infty$, the same is true of h_n. Moreover

$$a_{n+1} = \frac{a_n + h_n}{2}$$

and

$$h_{n+1} = \frac{k}{a_{n+1}} = \frac{2k}{a_n + \frac{k}{a_n}} = \frac{2}{\frac{a_n}{k} + \frac{1}{a_n}} = \frac{2}{\frac{1}{h_n} + \frac{1}{a_n}}.$$

Thus a_{n+1} and h_{n+1} are the arithmetic and harmonic means, respectively, of a_n and h_n, so that $h_{n+1} \leq a_{n+1}$ by Lemma 2.5.5. Note also that $a_n h_n = k$ for all n. Thus the geometric mean of a_n and h_n is $\sqrt{a_n h_n} = \sqrt{k}$, and both $\{a_n\}$ and $\{h_n\}$ converge to this number. The convergence of $\{h_n\}$ can also be confirmed by the observations that it is a nondecreasing sequence (because $\{a_n\}$ is nonincreasing) and bounded above by $\max\{k/a_1, a_2\}$: We have $h_1 = k/a_1$ and $h_n \leq a_n \leq a_2$ for all $n > 1$.

We next consider two sequences defined by arithmetic and geometric means. These rapidly converging sequences were first studied by Lagrange and Gauss independently in the eighteenth century and by Borchardt in 1888 in relation to the computation of elliptic integrals. For a good discussion of this approach, see [5].

Example 2.7.7. Let a_0 and g_0 be positive numbers, and for all $n \geq 0$ let

$$a_{n+1} = \frac{a_n + g_n}{2}$$

and

$$g_{n+1} = \sqrt{a_n g_n}.$$

If $g_0 = a_0$, then $a_1 = g_1 = a_0$ and it follows by induction that $a_n = g_n = a_0$ for all n. We assume therefore that $a_0 \neq g_0$. From Lemma 2.5.5 we have $a_n \geq g_n > 0$ for all positive n. Hence

$$a_n \geq a_{n+1} \geq g_{n+1} \geq g_n$$

for all $n > 0$. Thus $\{a_n\}$ is a nonincreasing sequence bounded below by $\min\{a_0, g_1\}$ and $\{g_n\}$ is a nondecreasing sequence bounded above by $\max\{g_0, a_1\}$. Both sequences therefore converge.

Now for all $n > 0$ we have

$$0 \le a_{n+1} - g_{n+1}$$
$$\le a_{n+1} - g_n$$
$$= \frac{a_n - g_n}{2}.$$

By induction we deduce that

$$0 \le a_n - g_n$$
$$\le \frac{a_1 - g_1}{2^{n-1}}. \qquad (2.10)$$

From the sandwich theorem we conclude that

$$\lim_{n \to \infty} a_n = \lim_{n \to \infty} g_n.$$

This common limit is denoted by $\mathrm{agm}(a_0, g_0)$ and called the **arithmetic–geometric mean** of a_0 and g_0. It is positive, since

$$\lim_{n \to \infty} a_n \ge g_1 > 0.$$

It cannot be written in closed form but may be expressed in terms of a certain elliptic integral. The number $1/\mathrm{agm}\left(\sqrt{2}, 1\right)$ is known as **Gauss's constant**. Its value is approximately 0.8346268.

Although inequality (2.10) provides a means of estimating the rate of convergence of the sequences, a better estimate may be obtained by comparing the values of $a_n - g_n$ and $a_{n+1} - g_{n+1}$. For each n define

$$\epsilon_n = a_n - g_n$$
$$= \left(\sqrt{a_n} - \sqrt{g_n}\right)\left(\sqrt{a_n} + \sqrt{g_n}\right).$$

Then

$$\epsilon_{n+1} = a_{n+1} - g_{n+1}$$
$$= \frac{a_n + g_n}{2} - \sqrt{a_n g_n}$$
$$= \frac{\left(\sqrt{a_n} - \sqrt{g_n}\right)^2}{2},$$

so that

$$\epsilon_n^2 = \left(\sqrt{a_n} + \sqrt{g_n}\right)^2 \left(\sqrt{a_n} - \sqrt{g_n}\right)^2$$
$$= 2\left(\sqrt{a_n} + \sqrt{g_n}\right)^2 \epsilon_{n+1}.$$

Therefore $\epsilon_{n+1} \neq 0$ if $\epsilon_n \neq 0$, and since $\epsilon_0 \neq 0$, we conclude inductively that $\epsilon_n \neq 0$ for all n. Thus

$$\frac{\epsilon_{n+1}}{\epsilon_n^2} = \frac{1}{2\left(\sqrt{a_n} + \sqrt{g_n}\right)^2},$$

so that

$$\lim_{n \to \infty} \frac{\epsilon_{n+1}}{\epsilon_n^2} = \frac{1}{8\mathrm{agm}(a_0, g_0)}.$$

We leave it to the reader to show that the sequences defined by the harmonic and geometric means also converge to the same limit. These sequences are given by

$$h_{n+1} = \frac{2}{\frac{1}{h_n} + \frac{1}{g_n}}$$

and

$$g_{n+1} = \sqrt{h_n g_n},$$

respectively, for some given positive numbers h_0 and g_0, respectively. Their common limit is denoted by $\mathrm{hgm}(h_0, g_0)$. In fact,

$$\mathrm{hgm}(h_0, g_0) = \frac{1}{\mathrm{agm}\left(\frac{1}{h_0}, \frac{1}{g_0}\right)}.$$

$$\triangle$$

Recall that an empty product (a product with no factors) is defined to be 1. We use this convention in the next example.

Example 2.7.8. Let

$$s_n = \left(1 + \frac{1}{n}\right)^n = \left(\frac{n+1}{n}\right)^n$$

for all $n > 0$. The sequence $\{s_n\}$ has been presented earlier as Example 2.1.6. We shall show that it is increasing and bounded above.

For all $n > 0$ the binomial theorem shows that

$$s_n = \sum_{j=0}^{n} \binom{n}{j} \left(\frac{1}{n}\right)^j$$

$$= \sum_{j=0}^{n} \frac{n(n-1)\cdots(n-j+1)}{j! \, n^j}$$

$$= \sum_{j=0}^{n} \frac{1}{j!} \left(1 - \frac{1}{n}\right)\left(1 - \frac{2}{n}\right)\cdots\left(1 - \frac{j-1}{n}\right)$$

$$\leq \sum_{j=0}^{n} \frac{1}{j!}$$

$$\leq 1 + \sum_{j=1}^{n} \frac{1}{2^{j-1}}$$

$$= 1 + \sum_{j=0}^{n-1} \frac{1}{2^j}$$

$$= 1 + \frac{1 - \frac{1}{2^n}}{1 - \frac{1}{2}}$$

$$< 1 + 2$$

$$= 3.$$

Hence the sequence $\{s_n\}$ is bounded above by 3. Moreover for all $n > 0$ we have

$$s_{n+1} = \sum_{j=0}^{n+1} \frac{1}{j!} \left(1 - \frac{1}{n+1}\right)\left(1 - \frac{2}{n+1}\right)\cdots\left(1 - \frac{j-1}{n+1}\right)$$

$$> \sum_{j=0}^{n} \frac{1}{j!} \left(1 - \frac{1}{n+1}\right)\left(1 - \frac{2}{n+1}\right)\cdots\left(1 - \frac{j-1}{n+1}\right)$$

$$\geq \sum_{j=0}^{n} \frac{1}{j!} \left(1 - \frac{1}{n}\right)\left(1 - \frac{2}{n}\right)\cdots\left(1 - \frac{j-1}{n}\right)$$

$$= s_n,$$

and so $\{s_n\}$ is an increasing bounded sequence. Therefore it converges. Its limit is denoted by e. Since $s_n \geq s_1 = 2$ for all n and the sequence is increasing and bounded above by 3, we find that $2 < e < 3$. Indeed we have $s_2 > s_1 = 2$, so that $e > 2$. Moreover for all $n \geq 3$ we have

$$s_n \le 1 + 1 + \frac{1}{2} + \frac{1}{6} + \sum_{j=3}^{n-1} \frac{1}{2^j}$$

$$= 1 + \sum_{j=0}^{n-1} \frac{1}{2^j} - \frac{1}{4} + \frac{1}{6}$$

$$< 1 + 2 - \frac{1}{12}$$

$$= \frac{35}{12}.$$

Thus $e \le 35/12 < 3$.

More meticulous calculations show that e is approximately 2.718. $\triangle$

We now consider the more general sequence

$$\left\{ \left(1 + \frac{x}{n}\right)^n \right\} = \left\{ \left(\frac{n+x}{n}\right)^n \right\},$$

where $x \in \mathbb{R}$. For each $x \ne 0$ we have

$$\left(1 + \frac{x}{n}\right)^n = \left(1 + \frac{1}{\frac{n}{x}}\right)^n = \left(1 + \frac{1}{m}\right)^{mx} = \left(\left(1 + \frac{1}{m}\right)^m\right)^x,$$

where $m = n/x$. In the case where m is a positive integer, we have

$$e = \lim_{m \to \infty} \left(1 + \frac{1}{m}\right)^m.$$

It therefore seems reasonable to define

$$e^x = \lim_{n \to \infty} \left(1 + \frac{x}{n}\right)^n$$

for all real x, but first we must show that the limit exists.

Proposition 2.7.3. *For all $x \in \mathbb{R}$ the sequence*

$$\{s_n\} = \left\{ \left(1 + \frac{x}{n}\right)^n \right\}$$

is convergent. Moreover if $x \ne 0$, then the sequence is increasing for large enough n.

Proof. The sequence converges to 1 if $x = 0$. Suppose therefore that $x \neq 0$. Letting $n > |x|$, define $a_1 = 1$ and

$$a_k = 1 + \frac{x}{n}$$

for each $k \in \{2, 3, \ldots, n + 1\}$. Thus $a_k \neq 1 = a_1$ for each $k > 1$. Moreover $n > |x| \geq -x$, so that $x/n > -1$; hence $a_k > 0$ for each k. By Lemma 2.5.5 we have

$$\left(\left(1 + \frac{x}{n} \right)^n \right)^{\frac{1}{n+1}} = (a_1 a_2 \cdots a_{n+1})^{\frac{1}{n+1}}$$

$$< \frac{1}{n+1} \left(1 + n \left(1 + \frac{x}{n} \right) \right)$$

$$= 1 + \frac{x}{n+1};$$

hence

$$\left(1 + \frac{x}{n} \right)^n < \left(1 + \frac{x}{n+1} \right)^{n+1}.$$

We conclude that the sequence $\{s_n\}$ is increasing for $n > |x|$.

It therefore suffices to show that $\{s_n\}$ is bounded. Choose an integer $M \geq |x|$. Then the sequence

$$\left\{ \left(1 + \frac{M}{n} \right)^n \right\}$$

is increasing for $n > M > 0$. As $nM \geq n$, we therefore have

$$\left| \left(1 + \frac{x}{n} \right)^n \right| = \left| 1 + \frac{x}{n} \right|^n$$

$$\leq \left(1 + \frac{|x|}{n} \right)^n$$

$$\leq \left(1 + \frac{M}{n} \right)^n$$

$$\leq \left(1 + \frac{M}{nM} \right)^{nM}$$

$$= \left(\left(1 + \frac{1}{n} \right)^n \right)^M$$

$$< e^M,$$

because the sequence $\{(1 + 1/n)^n\}$ is increasing and converges to e. Therefore $\{s_n\}$ is a bounded sequence, as required.   □

Since $\{(1 + x/n)^n\}$ is a nondecreasing sequence of positive terms for $n > |x|$, it follows that $e^x > 0$ for all $x \in \mathbb{R}$. Note also that $e^0 = 1$ and $e^1 = e$.

Example 2.7.9. For all positive integers n define

$$t_n = \sum_{j=0}^{n} \frac{x^j}{j!},$$

where $x > 0$. We shall show that the sequence $\{t_n\}$ converges to e^x.

For each $n \in \mathbb{N}$ define

$$s_n = \left(1 + \frac{x}{n}\right)^n.$$

A simple modification of the calculation in Example 2.7.8 gives

$$s_n = \sum_{j=0}^{n} \frac{x^j}{j!} \left(1 - \frac{1}{n}\right) \left(1 - \frac{2}{n}\right) \cdots \left(1 - \frac{j-1}{n}\right)$$

$$\leq \sum_{j=0}^{n} \frac{x^j}{j!}$$

$$= t_n.$$

Moreover $\{t_n\}$ is certainly an increasing sequence. We shall prove it convergent by showing that it is bounded above by e^x. Let m and n be positive integers with $n > m$. Then

$$\left(1 + \frac{x}{n}\right)^n = \sum_{j=0}^{n} \frac{x^j}{j!} \left(1 - \frac{1}{n}\right) \left(1 - \frac{2}{n}\right) \cdots \left(1 - \frac{j-1}{n}\right)$$

$$> \sum_{j=0}^{m} \frac{x^j}{j!} \left(1 - \frac{1}{n}\right) \left(1 - \frac{2}{n}\right) \cdots \left(1 - \frac{j-1}{n}\right).$$

Hence

$$e^x = \lim_{n \to \infty} \left(1 + \frac{x}{n}\right)^n$$

$$\geq \lim_{n \to \infty} \sum_{j=0}^{m} \frac{x^j}{j!} \left(1 - \frac{1}{n}\right) \left(1 - \frac{2}{n}\right) \cdots \left(1 - \frac{j-1}{n}\right)$$

$$= \sum_{j=0}^{m} \frac{x^j}{j!}$$

$$= t_m.$$

Thus we have $s_m \leq t_m \leq e^x$ for all positive integers m, and so it follows from the sandwich theorem that

$$\lim_{n \to \infty} t_n = e^x.$$

$\triangle$

Since $\{t_n\}$ is an increasing sequence when $x > 0$, we find that

$$e^x > 1 + x > 1$$

for all positive x.

It follows from the definition that $e^0 = 1$. The result of Example 2.7.9 therefore holds also when $x = 0$ if we adopt the convention that $0^0 = 1$. In fact, it holds for all real x. In order to prove this result, we need the following lemma.

Lemma 2.7.4. *Let $a_1, a_2, \ldots, a_n$ be nonzero numbers that are greater than or equal to -1 and have equal sign. Then*

$$\prod_{j=1}^{n}(1 + a_j) \geq 1 + \sum_{j=1}^{n} a_j.$$

Proof. The result certainly holds if $n = 1$. Suppose therefore that $n > 1$ and that the lemma holds for sets of fewer than n numbers satisfying the hypotheses. The hypotheses imply that $1 + a_1 \geq 0$ and $a_1 a_j > 0$ for all $j > 1$. Therefore, using the inductive hypothesis, we find that

$$\prod_{j=1}^{n}(1 + a_j) = (1 + a_1) \prod_{j=2}^{n}(1 + a_j)$$

$$\geq (1 + a_1)\left(1 + \sum_{j=2}^{n} a_j\right)$$

$$= 1 + a_1 + \sum_{j=2}^{n} a_j + \sum_{j=2}^{n} a_1 a_j$$

$$> 1 + \sum_{j=1}^{n} a_j,$$

as required. $\square$

Corollary 2.7.5. *For each $x \geq -1$ and each nonnegative integer n, we have*

$$(1 + x)^n \geq 1 + nx.$$

Proof. We have already seen in Example 1.3.1 that this inequality holds for all $x \geq 0$. Suppose that $-1 \leq x < 0$. The result holds for $n = 0$ if we take $0^0 = 1$. For all $n > 0$ it follows immediately from the lemma by taking $a_j = x$ for all $j \leq n$.

$\square$

Theorem 2.7.6. *For all positive integers n define*

$$t_n = \sum_{j=0}^{n} \frac{x^j}{j!},$$

where $x \in \mathbb{R}$. Then the sequence $\{t_n\}$ converges to e^x.

Proof. Letting

$$s_n = \left(1 + \frac{x}{n}\right)^n$$

for all $n \in \mathbb{N}$, using Lemma 2.7.4 and Theorem 1.5.4, and recalling that the empty product is defined as 1, we have

$$|t_n - s_n| = \left| \sum_{j=0}^{n} \frac{x^j}{j!} - \sum_{j=0}^{n} \frac{x^j}{j!} \left(1 - \frac{1}{n}\right)\left(1 - \frac{2}{n}\right)\cdots\left(1 - \frac{j-1}{n}\right) \right|$$

$$= \left| \sum_{j=2}^{n} \left(1 - \left(1 - \frac{1}{n}\right)\left(1 - \frac{2}{n}\right)\cdots\left(1 - \frac{j-1}{n}\right)\right) \frac{x^j}{j!} \right|$$

$$\leq \sum_{j=2}^{n} \left(1 - \prod_{k=1}^{j-1}\left(1 - \frac{k}{n}\right)\right) \frac{|x|^j}{j!}$$

$$\leq \sum_{j=2}^{n} \left(1 - \left(1 - \sum_{k=1}^{j-1}\frac{k}{n}\right)\right) \frac{|x|^j}{j!}$$

$$= \sum_{j=2}^{n} \left(\sum_{k=1}^{j-1}\frac{k}{n}\right) \frac{|x|^j}{j!}$$

$$= \sum_{j=2}^{n} \frac{j(j-1)}{2n} \cdot \frac{|x|^j}{j!}$$

$$= \frac{x^2}{2n} \sum_{j=2}^{n} \frac{|x|^{j-2}}{(j-2)!}$$

$$= \frac{x^2}{2n} \sum_{j=0}^{n-2} \frac{|x|^j}{j!}$$

for all $n \geq 2$. We deduce that

$$\lim_{n \to \infty} |t_n - s_n| = 0,$$

for it follows from Example 2.7.9 that the sequence

$$\left\{ \sum_{j=0}^{n-2} \frac{|x|^j}{j!} \right\}$$

converges. Therefore

$$\lim_{n \to \infty} (t_n - s_n) = 0$$

and so

$$\lim_{n \to \infty} t_n = \lim_{n \to \infty} (t_n - s_n + s_n) = \lim_{n \to \infty} s_n = e^x.$$

$\square$

We can also show that e is irrational. First let $x = 1$ and let m and n be positive integers with $n > m$. Then $e - t_m > e - t_n \geq 0$ and

$$t_n - t_m = \sum_{j=m+1}^{n} \frac{1}{j!}$$

$$= \frac{1}{(m+1)!} \sum_{j=m+1}^{n} \frac{1}{(m+2)(m+3)\cdots j}$$

$$= \frac{1}{(m+1)!} \sum_{j=0}^{n-m-1} \frac{1}{(m+2)(m+3)\cdots(m+j+1)}$$

$$\leq \frac{1}{(m+1)!} \sum_{j=0}^{n-m-1} \frac{1}{(m+1)^j}$$

$$< \frac{1}{(m+1)!} \cdot \frac{1}{1 - \frac{1}{m+1}}$$

$$= \frac{1}{(m+1)!} \cdot \frac{m+1}{m}$$

$$= \frac{1}{m!m}.$$

Thus

$$e - t_m = e - t_n + t_n - t_m < e - t_n + \frac{1}{m!m},$$

and taking the limit as $n \to \infty$ yields

$$0 < e - t_m \leq \frac{1}{m!m}. \tag{2.11}$$

If e were rational, then we could write $e = p/q$ for some positive integers p and q. Moreover e is not an integer since $2 < e < 3$, and so $q > 1$. Putting $m = q$ in (2.11), we obtain

$$0 < q!(e - t_q) \leq \frac{1}{q} < 1.$$

But

$$q!e = p(q - 1)!$$

by assumption, and

$$q!t_q = \sum_{j=0}^{q} \frac{q!}{j!},$$

which is also an integer. We therefore have the contradiction that $q!(e - t_q)$ is an integer between 0 and 1.

Example 2.7.10. Let $s_n = n!/n^n$ for all $n > 0$. Then

$$\frac{s_{n+1}}{s_n} = \frac{(n + 1)!}{(n + 1)^{n+1}} \cdot \frac{n^n}{n!}$$

$$= \left(\frac{n}{n + 1} \right)^n$$

$$= \frac{1}{\left(1 + \frac{1}{n} \right)^n}$$

$$\to \frac{1}{e}$$

as $n \to \infty$. Since $e > 1$, it follows from Theorem 2.5.4 that the sequence $\{s_n\}$ converges to 0. $\triangle$

Example 2.7.11. Since

$$\lim_{n\to\infty} \left(\frac{n+1}{n}\right)^n = e,$$

Theorem 2.5.6 implies that

$$e = \lim_{n\to\infty} \left(\prod_{j=1}^{n} \left(\frac{j+1}{j}\right)^j\right)^{1/n}.$$

But it is easy to show by induction that

$$\prod_{j=1}^{n} \left(\frac{j+1}{j}\right)^j = \frac{(n+1)^n}{n!}$$

for all positive integers n, and so

$$e = \lim_{n\to\infty} \left(\frac{(n+1)^n}{n!}\right)^{1/n}$$

$$= \lim_{n\to\infty} \frac{n+1}{(n!)^{1/n}}.$$

Thus $n!$ is close to $(n+1)^n/e^n$ for large values of n.   △

We now show that our definition of e^x satisfies at least one of the usual laws governing exponentiation. First we prove the following useful lemma.

Lemma 2.7.7. *For every $x \in \mathbb{R}$,*

$$\lim_{n\to\infty} \left(1 + \frac{x}{n^2}\right)^n = 1.$$

Proof. For every integer $n > \sqrt{|x|}$ we have $n^2 > |x| \geq -x$, so that $x/n^2 > -1$. Applying Corollary 2.7.5 and writing $m = n^2$, we therefore find that

$$1 + \frac{x}{n} \leq \left(1 + \frac{x}{n^2}\right)^n = \left(\left(1 + \frac{x}{m}\right)^m\right)^{1/n} \leq e^{1/n}.$$

Using Example 2.5.2, we have

$$\lim_{n\to\infty} \left(1 + \frac{x}{n}\right) = 1 = \lim_{n\to\infty} e^{1/n}.$$

The required result now follows from the sandwich theorem.   □

Corollary 2.7.8. *For every real x we have*

$$e^{-x} = \frac{1}{e^x}.$$

Proof. This result follows from the lemma by taking limits of both sides of the identity

$$\left(1 + \frac{x}{n}\right)^n \left(1 - \frac{x}{n}\right)^n = \left(1 - \frac{x^2}{n^2}\right)^n.$$

□

Thus $e^x e^{-x} = 1$. This equation is a special case of the following law governing exponents. Its proof is an adaptation of an argument due to Kemeny [9].

Theorem 2.7.9. *For each $x, y \in \mathbb{R}$ we have*

$$e^x e^y = e^{x+y}.$$

Proof. The result clearly holds if $x = 0$ or $y = 0$, since $e^0 = 1$. We may therefore assume that $xy \neq 0$.

Let $z = -x - y$ and, for all positive integers n, define

$$
\begin{aligned}
s_n &= \left(\frac{n+x}{n}\right)^n \left(\frac{n+y}{n}\right)^n \left(\frac{n+z}{n}\right)^n \\
&= \left(\frac{n^3 + n^2(x+y+z) + n(xy + xz + yz) + xyz}{n^3}\right)^n \\
&= \left(1 + \frac{xy + xz + yz}{n^2} + \frac{xyz}{n^3}\right)^n.
\end{aligned}
$$

Note that the sequence $\{s_n\}$ converges to $e^x e^y e^z$, and therefore so does the subsequence $\{s_{2n-1}\}$.

Suppose that $xyz > 0$. Then

$$\frac{xyz}{(2n-1)^2} \geq \frac{xyz}{(2n-1)^3},$$

and so

$$\left(1 + \frac{xy + xz + yz}{(2n-1)^2}\right)^{2n-1} < s_{2n-1} \leq \left(1 + \frac{xy + xz + yz + xyz}{(2n-1)^2}\right)^{2n-1},$$

since $2n - 1$ is odd. Therefore Lemma 2.7.7 and the sandwich theorem show that

$$\lim_{n \to \infty} s_{2n-1} = 1.$$

Consequently,

$$e^x e^y e^z = 1,$$

and so

$$e^x e^y = \frac{1}{e^z} = e^{-z} = e^{x+y}.$$

Thus we have confirmed the theorem if $xyz > 0$.

We now distinguish three cases.

Case 1: Suppose $x < 0$ and $y < 0$. Then $z > 0$, so that $xyz > 0$. We conclude that

$$e^x e^y = e^{x+y}$$

in this case.

Case 2: If $x > 0$ and $y > 0$, then by case 1 we have

$$e^{-x} e^{-y} = e^{-(x+y)}.$$

Thus

$$\frac{1}{e^x e^y} = \frac{1}{e^{x+y}},$$

and the result follows.

Case 3: If $xy < 0$, then we may assume without loss of generality that $x < 0 < y$.

Case 3.1: If $-x < y$, then $z < 0$, so that $xyz > 0$.

Case 3.2: If $-x = y$, then

$$e^x e^y = e^x e^{-x} = 1 = e^0 = e^{x+y}.$$

Case 3.3: If $-x > y$, then $z > 0$. It therefore follows from case 1 that

$$e^{-z} e^{-y} = e^x,$$

and so

$$e^x e^y = e^{-z} = e^{x+y}.$$

□

The function given by e^x for all real x is called the **exponential** function and often denoted by exp. Here are some of its properties.

Theorem 2.7.10. *Let $\{a_n\}$ be a sequence of real numbers.*

1. If $\lim_{n\to\infty} a_n = \infty$, then

$$\lim_{n\to\infty} e^{a_n} = \infty.$$

2. If $\lim_{n\to\infty} a_n = a$, then

$$\lim_{n\to\infty} e^{a_n} = e^a.$$

3. If $a_n \neq 0$ for all n and $\lim_{n\to\infty} a_n = 0$, then

$$\lim_{n\to\infty} \frac{e^{a_n} - 1}{a_n} = 1.$$

Proof. 1. We have already seen that

$$e^x > 1 + x$$

for all $x > 0$. Choose $M > 1$. Since $\lim_{n\to\infty} a_n = \infty$, there exists N such that $a_n > M - 1 > 0$ for each integer $n \geq N$. Thus

$$e^{a_n} > 1 + a_n > M$$

for all such n, and the result follows.

2. We first prove the result for $a = 0$, in which case we need to show that

$$\lim_{n\to\infty} e^{a_n} = 1.$$

Since $\{a_n\}$ is convergent, it is bounded. Let $|a_n| < M$ for all n. For all positive integers m and n we have

$$\left| \sum_{j=0}^{m} \frac{a_n^j}{j!} - 1 \right| = \left| \sum_{j=1}^{m} \frac{a_n^j}{j!} \right|$$

$$\leq \sum_{j=1}^{m} \frac{|a_n|^j}{j!}$$

$$= |a_n| \sum_{j=1}^{m} \frac{|a_n|^{j-1}}{j!}$$

$$\leq |a_n| \sum_{j=1}^{m} \frac{M^{j-1}}{(j-1)!}$$

$$= |a_n| \sum_{j=0}^{m-1} \frac{M^j}{j!}.$$

Taking limits as $m \to \infty$, we therefore conclude from Proposition 2.3.2(1) that

$$0 \leq |e^{a_n} - 1| \leq |a_n| e^M.$$

Thus the sandwich theorem yields

$$\lim_{n\to\infty} |e^{a_n} - 1| = 0,$$

whence

$$\lim_{n\to\infty} (e^{a_n} - 1) = 0$$

and the desired result follows.

In the general case, where $\lim_{n\to\infty} a_n = a \neq 0$, we use the result just established:

$$\lim_{n\to\infty} e^{a_n} = \lim_{n\to\infty} e^a (e^{a_n - a}) = e^a \lim_{n\to\infty} e^{a_n - a} = e^a.$$

3. As $\{a_n\}$ converges, there exists M such that $|a_n| < M$ for all n. Thus for all integers $m > 1$ and $n > 0$ we have

$$\left| \frac{1}{a_n} \left(\sum_{j=0}^{m} \frac{a_n^j}{j!} - 1 \right) - 1 \right| = \left| \frac{1}{a_n} \sum_{j=1}^{m} \frac{a_n^j}{j!} - 1 \right|$$

$$= \left| \sum_{j=1}^{m} \frac{a_n^{j-1}}{j!} - 1 \right|$$

$$= \left| \sum_{j=0}^{m-1} \frac{a_n^j}{(j+1)!} - 1 \right|$$

$$= \left| \sum_{j=1}^{m-1} \frac{a_n^j}{(j+1)!} \right|$$

$$\leq |a_n| \sum_{j=1}^{m-1} \frac{|a_n|^{j-1}}{(j+1)!}$$

$$< |a_n| \sum_{j=1}^{m-1} \frac{M^{j-1}}{(j-1)!}$$

$$= |a_n| \sum_{j=0}^{m-2} \frac{M^j}{j!}.$$

We now complete the proof by an argument similar to that used in part (2): Taking limits as $m \to \infty$ yields

$$0 \leq \left| \frac{e^{a_n} - 1}{a_n} - 1 \right| \leq |a_n| e^M,$$

from which the desired result is obtained by taking limits as $n \to \infty$. $\quad\square$

Corollary 2.7.11. *If $\{a_n\}$ is a null sequence of nonzero terms, then*

$$\lim_{n \to \infty} \frac{e^{x+a_n} - e^x}{a_n} = e^x.$$

Proof. Part (3) of the theorem shows that

$$\lim_{n \to \infty} \frac{e^{x+a_n} - e^x}{a_n} = \lim_{n \to \infty} \frac{e^x(e^{a_n} - 1)}{a_n}$$

$$= e^x.$$

$\quad\square$

Theorem 2.7.12. *The exponential function is increasing.*

Proof. If $y > x$, then $y - x > 0$, so that $e^{y-x} > 1$; hence

$$e^y = e^{y-x} e^x > e^x$$

as $e^x > 0$. $\quad\square$

We have already shown that every bounded sequence has a convergent subsequence. We conclude this section by demonstrating that such a subsequence can be chosen to be monotonic.

A positive integer m is called a **peak index** for a real sequence $\{s_n\}$ if $s_n \le s_m$ for all $n \ge m$. For example, consider the sequence $\{1 + (-1)^n/n\}$. The number 1 is not a peak index since $s_1 = 0$ and $s_2 = 3/2$. Since

$$s_n \le 1 + \frac{1}{n} \le \frac{3}{2}$$

for all $n \ge 2$, we see that 2 is a peak index. Similarly, 3 is not a peak index but 4 is.

Theorem 2.7.13. *Every real sequence has a monotonic subsequence.*

Proof. Let P be the set of peak indices for a real sequence $\{s_n\}$.

Case 1: If P is finite, it must have an upper bound, N. We shall find an increasing subsequence $\{s_{k_n}\}$ of $\{s_n\}$. Let $k_1 = N + 1$. Then k_1 is not a peak index, and so there exists $k_2 > k_1$ such that $s_{k_2} > s_{k_1}$. For some integer $n > 1$ we may now assume the existence of integers $k_1, k_2, \ldots, k_n$ such that $k_j < k_l$ and $s_{k_j} < s_{k_l}$ whenever $j < l$. Since $k_n > N$, k_n is not a peak index and so there exists $k_{n+1} > k_n$ such that $s_{k_{n+1}} > s_{k_n}$. This observation completes the inductive definition of the increasing subsequence $\{s_{k_n}\}$ of $\{s_n\}$.

Case 2: If P is infinite, then for each positive integer n we may define k_n to be the nth peak index. It follows that $s_{k_{n+1}} \le s_{k_n}$ for all $n \in \mathbb{N}$, so that $\{s_{k_n}\}$ is a nonincreasing subsequence of $\{s_n\}$.

$\square$

Exercises 2.7.

1. Let $\{a_n\}$ and $\{b_n\}$ be sequences of positive numbers, and for all positive integers m and n define

$$p_n = \sum_{j=0}^{m} a_j n^j$$

and

$$q_n = \sum_{j=0}^{m} b_j n^j.$$

(a) Show that

$$\lim_{n \to \infty} \frac{p_n}{q_n} = \frac{a_m}{b_m}.$$

(b) Show that if $b_m > a_m$, then there exists N such that $p_n/q_n < 1$ for all $n \ge N$.

(c) Show that if $b_m < a_m$, then there exists N such that $p_n/q_n > 1$ for all $n \ge N$.

2. Show that the following sequences are monotonic for large enough n:

   (a) $\left\{\frac{(n-1)(n+2)}{(n+1)(n-3)}\right\}$; (b) $\left\{\frac{n^2-5}{n+1}\right\}$; (c) $\left\{\frac{5^n}{n!}\right\}$.

3. Show that a monotonic sequence is convergent if it has a convergent subsequence.

4. Show that the following sequences are nondecreasing and bounded above, and find their limits:

   (a) $x_0 = 0,\ x_{n+1} = \frac{3x_n+1}{4}$; (c) $x_0 = 1,\ x_{n+1} = \sqrt{3x_n + 1}$;

   (b) $x_0 = 0,\ x_{n+1} = \frac{2x_n+4}{3}$; (d) $x_0 = 1,\ x_{n+1} = \sqrt{2x_n + 7}$.

5. Consider a sequence given by

$$x_{n+1} = \frac{x_n^2 + 3}{4}$$

   for all $n > 0$.

   (a) Show that the sequence is nonincreasing if $1 < x_1 < 3$.

   (b) Show that the sequence is nondecreasing if $0 \le x_1 < 1$ or $x_1 > 3$.

   (c) For each of the sequences in (a) or (b), find its limit if it converges.

   (d) Study the convergence of the sequence for the following cases: $x_1 = 1$, $x_1 = 3$, $x_1 < 0$.

6. Let $x_1 = b$ and $x_{n+1} = \sqrt{ax_n}$ for all $n > 0$, where $a > 0$ and $b > 0$. Show that $\{x_n\}$ converges and find its limit.

7. Let $x_1 = a$ and $x_{n+1} = a + x_n^2$ for all $n > 0$, where $a \ge 0$. Discuss the convergence of the sequence $\{x_n\}$.

8. Let $\{a_n\}$ be increasing and $\{b_n\}$ decreasing, and suppose that

$$0 \le b_n - a_n \le \frac{1}{2^n}$$

   for all n. Show that $\{a_n\}$ and $\{b_n\}$ converge to the same limit.

9. Let $x_{n+1} = \sqrt{a + x_n}$ for all $n > 0$, where $a > 0$. Discuss the convergence of the sequence for each $x_1 > 0$.

10. Let $x_1 = c > 0$ and

$$x_{n+1} = \frac{6(1 + x_n)}{7 + x_n}$$

   for all $n > 0$. Discuss the convergence of the sequence. (Consider the cases $c \ge 2$ and $0 < c < 2$ separately.)

11. Consider a sequence given by

$$x_{n+1} = \frac{3x_n + 1}{x_n + 3}$$

   for all $n > 0$. Study the convergence of the sequence for the cases $x_1 \le -1$, $-1 < x_1 < 1$, $x_1 = 1$, and $x_1 > 1$.

12. Let $a \geq 1, 0 < x_1 < a^2$, and

$$x_{n+1} = a - \sqrt{a^2 - x_n}$$

for all $n > 0$.

(a) Show that $0 < x_n < a^2$ for all n and that $\{x_n\}$ is nonincreasing.
(b) Show that $\{x_n\}$ converges to 0.
(c) Show that

$$\lim_{n \to \infty} \frac{x_n}{x_{n+1}} = 2a.$$

13. For all $n \geq 0$ let

$$x_n = \sum_{k=n+1}^{2n} \frac{1}{k}.$$

Show that $\{x_n\}$ is nondecreasing and converges to a limit between $1/2$ and 1.

14. Let $x_1 = 1$ and

$$x_{n+1} = \frac{4 + 3x_n}{3 + 2x_n}$$

for all $n > 0$. Show that the sequence $\{x_n\}$ is nondecreasing and bounded above by $3/2$, and find its limit.

15. Let $x_1 = \sqrt{2}$ and

$$x_{n+1} = \sqrt{2 + \sqrt{x_n}}$$

for all $n > 0$. Prove that the sequence converges.

16. Show that the sequences

$$\left\{ \frac{2^n (n-1)! n!}{(2n)!} \right\}$$

and

$$\left\{ \frac{2^{2n-1} (n-1)! n!}{(2n)!} \right\}$$

are monotonic and find their limits.

17. Let $\sqrt{k} \leq x_1 < 3\sqrt{k}$, where $k > 0$, and for each $n > 0$ define

$$x_{n+1} = \frac{1}{2} \left(x_n + \frac{k}{x_n} \right).$$

Show that

$$\left| x_{n+1} - \sqrt{k} \right| \leq 2\sqrt{k} \left(\frac{x_1 - \sqrt{k}}{2\sqrt{k}} \right)^{2^n}$$

and hence that the sequence converges to $\sqrt{k}$.

18. Let h_0 and g_0 be positive numbers and let $a_0 = 1/h_0$ and $b_0 = 1/g_0$. For all $n \geq 0$ define

$$h_{n+1} = \frac{2}{\frac{1}{h_n} + \frac{1}{g_n}},$$

$$g_{n+1} = \sqrt{h_n g_n},$$

$$a_{n+1} = \frac{a_n + b_n}{2},$$

and

$$b_{n+1} = \sqrt{a_n b_n}.$$

Show that $h_n = 1/a_n$ and $g_n = 1/b_n$ for all n and hence that

$$\lim_{n \to \infty} h_n = \lim_{n \to \infty} g_n = \frac{1}{\text{agm} \left(\frac{1}{h_0}, \frac{1}{g_0} \right)}.$$

19. Let $x_1 = 1/2$ and $y_1 = 1$, and define

$$x_{n+1} = \sqrt{x_n y_n}$$

and

$$y_{n+1} = \frac{2}{\frac{1}{x_{n+1}} + \frac{1}{y_n}}$$

for all $n \geq 1$. Prove that $x_n < x_{n+1} < y_{n+1} < y_n$ for all n and deduce that both sequences $\{x_n\}$ and $\{y_n\}$ converge to the same limit L, where $1/2 < L < 1$.

20. Let $x_1 > 0$ and $y_1 > 0$, and define

$$x_{n+1} = \frac{x_n + y_n}{2}$$

and

$$y_{n+1} = \sqrt{x_{n+1} y_n}$$

for all $n \geq 1$. Prove that the sequences $\{x_n\}$ and $\{y_n\}$ are monotonic and converge to the same limit.

21. Show that the sequence

$$\left\{ \left(1 + \frac{1}{n} \right)^{n+1} \right\}$$

is nonincreasing and bounded below, and find its limit.

2.8 Unbounded Sequences

In the previous section we saw that a monotonic sequence is convergent if and only if it is bounded. Specifically, if the sequence $\{s_n\}$ is nondecreasing and unbounded, then $\lim_{n \to \infty} s_n = \infty$, and if it is nonincreasing and unbounded, then $\lim_{n \to \infty} s_n = -\infty$. However it may be that a divergent sequence is bounded. An example is furnished by the sequence $\{(-1)^n\}$. It may also be that $\lim_{n \to \infty} s_n = \pm\infty$, but $\{s_n\}$ is not monotonic.

Example 2.8.1. Let

$$s_n = n + 2(-1)^n$$

for all n. Then

$$s_{2n} = 2n + 2,$$

$$s_{2n+1} = 2n + 1 - 2 = 2n - 1,$$

and

$$s_{2n+2} = 2n + 2 + 2 = 2n + 4.$$

Hence $\{s_n\}$ is not monotonic. But for each n we have

$$s_n \geq n - 2,$$

and so

$$\lim_{n \to \infty} s_n = \infty.$$

$\triangle$

The next proposition is easy to establish.

Proposition 2.8.1. *Suppose $\{s_n\}$ and $\{t_n\}$ are sequences and there exists N such that $t_n \geq s_n$ for all $n \geq N$.*

1. If $s_n \to \infty$ as $n \to \infty$, then $t_n \to \infty$ as $n \to \infty$.
2. If $t_n \to -\infty$ as $n \to \infty$, then $s_n \to -\infty$ as $n \to \infty$.

Proof. The first part is immediate from Definition 2.7.1 and the fact that if, given any number M, there exists N_1 such that $s_n > M$ for all $n \geq N_1$, then $t_n \geq s_n > M$ for all $n \geq \max\{N, N_1\}$. The proof of the second part is similar. □

At this point the reader may be wondering to what extent the symbols ∞ and $-\infty$ may be treated as if they were numbers. We now attempt to answer that question by studying the properties of limits of sums and products of unbounded sequences.

For sums we have the following theorem.

Theorem 2.8.2. *Let $\{s_n\}$ and $\{t_n\}$ be sequences and L a number. If $s_n \to \infty$ and either $t_n \to \infty$ or $t_n \to L$ as $n \to \infty$, then $s_n + t_n \to \infty$.*

Proof. Choose a number M, and suppose first that $t_n \to \infty$. There exists N such that $s_n > M$ and $t_n > M$ for all $n \geq N$. Then $s_n + t_n > 2M$ for all such n, as required.

If $t_n \to L$, then there exists N_1 such that $|t_n - L| < 1$ for all $n \geq N_1$. For each such n it follows that

$$t_n > L - 1.$$

Moreover there exists N such that

$$s_n > M - L + 1$$

for all $n \geq N$. For all $n \geq \max\{N, N_1\}$ we deduce that $s_n + t_n > M$. □

Thus Theorem 2.3.7(1) may be extended if we write $\infty + \infty = \infty + L = \infty$. Because of the commutativity of addition of real numbers, we also write $L + \infty = \infty$. A corresponding theorem may be proved in which s_n approaches $-\infty$ and t_n approaches L or $-\infty$, and so we may also write $-\infty + (-\infty) = -\infty + L = L + (-\infty) = -\infty$. Extending the rule that $x - y = x + (-y)$, we simplify the left-hand side of this equation to $-\infty - \infty$. Similarly, $L + (-\infty)$ may be simplified to $L - \infty$. On the other hand, we cannot ascribe any meaning to $\infty - \infty$. We may be tempted to set it equal to 0. However, suppose $s_n = 2n$ and $t_n = n$ for all n. Then $s_n \to \infty$ and $t_n \to \infty$ as $n \to \infty$, but the sequence $\{s_n - t_n\}$ does not approach 0 as $n \to \infty$ since $s_n - t_n = 2n - n = n \to \infty$.

Let us move on to products.

Theorem 2.8.3. *Let $\{s_n\}$ and $\{t_n\}$ be sequences and L a positive number. If $s_n \to \infty$ and either $t_n \to \infty$ or $t_n \to L$ as $n \to \infty$, then $s_n t_n \to \infty$.*

Proof. Choose $M > 0$, and suppose that $t_n \to \infty$. There exists N such that $s_n > M$ and $t_n > 1$ for all $n \geq N$. For all such n we deduce that $s_n t_n > M$, as required.

Suppose on the other hand that $t_n \to L$. There exists N_1 such that $t_n > L/2$ for all $n \geq N_1$. There also exists N such that $s_n > 2M/L$ for all $n \geq N$. For all $n \geq \max\{N, N_1\}$ it therefore follows that

$$s_n t_n > \frac{2M}{L} \cdot \frac{L}{2} = M.$$

$\square$

Again we use this theorem as a pretext for extending Theorem 2.3.7, this time by writing $\infty \cdot \infty = \infty \cdot L = L \cdot \infty = \infty$, where $L > 0$. A corresponding result may be proved in which $s_n \to -\infty$: Simply apply the theorem to the sequence $\{-s_n\}$. Thus we also write $(-\infty) \cdot \infty = \infty \cdot (-\infty) = (-\infty) \cdot L = L \cdot (-\infty) = -\infty$. Similarly, replacing t_n by $-t_n$ gives the additional equations $\infty \cdot (-L) = (-L) \cdot \infty = -\infty$ and $(-\infty) \cdot (-\infty) = (-\infty) \cdot (-L) = (-L) \cdot (-\infty) = \infty$.

On the other hand, we cannot ascribe meanings to $\infty \cdot 0$ or $(-\infty) \cdot 0$. For example, let $k \neq 0$ and for all $n > 0$ define $s_n = kn$ and $t_n = 1/n$. Then s_n approaches ∞ or $-\infty$ and t_n approaches 0, but $s_n t_n = k$.

We prepare ourselves for the incorporation of division into this framework by proving the following theorem.

Theorem 2.8.4. *Let $\{s_n\}$ be a sequence of positive numbers.*

1. We have $s_n \to 0$ as $n \to \infty$ if and only if $1/s_n \to \infty$ as $n \to \infty$.
2. Similarly, $s_n \to \infty$ as $n \to \infty$ if and only if $1/s_n \to 0$ as $n \to \infty$.

Proof. 1. Suppose first that $s_n \to 0$ as $n \to \infty$, and choose $M > 0$. There exists N such that $s_n < 1/M$ for all $n \geq N$. Thus $1/s_n > M$ for all such n, and we have proved that $1/s_n \to \infty$ as $n \to \infty$.

Conversely, suppose that $1/s_n \to \infty$ as $n \to \infty$, and choose $\varepsilon > 0$. There exists N such that $1/s_n > 1/\varepsilon$ for all $n \geq N$. Hence $s_n < \varepsilon$ for all such n, as required.

2. Apply part (1) to the sequence $\{1/s_n\}$.

$\square$

The following corollary is immediate.

Corollary 2.8.5. *Let $\{s_n\}$ be a sequence of nonzero terms. Then $|s_n| \to \infty$ as $n \to \infty$ if and only if $1/|s_n| \to 0$ as $n \to \infty$.*

Example 2.8.2. Let $\{a_n\}$ be a sequence of real numbers, and suppose that

$$\lim_{n \to \infty} a_n = -\infty.$$

Then

$$\lim_{n \to \infty} (-a_n) = \infty.$$

It therefore follows from Theorems 2.7.10(1) and 2.8.4(2) that

$$\lim_{n\to\infty} e^{a_n} = \lim_{n\to\infty} \frac{1}{e^{-a_n}} = 0.$$

$$\triangle$$

Example 2.8.3. Let $s_n = c^n$ for all positive integers n, where c is a fixed real number. By Example 2.6.3 the sequence $\{s_n\}$ is convergent if and only if $|c| < 1$ or $c = 1$. In fact,

$$\lim_{n\to\infty} s_n = \infty$$

if $c > 1$, for in that case we have $0 < 1/c < 1$ so that

$$\lim_{n\to\infty} \frac{1}{c^n} = 0.$$

$$\triangle$$

Example 2.8.4. Let $\{s_n\}$ be a sequence of nonzero real numbers. Theorem 2.5.4 shows that if

$$\lim_{n\to\infty} \left| \frac{s_{n+1}}{s_n} \right| = L < 1,$$

then

$$\lim_{n\to\infty} s_n = 0.$$

We turn now to the case where $L > 1$. By Proposition 2.3.3 there exist numbers $k > 1$ and N such that

$$\left| \frac{s_{n+1}}{s_n} \right| > k$$

for all $n \geq N$. Hence

$$|s_{n+1}| > k|s_n|$$

for all $n \geq N$. By induction it follows that

$$|s_{N+p}| > k^p |s_N|$$

for all positive integers p. Since $k > 1$, we deduce from the previous example that $k^p \to \infty$ as $p \to \infty$, and so

$$\lim_{p \to \infty} |s_{N+p}| = \infty$$

by Proposition 2.8.1. It follows that $|s_n| \to \infty$ as $n \to \infty$. $\triangle$

Theorem 2.8.4(2) motivates the equation $1/\infty = 0$. A similar theorem may be proved in which s_n approaches $-\infty$, where $\{s_n\}$ is a sequence of negative terms, and so we also write $1/(-\infty) = 0$. However, we cannot use part (1) to justify writing $1/0 = \infty$ because of the requirement that $\{s_n\}$ be a sequence of positive terms. Had $\{s_n\}$ been a sequence of negative terms, we would have been equally tempted to write $1/0 = -\infty$! By extending the rule that $x/y = x(1/y)$ whenever $y \neq 0$, we may write $L/\infty = L/(-\infty) = 0$ for each number L. Furthermore, if $L > 0$, then we may write $\infty/L = \infty \cdot (1/L) = \infty$ and, similarly, $(-\infty)/L = -\infty$. Likewise we write $\infty/(-L) = -\infty$ and $(-\infty)/(-L) = \infty$. Note that no meaning is ascribed to such forms as $0/0$ or ∞/∞. This question will be explored later.

Exercises 2.8.

1. Suppose that the sequence $\{x_n\}$ is increasing and that $x_n^{1/n} > 1$ for all $n > 0$. Show that

$$\lim_{n \to \infty} x_n = \infty.$$

   (Hint: Prove it by contradiction.)
2. Show that if $0 < a < x_n$ and $\lim_{n \to \infty} y_n = \infty$, then $\lim_{n \to \infty} x_n y_n = \infty$. Does the result remain true if we replace the inequalities by $x_n > 0$?
3. Find the following limits:

   (a) $\lim_{n \to \infty} (n^2 - n^3)$;
   (b) $\lim_{n \to \infty} \frac{5^n}{2^n + 3^n}$.

4. Let $\{x_n\}$ and $\{y_n\}$ be sequences such that

$$\lim_{n \to \infty} x_n = \lim_{n \to \infty} y_n = \infty.$$

   Give examples to show that

$$\lim_{n \to \infty} (x_n - y_n)$$

   may be any number or $\pm\infty$.

Chapter 3
Series

3.1 Introduction

The theory of sequences can be combined with the familiar notion of a finite sum to produce the theory of infinite series. The concept of a series is an attempt to encapsulate the idea of a sum of infinitely many real or complex numbers. Applications of series appear in many areas of pure and applied mathematics, and the study of their properties forms a major part of analysis.

The idea of a series disturbed the ancients. In the fifth century BC, for example, Zeno argued that it is impossible to walk from one place to another. For the walker must first travel half the distance, then half the remaining distance, then half the distance left after that, and so forth. The journey can never be completed, because after each stage there is still some distance to go. The inference is that motion is impossible!

Where did Zeno go wrong? He argued that in the first stage of the walk 1/2 of the total distance must be covered, in the next stage 1/4 of the total distance, in the third stage 1/8 of the distance, and so on. Zeno was thus attempting to add infinitely many numbers, and concluded that the sum would be infinite. The absurdity of his conclusion suggests that the sum of infinitely many numbers should not necessarily be infinite. But at this juncture it is not really clear precisely what is meant by the sum of infinitely many numbers. For us to be able to make progress, this notion must be clarified. Specifically, to resolve Zeno's paradox, we need a definition that enables us to conclude that

$$\frac{1}{2} + \frac{1}{4} + \frac{1}{8} + \cdots = 1. \tag{3.1}$$

This problem is resolved through what is called the sequence of partial sums. In other words, we construct a sequence whose first term is the first of the infinite set

© Springer Science+Business Media New York 2015
C.H.C. Little et al., *Real Analysis via Sequences and Series*, Undergraduate
Texts in Mathematics, DOI 10.1007/978-1-4939-2651-0_3

of numbers to be added up, whose second term is the sum of the first two numbers to be added up, and so forth. In general, the sum of the first n numbers in our infinite set gives the nth term of the sequence of partial sums. The infinite sum, which is called a series, is then defined as the limit of the sequence of partial sums, provided of course that the limit exists.

For example, let us return to Zeno's paradox. The sum of the first n terms on the left-hand side of Eq. (3.1) is $\sum_{j=1}^{n} 1/2^j$. In this case the sequence of partial sums is therefore $\{\sum_{j=1}^{n} 1/2^j\}_{n \geq 1}$. We now show that this sequence indeed converges to 1. Setting $a = 1/2$ in Corollary 1.5.6, we have

$$\sum_{j=1}^{n} \frac{1}{2^j} = \sum_{j=0}^{n-1} \frac{1}{2^{j+1}}$$

$$= \frac{1}{2} \sum_{j=0}^{n-1} \frac{1}{2^j}$$

$$= \frac{1}{2} \left(\frac{1 - \frac{1}{2^n}}{1 - \frac{1}{2}} \right)$$

$$= 1 - \frac{1}{2^n}.$$

Thus the sequence $\{\sum_{j=1}^{n} 1/2^j\}$ converges to 1, as desired.

The use of series troubled many mathematicians in the 18th century. The construction of an acceptable, rigorous theory took many decades and involved such mathematicians as Weierstrass, Bolzano, Fourier, Cauchy, Dirichlet, Riemann, and Dedekind. The theory did not reach its current form until the end of the 19th century.

3.2   Definition of a Series

Let us now formalize the concept of a series. Let $\{z_n\}_{n \geq 0}$ be a sequence of real or complex numbers, and for each n let

$$S_n = \sum_{j=0}^{n} z_j.$$

The sequence $\{S_n\}$ is called a **series** and is denoted by

$$\sum_{j=0}^{\infty} z_j \qquad\qquad (3.2)$$

or

$$z_0 + z_1 + \cdots .$$

We refer to $z_0, z_1, \ldots$ as the **terms** of the series. We consider them to be in the order $z_0, z_1, \ldots$, and so we may refer to z_j as the $(j+1)$th term, for each $j \geq 0$. A series is **real** if all its terms are real. For every series, whether real or not, the numbers $S_0, S_1, \ldots$ are the **partial sums**. More particularly, for each n we may describe S_n as the partial sum **corresponding** to z_n, or the $(n+1)$th partial sum. If $\{S_n\}$ converges to some number S, then we write

$$\sum_{j=0}^{\infty} z_j = S.$$

In other words,

$$\sum_{j=0}^{\infty} z_j = \lim_{n \to \infty} \sum_{j=0}^{n} z_j .$$

As in the case of finite sums, we note that the index j is a dummy variable in the expression (3.2). Thus, if k is another index, then

$$\sum_{j=0}^{\infty} z_j = \sum_{k=0}^{\infty} z_k .$$

Motivated by Proposition 1.5.1, we also write

$$\sum_{j=m}^{\infty} z_j = \sum_{j=0}^{\infty} z_{j+m}$$

for each integer m for which $z_m, z_{m+1}, \ldots$ are defined. It is also evident that if $m \leq r$, then

$$\sum_{j=m}^{\infty} z_j = \sum_{j=m}^{r} z_j + \sum_{j=r+1}^{\infty} z_j .$$

In writing this equation we admit the possibility that both series diverge. However, if one of them converges, then so does the other. When testing these series for convergence, it therefore suffices to test just one of them.

It is seldom easy to determine the number, if any, to which a given series converges. However, we do have the following theorem, which follows easily from our earlier work on finite sums.

Theorem 3.2.1. *If z is a complex number such that $|z| < 1$, then*

$$\sum_{j=0}^{\infty} z^j = \frac{1}{1-z},$$

where $0^0 = 1$, but if $|z| \geq 1$, then the series diverges.

Proof. Let n be a nonnegative integer. If $z = 1$, then

$$\sum_{j=0}^{n} z^j = \sum_{j=0}^{n} 1 = n + 1,$$

and so the series diverges. Suppose therefore that $z \neq 1$. Then

$$\sum_{j=0}^{n} z^j = \frac{1 - z^{n+1}}{1 - z},$$

by Corollary 1.5.6. Using the results of Examples 2.2.4 and 2.6.3, we find that the series converges to $1/(1-z)$ if $|z| < 1$ but diverges if $|z| \geq 1$.   □

The series considered in Theorem 3.2.1 is said to be **geometric**.

Example 3.2.1. The repeating decimal $0.22\ldots$ can be written as

$$0.2 + 0.02 + 0.002 + \ldots = 2 \sum_{j=1}^{\infty} a^j$$

$$= 2 \left(\sum_{j=0}^{\infty} a^j - 1 \right),$$

where $a = 0.1 = 1/10$. By Theorem 3.2.1 this series converges to

$$2 \left(\frac{1}{1 - \frac{1}{10}} - 1 \right) = 2 \left(\frac{10}{9} - 1 \right) = \frac{2}{9}.$$

△

The telescoping property can also sometimes be used to deduce the number to which a given series converges. Specifically, we have the following result.

Theorem 3.2.2. *The series*

$$\sum_{j=0}^{\infty} (z_{j+1} - z_j)$$

converges if and only if the sequence $\{z_n\}$ converges, and in that case

$$\sum_{j=0}^{\infty}(z_{j+1} - z_j) = \lim_{n\to\infty} z_n - z_0.$$

Proof. By the telescoping property,

$$\sum_{j=0}^{n}(z_{j+1} - z_j) = z_{n+1} - z_0$$

for every n, and the result follows immediately.   □

A series of the type contemplated in Theorem 3.2.2 is said to be **telescoping** or to **telescope**. Note that

$$\sum_{j=0}^{\infty}(z_j - z_{j+1}) = -\sum_{j=0}^{\infty}(z_{j+1} - z_j) = z_0 - \lim_{n\to\infty} z_n$$

if the series converges.

Example 3.2.2. Since

$$\frac{1}{j(j+1)} = \frac{1}{j} - \frac{1}{j+1}$$

for all $j > 0$, the series

$$\sum_{j=1}^{\infty}\frac{1}{j(j+1)}$$

telescopes. It converges to $1 - 0 = 1$.   △

Example 3.2.3. Let a and b be complex numbers, and let $z_0 = a$, $z_1 = b$, and

$$z_n = \frac{z_{n-1} + z_{n-2}}{2}$$

for all $n \geq 2$. In Example 2.6.1 we showed that this sequence is Cauchy and converges to $(a + 2b)/3$. We now give another way to find its limit.

Note first that

$$z_n - z_{n-1} = -\frac{1}{2}(z_{n-1} - z_{n-2})$$

for all $n \geq 2$. By induction it follows that

$$z_n - z_{n-1} = \left(-\frac{1}{2}\right)^{n-1} (b - a),$$

for all $n \geq 1$, so that

$$\sum_{n=1}^{\infty}(z_n - z_{n-1}) = \sum_{n=0}^{\infty}\left(-\frac{1}{2}\right)^n (b - a) = \frac{b - a}{1 - \left(-\frac{1}{2}\right)} = \frac{2(b - a)}{3}.$$

But the series on the left-hand side is telescoping, and so

$$\lim_{n\to\infty} z_n - z_0 = \frac{2(b - a)}{3},$$

whence

$$\lim_{n\to\infty} z_n = a + \frac{2(b - a)}{3} = \frac{a + 2b}{3}.$$

△

Suppose a series $\sum_{j=1}^{\infty} z_j$ converges to some number S. Then, by definition, the sequence $\{S_n\}$ also converges to S, where

$$S_n = \sum_{j=1}^{n} z_j$$

for each $n > 0$. So does the subsequence $\{S_{t_n}\}$, where $\{t_n\}$ is any increasing sequence of positive integers. (See Theorem 2.4.1.) Now

$$S_{t_n} = \sum_{j=1}^{t_n} z_j = \sum_{k=0}^{n-1} \sum_{j=t_k+1}^{t_{k+1}} z_j,$$

where we define $t_0 = 0$. Thus

$$\sum_{j=1}^{\infty} z_j = \sum_{k=0}^{\infty} \sum_{j=t_k+1}^{t_{k+1}} z_j. \tag{3.3}$$

This result shows that the terms of a convergent series may be grouped together by means of parentheses without affecting the convergence of the series. In Eq. (3.3) we group together first the terms

$$z_1, z_2, \ldots, z_{t_1},$$

followed by the terms

$$z_{t_1+1}, z_{t_1+2}, \ldots, z_{t_2},$$

and so forth, giving

$$(z_1 + z_2 + \cdots + z_{t_1}) + (z_{t_1+1} + z_{t_1+2} + \cdots + z_{t_2}) + \cdots .$$

On the other hand, if some terms are already grouped together, then removal of the parentheses used to group them may change a convergent series into a divergent one. For example,

$$\sum_{j=0}^{\infty}(1 - 1) = \sum_{j=0}^{\infty}0 = 0,$$

but the series obtained by removing the parentheses is

$$\sum_{j=0}^{\infty}(-1)^j,$$

which diverges because its partial sums alternate between 1 and 0. However, if the series with the parentheses removed does converge to a number S, then the original series must also converge to S, as we have just seen.

There is one important circumstance under which the removal of parentheses does not alter the convergence of the series: let us suppose that $a_j \geq 0$ for each j. Assuming that the series on the right-hand side of (3.3) converges when $z_j = a_j$ for all $j > 0$, by definition the sequence $\{S_{t_n}\}$ also converges to some number L, where

$$S_n = \sum_{j=1}^{n} a_j$$

for each $n > 0$. Hence $\{S_{t_n}\}$ is bounded above. Now the sequences $\{S_n\}$ and $\{S_{t_n}\}$ are nondecreasing, since $a_j \geq 0$ for each j. Therefore both sequences are bounded above by L because for each n there exists an integer k such that $n \leq t_k$, and so $S_n \leq S_{t_k} \leq L$. Being nondecreasing, the sequence $\{S_n\}$ converges to some number $M \leq L$. In other words, the series $\sum_{j=1}^{\infty} a_j$ converges to M. But we also have $L \leq M$ because $S_{t_k} \leq S_n \leq M$ for all $n \geq t_k$. Thus we have proved the following fact.

Theorem 3.2.3. *Suppose that $a_j \geq 0$ for each j. If the series*

$$\sum_{k=0}^{\infty} \sum_{j=t_k+1}^{t_{k+1}} a_j$$

converges to some number L, then so does $\sum_{j=1}^{\infty} a_j$.

Exercises 3.1. 1. Find the limits of the following series:

(a) $\sum_{j=0}^{\infty} \frac{2}{3^j}$

(e) $\sum_{j=0}^{\infty} \frac{1}{\sqrt{j}+\sqrt{j+1}}$

(b) $\sum_{j=2}^{\infty} \frac{j-1}{j!}$

(f) $\sum_{j=3}^{\infty} \frac{4}{j^2-4}$

(c) $\sum_{j=1}^{\infty} \frac{1}{j(j+2)}$

(g) $\sum_{j=0}^{\infty} \frac{2}{j^2+4j+3}$

(d) $\sum_{j=1}^{\infty} \frac{2j+1}{j^2(j+1)^2}$

(h) $\sum_{j=0}^{\infty} \frac{6^j}{3^{j+1}-2^{j+1}+3^j-2^j}$.

2. Show that if $\lim_{n\to\infty} x_n = \infty$, then

$$\sum_{j=0}^{\infty} \left(\frac{1}{x_j} - \frac{1}{x_{j+1}} \right)$$

converges.

3. A decimal number is said to be repeating if there is a finite sequence of digits that is repeated indefinitely.

(a) Express the repeating decimal $1.2323\ldots$ as a fraction.

(b) Show that every repeating decimal number is rational.

4. For all $n \in \mathbb{N}$ let

$$S_n = \sum_{j=1}^{n} \frac{1}{j}.$$

Show that

$$S_{2n} - S_n \geq \frac{1}{2}$$

and hence that $\{S_n\}$ is not Cauchy.

5. Let $\{x_n\}$ be a sequence of positive terms, and suppose that $\{S_n\}$ diverges, where $S_n = \sum_{j=0}^{n} x_j$ for all n.

(a) Show that

$$\left\{ \sum_{j=0}^{n} \frac{x_j}{S_j^2} \right\}$$

is Cauchy and hence converges.

(b) Is the above true for

$$\left\{ \sum_{j=0}^{n} \frac{x_j}{S_j} \right\}?$$

6. Show that

$$\sum_{j=1}^{\infty} \frac{1}{j(j+1)\cdots(j+k)},$$

where $k \in \mathbb{N}$, is telescoping and hence find the sum.

7. Find the sum

$$\sum_{j=0}^{\infty} \left(\frac{1}{j+i+1} - \frac{1}{j+i} \right).$$

8. Let $\{x_n\}$ be a sequence of positive terms. Show that $\sum_{j=0}^{\infty} x_j$ converges if and only if the sequence $\{\sum_{j=0}^{n} x_j\}$ is bounded.

3.3 Elementary Properties of Series

Since the behavior of a series is determined by its sequence of partial sums, many theorems about series can be derived from the analogous theorems about sequences. For instance, our next theorem follows immediately from Theorem 1.5.2.

Theorem 3.3.1. *Let $\sum_{j=0}^{\infty} w_j$ and $\sum_{j=0}^{\infty} z_j$ be convergent series. Then, for all numbers s and t,*

$$\sum_{j=0}^{\infty} (sw_j + tz_j) = s \sum_{j=0}^{\infty} w_j + t, \sum_{j=0}^{\infty} z_j. \tag{3.4}$$

Note that the convergence of the series on the left-hand side of Eq. (3.4) constitutes part of the conclusion of the theorem. However, as for sequences, the series on the left-hand side may converge while the series on the right do not. For instance, for $s = t = 1$ the series on the right-hand side both diverge if $w_j = 1$ and $z_j = -1$ for all j, but the series on the left-hand side converges to 0.

The following theorem also is immediate from Theorem 2.3.11 and the definition of series in terms of sequences of partial sums.

Theorem 3.3.2. *A series $\sum_{j=0}^{\infty} z_j$ converges if and only if the series $\sum_{j=0}^{\infty} \mathrm{Re}\,(z_j)$ and $\sum_{j=0}^{\infty} \mathrm{Im}\,(z_j)$ both converge.*

The following result gives a necessary condition for a series to be convergent.

Theorem 3.3.3. *If the series $\sum_{j=0}^{\infty} z_j$ converges, then $\lim_{n\to\infty} z_n = 0$.*

Proof. For each $n \geq 0$ let

$$S_n = \sum_{j=0}^{n} z_j.$$

Then $z_n = S_n - S_{n-1}$ for each $n > 0$. If

$$\sum_{j=0}^{\infty} z_j = S,$$

then

$$\lim_{n \to \infty} S_{n-1} = \lim_{n \to \infty} S_n = S,$$

so that

$$\lim_{n \to \infty} z_n = \lim_{n \to \infty} (S_n - S_{n-1}) = S - S = 0.$$

$\square$

The contrapositive of the theorem above is particularly useful in establishing the divergence of a series. It is known as the *n*th-term test.

Corollary 3.3.4 (*n***th-Term Test**). *If* $\lim_{n \to \infty} z_n$ *does not exist or is nonzero, then the series* $\sum_{j=0}^{\infty} z_j$ *diverges.*

For example, if $|z| \geq 1$, then the series $\sum_{j=0}^{\infty} z^j$ must diverge since

$$\lim_{n \to \infty} |z^n| = \lim_{n \to \infty} |z|^n \neq 0.$$

This argument gives an alternative proof of part of Theorem 3.2.1.

However, convergence of the sequence $\{z_n\}$ to 0 does not imply convergence of the series $\sum_{j=0}^{\infty} z_j$. This point is illustrated in the following example. The series in this example is called the **harmonic series**.

Example 3.3.1. The series

$$\sum_{j=1}^{\infty} \frac{1}{j}$$

diverges. A proof has been given in Example 2.6.2. We now present an alternative proof.

Let $S_n = \sum_{j=1}^{n} 1/j$ for all positive integers n. We use induction to show that

$$S_{2^n} > \frac{n+1}{2}$$

for all nonnegative integers n. It will then follow that $\{S_n\}$ diverges, as it has a divergent subsequence. Since $S_1 = 1 > 1/2$, the required inequality certainly holds for $n = 0$. We may therefore assume that $n > 0$ and that the result holds for $n - 1$. Then

$$S_{2^n} = \sum_{j=1}^{2^n} \frac{1}{j}$$

$$= \sum_{j=1}^{2^{n-1}} \frac{1}{j} + \sum_{j=2^{n-1}+1}^{2^n} \frac{1}{j}$$

$$\geq S_{2^{n-1}} + \sum_{j=2^{n-1}+1}^{2^n} \frac{1}{2^n}$$

$$> \frac{n}{2} + \frac{1}{2^n} \sum_{j=2^{n-1}+1}^{2^n} 1$$

$$= \frac{n}{2} + \frac{2^n - 2^{n-1}}{2^n}$$

$$= \frac{n+1}{2}.$$

$\triangle$

Remark. For every positive integer n we have

$$S_{2^n} = \sum_{j=1}^{2^n} \frac{1}{j}$$

$$= 1 + \sum_{j=2}^{2^n} \frac{1}{j}$$

$$= 1 + \sum_{k=0}^{n-1} \sum_{j=2^k+1}^{2^{k+1}} \frac{1}{j}$$

$$< 1 + \sum_{k=0}^{n-1} \sum_{j=2^k+1}^{2^{k+1}} \frac{1}{2^k}$$

$$= 1 + \sum_{k=0}^{n-1} \frac{1}{2^k} \sum_{j=2^k+1}^{2^{k+1}} 1$$

$$= 1 + \sum_{k=0}^{n-1} \frac{2^k}{2^k}$$

$$= 1 + \sum_{k=0}^{n-1} 1$$

$$= 1 + n.$$

Thus the divergence of the harmonic series is extraordinarily slow.

Exercises 3.2. 1. Can $\sum_{j=0}^{\infty}(x_j + y_j)$ converge when at least one of $\sum_{j=0}^{\infty} x_j$ and $\sum_{j=0}^{\infty} y_j$ diverges?

2. Show that if $\sum_{j=0}^{\infty} z_j$ converges, where $z_j \neq 0$ for all j, then

$$\sum_{j=0}^{\infty} \frac{1}{z_j}$$

diverges.

3. Is

$$\sum_{j=1}^{\infty}(-1)^j 2^{1/j}$$

convergent?

4. Show that $\sum_{j=1}^{\infty} aj$ converges if and only if $a = 0$.

5. Prove that

$$\sum_{j=0}^{\infty} \frac{j^2}{7j^2 + 11}$$

diverges.

6. Evaluate the following sums:

(a) $\sum_{j=0}^{\infty} \frac{(-1)^j + 2^j}{3^j}$;

(b) $\sum_{j=1}^{\infty} \left(\frac{2}{5^j} - \frac{1}{j(j+1)} \right)$.

7. Let $\{x_n\}$ be a decreasing sequence of positive terms and suppose that $\sum_{j=0}^{\infty} x_j$ converges. Show that

$$\lim_{n \to \infty} n x_n = 0.$$

This result is known as Pringsheim's theorem. [Hint: Define $m = n/2$ for even n and $m = (n + 1)/2$ for odd n. Then

$$S_n - S_m \geq n x_n.]$$

8. Show that if $an + b \neq 0$ for all nonnegative integers n, then

$$\sum_{j=0}^{\infty} \frac{1}{aj + b}$$

diverges. (Use question 7.)

3.4 The Comparison Test

Often it is difficult or impossible to compute the partial sums for a series, and so it is necessary to have some tests for convergence that do not depend on knowledge of the partial sums. Many such tests require the series to contain only nonnegative terms (which of course must be real). We devote the next few sections to such tests. One of them is the comparison test, in which we compare the terms of a given series with the corresponding terms of a series whose behavior is already known.

Theorem 3.4.1 (Comparison Test). *Let $\sum_{j=0}^{\infty} a_j$ and $\sum_{j=0}^{\infty} b_j$ be two series of nonnegative terms, and suppose that $a_j \leq b_j$ for all j greater than or equal to some nonnegative integer N. If the latter series converges, then so does the former.*

Proof. We may assume that $N = 0$, since the series $\sum_{j=0}^{\infty} b_j$ converges if and only if $\sum_{j=N}^{\infty} b_j$ does so, and similarly for $\sum_{j=0}^{\infty} a_j$. Since the series $\sum_{j=0}^{\infty} b_j$ converges, so does the sequence $\{\sum_{j=0}^{n} b_j\}$ of partial sums. This is a nondecreasing sequence since $b_j \geq 0$ for all j. If its limit is L then, since $a_j \leq b_j$ for all j, it follows that

$$\sum_{j=0}^{n} a_j \leq \sum_{j=0}^{n} b_j \leq L$$

for each n. Thus the nondecreasing sequence $\{\sum_{j=0}^{n} a_j\}$ is bounded above and hence converges. We conclude that the series $\sum_{j=0}^{\infty} a_j$ converges. $\square$

Corollary 3.4.2. *Let $\sum_{j=0}^{\infty} a_j$ and $\sum_{j=0}^{\infty} b_j$ be series of nonnegative terms, and suppose that $a_j \leq b_j$ whenever $j \geq N \geq 0$. If the former series diverges, then so does the latter.*

Remark. If a series of nonnegative terms converges, then the corresponding sequence of partial sums is convergent and therefore bounded. Conversely, if the sequence of partial sums is bounded, then, being nondecreasing, it converges and so the series is convergent.

Example 3.4.1. Let us test the series

$$\sum_{j=1}^{\infty} \frac{1}{3^j j}.$$

Set $a_j = 1/(3^j j)$ and $b_j = 1/3^j$ for each $j \geq 1$. Then $a_j \leq b_j$ for each j. But $\sum_{j=0}^{\infty} b_j$ is the convergent geometric series

$$\sum_{j=0}^{\infty} \left(\frac{1}{3}\right)^j.$$

Thus $\sum_{j=1}^{\infty} b_j$ converges, and therefore so does $\sum_{j=1}^{\infty} a_j$ by the comparison test. $\triangle$

Exercises 3.3.   1. Show that

$$\sum_{j=1}^{\infty} \frac{1}{e^j \sqrt{j}}$$

converges.

2. Show that

$$\sum_{j=1}^{\infty} \frac{1}{\sqrt{j}}$$

diverges.

3. Let $\{x_n\}$ be a sequence of positive terms and suppose that $\sum_{j=0}^{\infty} x_j$ converges. Show that $\sum_{j=0}^{\infty} x_j^2$ converges.

4. Let $\{x_n\}$ be a sequence of positive terms. Show that $\sum_{j=0}^{\infty} x_j$ converges if and only if

$$\sum_{j=0}^{\infty} \frac{x_j}{1 + x_j}$$

converges.

5. Show that if $\sum_{j=0}^{\infty} x_j^2$ converges, then so does

$$\sum_{j=1}^{\infty} \frac{|x_j|}{j}.$$

6. Suppose that $0 < a_{n+1} \leq \alpha a_n$ for all n, where $\alpha < 1$. Show that

$$\sum_{j=0}^{\infty} a_j \leq \frac{a_0}{1 - \alpha}.$$

7. Let $\{x_n\}$ be a sequence of positive terms. Show that if $\sum_{j=0}^{\infty} x_j$ converges, then so does

$$\sum_{j=0}^{\infty} \sqrt{x_j x_{j+1}}.$$

(Hint: Use Lemma 2.5.5.)

8. Let $\{x_n\}$ be a sequence of positive terms, and suppose that $\{S_n\}$ diverges, where $S_n = \sum_{j=0}^{n} x_j$ for all n. Show that

$$\sum_{j=0}^{\infty} \frac{x_j}{S_j^m}$$

converges if and only if $m > 1$. (See question 5 at the end of Sect. 3.2.)

9. Let $\{a_n\}$ be an increasing sequence of positive terms. Show that

$$\sum_{j=0}^{\infty} \left(1 - \frac{a_j}{a_{j+1}}\right)$$

converges if and only if $\{a_n\}$ is bounded. (Apply question 5 with

$$x_n = a_{n+1} - a_n.)$$

10. Show that if $m > 0$ and $p > 0$, then

$$\lim_{n \to \infty} \left(\sum_{j=1}^{n} \frac{j^m}{1 + j^{m+1}} \bigg/ \sum_{j=1}^{n} \frac{j^p}{1 + j^{p+1}}\right) = 1.$$

11. Let $\{y_n\}$ be a bounded sequence of positive terms. Suppose that $\sum_{j=0}^{\infty} x_j$ is a convergent series of nonnegative terms. Show that the series $\sum_{j=0}^{\infty} x_j y_j$ is convergent.

3.5 Cauchy's Condensation Test

Given the series $\sum_{j=0}^{\infty} a_j$, the series

$$\sum_{k=0}^{\infty} 2^k a_{2^k}$$

is called the corresponding **condensed series**. We may write it as

$$a_1 + (a_2 + a_2) + (a_4 + a_4 + a_4 + a_4) + \cdots.$$

The importance of the condensed series is revealed by the following theorem. It is remarkable in that it enables us to settle the convergence of a series by considering only a very thin sample of its terms.

Theorem 3.5.1 (Cauchy's Condensation Test). *If $\{a_n\}$ is a nonincreasing sequence of nonnegative terms, then the series $\sum_{j=0}^{\infty} a_j$ converges if and only if its condensed series does so.*

Proof. It suffices to check the convergence of the series $\sum_{j=1}^{\infty} a_j$ and its condensed series. We may write the sum of the first $2^{n+1} - 1$ terms of the former series as

$$a_1 + (a_2 + a_3) + (a_4 + a_5 + a_6 + a_7) + \ldots + (a_{2^n} + a_{2^n+1} + \ldots + a_{2^{n+1}-1}).$$

Thus

$$\sum_{j=1}^{2^{n+1}-1} a_j = \sum_{k=0}^{n} \sum_{j=2^k}^{2^{k+1}-1} a_j.$$

Since $a_j \leq a_{2^k}$ whenever $2^k \leq j < 2^{k+1}$, we find that

$$\sum_{j=2^k}^{2^{k+1}-1} a_j \leq \sum_{j=2^k}^{2^{k+1}-1} a_{2^k}$$

$$= a_{2^k} \sum_{j=2^k}^{2^{k+1}-1} 1$$

$$= 2^k a_{2^k}.$$

It therefore follows from the comparison test and Theorem 3.2.3, because of the assumption that $a_j \geq 0$ for all j, that if the condensed series converges, then so does $\sum_{j=1}^{\infty} a_j$.

On the other hand, for all $n > 0$ we can also write

$$\sum_{j=1}^{2^n} a_j = a_1 + a_2 + (a_3 + a_4) + \ldots + (a_{2^{n-1}+1} + a_{2^{n-1}+2} + \ldots + a_{2^n})$$

$$= a_1 + \sum_{k=1}^{n} \sum_{j=2^{k-1}+1}^{2^k} a_j.$$

As $a_j \geq a_{2^k}$ whenever $2^{k-1} < j \leq 2^k$, it follows that

$$\sum_{j=2^{k-1}+1}^{2^k} a_j \geq \sum_{j=2^{k-1}+1}^{2^k} a_{2^k}$$

$$= 2^{k-1} a_{2^k}$$
$$= \frac{1}{2} \cdot 2^k a_{2^k},$$

and so

$$\sum_{k=1}^{n} 2^k a_{2^k} \leq 2 \sum_{k=1}^{n} \sum_{j=2^{k-1}+1}^{2^k} a_j.$$

If the series $\sum_{j=1}^{\infty} a_j$ converges, then so does

$$\sum_{k=1}^{\infty} \sum_{j=2^{k-1}+1}^{2^k} 2a_j,$$

and consequently the comparison test shows that the condensed series also converges. □

Remark. We observe from the proof above that if

$$\sum_{j=1}^{\infty} a_j = S$$

and

$$\sum_{j=0}^{\infty} 2^j a_{2^j} = T,$$

then $S \leq T \leq 2S$.

The next theorem, which generalizes Example 3.3.1, is an application of the condensation test.

Theorem 3.5.2. *For every rational number p, the series*

$$\sum_{j=1}^{\infty} \frac{1}{j^p}$$

converges if $p > 1$ and diverges otherwise.

Proof. If $p \leq 0$, then $1/j^p \geq 1$ for all positive integers j. Thus the sequence $\{1/n^p\}$ does not converge to 0 and the series diverges by the nth-term test.

For each $p > 0$ the sequence $\{1/n^p\}$ is decreasing and Cauchy's condensation test can be applied. The condensed series is

$$\sum_{j=0}^{\infty} 2^j \frac{1}{2^{jp}} = \sum_{j=0}^{\infty} \frac{1}{2^{j(p-1)}} = \sum_{j=0}^{\infty} \left(\frac{1}{2^{p-1}} \right)^j.$$

This is a geometric series and converges if and only if $1/2^{p-1} < 1$. Hence the condensed series, and therefore the given series, converges if and only if $p - 1 > 0$, as required.   □

The series given in Theorem 3.5.2 is generally referred to as the p-series.

Exercises 3.4. 1. Using the knowledge gained so far, discuss the convergence of the following series:

(a) $\sum_{j=2}^{\infty} \frac{1}{j\sqrt{j-1}}$;

(b) $\sum_{j=0}^{\infty} \frac{1}{aj^2+b}$, where $a > 0$ and $b > 0$;

(c) $\sum_{j=0}^{\infty} \frac{1}{j!}$.

2. Let $\{a_n\}$ be a nonincreasing sequence of nonnegative terms. Prove that the series $\sum_{j=0}^{\infty} a_j$ converges if and only if the series

$$\sum_{j=0}^{\infty} 3^j a_{3^j}$$

converges. (This result is in fact true if we replace 3 by any integer greater than 2 [12].)

3. (a) The following result is a special case of a theorem in [1]. Let $\sum_{j=1}^{\infty} a_j$ be a series of positive terms. Suppose that the set

$$\left\{ \frac{\max\{a_{2n}, a_{2n+1}\}}{a_n} \,\middle|\, n \in \mathbb{N} \right\}$$

is bounded below and above by l and L, respectively. Then the series converges if $L < 1/2$ and diverges if $l > 1/2$.

Fill in the details in the following brief sketch of the proof. Suppose that $L < 1/2$. As in Cauchy's condensation test,

$$S_{2^{n+1}-1} = \sum_{j=1}^{2^{n+1}-1} a_j = a_1 + \sum_{k=1}^{n} T_k,$$

where

$$T_k = \sum_{j=2^k}^{2^{k+1}-1} a_j$$

$$= \sum_{j=2^{k-1}}^{2^k-1} (a_{2j} + a_{2j+1})$$

$$\leq 2L T_{k-1}$$

$$\leq (2L)^k a_1.$$

Deduce that the sequence $\{S_{2^n+1-1}\}$ converges, and hence that $\{S_n\}$ does so. Similarly, prove the theorem for the case where $l > 1/2$.

(b) Use the theorem of part (a) to prove the convergence of the following series:

i. $\sum_{j=1}^{\infty} \frac{1}{j^p}$ for all rational $p > 1$;

ii. $\sum_{j=1}^{\infty} \frac{(2j-1)!}{2^{2j-1}(j-1)!(j+1)!}$.

[Hint: If a_n denotes the nth term of the series, then

$$\frac{a_{2n}}{a_n} \le \frac{1}{2}\left(1 - \frac{1}{4n}\right)^{n-1}.]$$

4. Test the convergence of the series

$$\sum_{j=1}^{\infty} \frac{(2j-1)!}{2^{2j-1}(j-1)!j!}.$$

[Hint: If a_n denotes the nth term of the series, then $a_n > 1/(2n)$.]

3.6   The Limit Comparison Test

Our next test for convergence is often easier to apply than the comparison test. First, however, we require some new notation.

Definition 3.6.1. Let $\{a_n\}$ and $\{b_n\}$ be sequences of positive terms. Then a_n and b_n are said to be of the **same order of magnitude** if there is a positive number L such that

$$\lim_{n \to \infty} \frac{a_n}{b_n} = L.$$

In this case we write $a_n \sim Lb_n$. We say that a_n is of a **lesser order of magnitude** than b_n, and write $a_n << b_n$, if

$$\lim_{n \to \infty} \frac{a_n}{b_n} = 0.$$

Finally, a_n is of a **greater order of magnitude** than b_n if

$$\lim_{n \to \infty} \frac{a_n}{b_n} = \infty.$$

We then write $a_n >> b_n$.

The basic intuitive idea is that if $a_n << b_n$, then b_n increases with n much faster (or decreases much more slowly) than does a_n. Theorem 2.8.4 shows that $a_n << b_n$ if and only if $b_n >> a_n$. Moreover if $\{c_n\}$ is another sequence of positive terms and $a_n << b_n << c_n$, then $a_n << c_n$. This result follows immediately from the observation that

$$\frac{a_n}{c_n} = \frac{a_n}{b_n} \cdot \frac{b_n}{c_n}.$$

Observe also that if $a_n \sim Lb_n$, then $b_n \sim a_n/L$.

The following example is worth noting.

Example 3.6.1. If p is a rational number and $c > 1$, then

$$n^p << c^n << n! << n^n.$$

Proof. These results are immediate from Examples 2.5.5, 2.5.6, and 2.7.10.   △

We are now ready for our next convergence test, which is known as the limit comparison test.

Theorem 3.6.1 (Limit Comparison Test). *Let $\sum_{j=0}^{\infty} a_j$ and $\sum_{j=0}^{\infty} b_j$ be series of positive terms.*

1. *If $a_n \sim Lb_n$ for some $L > 0$, then both series converge or both diverge.*
2. *If $a_n << b_n$ and $\sum_{j=0}^{\infty} b_j$ converges, then so does $\sum_{j=0}^{\infty} a_j$.*
3. *If $a_n >> b_n$ and $\sum_{j=0}^{\infty} b_j$ diverges, then so does $\sum_{j=0}^{\infty} a_j$.*

Proof. 1. We are given that

$$\lim_{n\to\infty} \frac{a_n}{b_n} = L > 0.$$

Proposition 2.3.5 shows the existence of a number N_1 such that $a_n/b_n > L/2$ for all $n \geq N_1$. Similarly, there exists N_2 such that $a_n/b_n < L + L/2 = 3L/2$ for all $n \geq N_2$. Take $N = \max\{N_1, N_2\}$ and choose $n \geq N$. Then

$$\frac{L}{2} < \frac{a_n}{b_n} < \frac{3L}{2};$$

hence

$$a_n < \frac{3Lb_n}{2}.$$

We now apply the comparison test. Suppose $\sum_{j=0}^{\infty} b_j$ is convergent. By Theorem 3.3.1 the series

$$\sum_{j=0}^{\infty} \frac{3Lb_j}{2}$$

also converges, whence $\sum_{j=0}^{\infty} a_j$ converges by the comparison test. Similarly, if $\sum_{j=0}^{\infty} a_j$ converges, then so does $\sum_{j=0}^{\infty} b_j$.

2. Suppose

$$\lim_{n \to \infty} \frac{a_n}{b_n} = 0.$$

Then there exists N such that

$$\frac{a_n}{b_n} < 1$$

for all $n \geq N$. Hence $a_n < b_n$ for all $n \geq N$, and we conclude from the comparison test that $\sum_{j=0}^{\infty} a_j$ converges if $\sum_{j=0}^{\infty} b_j$ does so.

3. This statement is equivalent to the previous one.   □

Example 3.6.2. Test the series

$$\sum_{j=0}^{\infty} \frac{1}{(j+1)(j+2)}$$

for convergence.

Solution. Let

$$a_n = \frac{1}{(n+1)(n+2)}$$

for all $n > 0$. We show that a_n has the same order of magnitude as $1/n^2$. Accordingly, we put

$$b_n = \frac{1}{n^2}$$

for all $n > 0$. Then

$$\frac{a_n}{b_n} = \frac{n^2}{(n+1)(n+2)} \to 1$$

as $n \to \infty$. Thus $a_n \sim b_n$. As $\sum_{j=1}^{\infty} b_j$ converges by Theorem 3.5.2, so does $\sum_{j=0}^{\infty} a_j$ by the limit comparison test.   △

Example 3.6.3. Test the series

$$\sum_{j=0}^{\infty}(\sqrt{j+1}-\sqrt{j})^2$$

for convergence.

Solution. For all $n \geq 0$ let

$$a_n = (\sqrt{n+1}-\sqrt{n})^2$$

$$= \left(\frac{(\sqrt{n+1}-\sqrt{n})(\sqrt{n+1}+\sqrt{n})}{\sqrt{n+1}+\sqrt{n}}\right)^2$$

$$= \frac{1}{(\sqrt{n+1}+\sqrt{n})^2},$$

and let

$$b_n = \frac{1}{n}$$

for all $n > 0$. Since

$$\frac{b_n}{a_n} = \frac{(\sqrt{n+1}+\sqrt{n})^2}{n}$$

$$= \frac{2n+1+2\sqrt{n(n+1)}}{n}$$

$$= 2 + \frac{1}{n} + 2\sqrt{1+\frac{1}{n}}$$

$$\to 4$$

as $n \to \infty$, it follows that $a_n \sim b_n/4$. As $\sum_{j=1}^{\infty} b_j$ is the divergent harmonic series, $\sum_{j=0}^{\infty} a_j$ diverges by the limit comparison test.   △

Example 3.6.4. Test the series

$$\sum_{j=1}^{\infty}\frac{1}{j^{1+\frac{1}{j}}}.$$

Solution. Taking

$$a_n = \frac{1}{n^{1+\frac{1}{n}}}$$

and

$$b_n = \frac{1}{n}$$

for all $n > 0$, we have

$$\frac{b_n}{a_n} = \frac{n^{1+\frac{1}{n}}}{n} = n^{1/n} \to 1$$

as $n \to \infty$. Since $\sum_{j=1}^{\infty} b_j$ diverges, so does $\sum_{j=1}^{\infty} a_j$.
  The reader should compare this result with that of Theorem 3.5.2. $\triangle$

Exercises 3.5. 1. Test the convergence of each of the following series:

(a) $\sum_{j=0}^{\infty} \frac{2^{j+1}}{5^j - j}$;

(b) $\sum_{j=1}^{\infty} \frac{\sqrt{j+1} - \sqrt{j}}{j}$;

(c) $\sum_{j=1}^{\infty} \frac{e^{1/j}}{j^2}$;

(d) $\sum_{j=3}^{\infty} \frac{1}{\sqrt{j-3} + \sqrt{j+3}}$;

(e) $\sum_{j=1}^{\infty} (e^{1/j^p} - 1)$ for every rational p [use Theorem 2.7.10(3)];

(f) $\sum_{j=1}^{\infty} (e^{1/j} - 1)^p$ for every rational p;

(g) $\sum_{j=0}^{\infty} (\sqrt{j^2 + 1} - j)$;

(h) $\sum_{j=1}^{\infty} \frac{\sqrt{j+1} - \sqrt{j}}{\sqrt{j(j+1)}}$;

(i) $\sum_{j=1}^{\infty} \frac{1}{j^{2-1/j}}$.

2. Find all integers t such that

$$\sum_{j=1}^{\infty} \frac{1}{j^{1+t-\sqrt{t}}}$$

is convergent.

3. Let $a > 0$ and $b > 0$. Find all rational p such that

$$\sum_{j=0}^{\infty} \frac{1}{(aj+b)^p}$$

is convergent.

4. Test the convergence of the series

$$\sum_{j=0}^{\infty} \frac{1}{aj^2 + bj + c},$$

where a, b, c are all positive.

5. Prove that

$$\lim_{n \to \infty} (n!)^{1/n} = \infty.$$

3.7   The Ratio Test

In order to use the comparison or limit comparison test effectively, we need a supply of series whose convergence or divergence has already been established. We shall therefore develop another test which involves only the terms of the series being tested. Its proof uses the comparison test.

Theorem 3.7.1. *Let $\sum_{j=0}^{\infty} a_j$ be a series of positive terms and suppose there exist numbers r and N such that*

$$\frac{a_{n+1}}{a_n} \le r < 1$$

for all $n \ge N$. Then the series converges. On the other hand, if

$$\frac{a_{n+1}}{a_n} \ge 1$$

for all $n \ge N$, then the series diverges.

Proof. In the first case we have $a_{n+1} \le r a_n$ for all $n \ge N$. It follows by induction that

$$a_{N+j} \le r^j a_N$$

for each positive integer j. Now the series $\sum_{j=0}^{\infty} a_N r^j$ converges since $0 < r < 1$, and so $\sum_{j=0}^{\infty} a_{N+j}$ converges also, by the comparison test. Hence $\sum_{j=0}^{\infty} a_j$ converges.

   In the second case we have $a_{n+1} \ge a_n$ for all $n \ge N$ and the series diverges by the nth-term test.   □

Corollary 3.7.2 (Ratio Test). *Let $\sum_{j=0}^{\infty} a_j$ be a series of positive terms and let*

$$\lim_{n \to \infty} \frac{a_{n+1}}{a_n} = L$$

for some number L. Then the series converges if $L < 1$ and diverges if $L > 1$.

Proof. If $L < 1$, then by Proposition 2.3.4 there exist numbers N_1 and $r < 1$ such that

$$\frac{a_{n+1}}{a_n} < r$$

for all $n \geq N_1$. The result therefore follows immediately from Theorem 3.7.1 in this case.

On the other hand, suppose $L > 1$. There exists N_2 such that $a_{n+1}/a_n > 1$ for all $n \geq N_2$, and again the result follows from Theorem 3.7.1. $\qquad\square$

Remark. No conclusion can be drawn when $L = 1$ in the ratio test. For example, this is the case for both the series $\sum_{j=1}^{\infty} 1/j$ and $\sum_{j=1}^{\infty} 1/j^2$, but the former series diverges whereas the latter converges.

Example 3.7.1. Test the series

$$\sum_{j=0}^{\infty} \frac{(j+3)2^{j+1}}{3^j}.$$

Solution. Putting

$$a_n = \frac{(n+3)2^{n+1}}{3^n}$$

for all $n \geq 0$, we have

$$\begin{aligned}
\frac{a_{n+1}}{a_n} &= \frac{(n+4)2^{n+2}}{3^{n+1}} \cdot \frac{3^n}{(n+3)2^{n+1}} \\
&= \frac{2(n+4)}{3(n+3)} \\
&\to \frac{2}{3}
\end{aligned}$$

as $n \to \infty$. Hence the series converges by the ratio test. $\qquad\triangle$

Example 3.7.2. Test the series

$$\sum_{j=0}^{\infty} \frac{x^{2j}}{(2j)!},$$

where $x \neq 0$.

Solution. With

$$a_n = \frac{x^{2n}}{(2n)!}$$

for all n, we have

$$\frac{a_{n+1}}{a_n} = \frac{x^{2n+2}}{(2n+2)!} \cdot \frac{(2n)!}{x^{2n}}$$

$$= \frac{x^2}{(2n+2)(2n+1)}$$

$$\to 0,$$

and so the series converges.

A similar argument shows that

$$\sum_{j=0}^{\infty} \frac{x^{2j+1}}{(2j+1)!}$$

also converges for all $x > 0$.   △

Example 3.7.3. The results of Example 3.6.1 can also be achieved by using the ratio test. For instance, let us show that $n! << n^n$. If we set $a_n = n!/n^n$ for all $n > 0$, then the calculation in Example 2.7.10 shows that

$$\frac{a_{n+1}}{a_n} = \frac{1}{\left(1 + \frac{1}{n}\right)^n} \to \frac{1}{e} < 1.$$

Hence the series

$$\sum_{j=1}^{\infty} \frac{j!}{j^j}$$

converges by the ratio test. We deduce that

$$\lim_{n \to \infty} \frac{n!}{n^n} = 0,$$

as required. $\triangle$

Exercises 3.6. 1. Let $\sum_{j=0}^{\infty} a_j$ be a series of positive terms. Show that if $a_{n+1} \geq a_n$ for all n, then the series diverges.

2. Test the convergence of the following series:

(a) $\sum_{j=1}^{\infty} \frac{j! e^j}{j^j}$;   (d) $\sum_{j=1}^{\infty} \frac{(2j)!}{j^j}$;

(b) $\sum_{j=1}^{\infty} \frac{j^2}{j!}$;   (e) $\sum_{j=0}^{\infty} \frac{(j!)^2}{(2j)!}$;

(c) $\sum_{j=1}^{\infty} \frac{2^j j^2}{j!}$;   (f) $\sum_{j=0}^{\infty} \frac{(j!)^2 3^j}{(2j)!}$.

3. (a) Suppose that $a_n > 0$, $b_n > 0$, and $a_{n+1}/a_n \leq b_{n+1}/b_n$ for all n. Show that $\sum_{j=0}^{\infty} a_j$ converges if $\sum_{j=0}^{\infty} b_j$ does. (Hint: Show that the sequence $\{a_n/b_n\}$ is decreasing and use question 11 in the exercises at the end of Sect. 3.4.)

   (b) Use part (a) to test the convergence of the following series:

   i. $\sum_{j=1}^{\infty} \frac{j^j}{j e^j j!}$;

   ii. $\sum_{j=1}^{\infty} \frac{j^j}{e^j j!}$.

   Note that the ratio test yields no conclusion in these examples.

4. Test the convergence of the series

$$\sum_{j=1}^{\infty} \frac{j! x^j}{j^j}$$

for all $x > 0$.

5. The result in question 3 at the end of Sect. 3.5 is known as the second ratio test. Here is a special case that can be proved by using Cauchy's condensation test. Let $\{a_n\}$ be a decreasing sequence of positive terms. By applying the ratio test to the condensed series, show that $\sum_{j=0}^{\infty} a_j$ converges if

$$\lim_{n \to \infty} \frac{a_{2n}}{a_n} < \frac{1}{2}$$

and diverges if

$$\lim_{n \to \infty} \frac{a_{2n+1}}{a_n} > \frac{1}{2}.$$

3.8   The Root Test

A somewhat different application of the comparison test gives us another test called the root test. Again, it is applicable only to series of nonnegative terms. Its proof is similar to that of the ratio test.

Theorem 3.8.1. *Let $\sum_{j=0}^{\infty} a_j$ be a series of nonnegative terms, and suppose that there exist numbers r and $N > 0$ such that $a_n^{1/n} \leq r < 1$ for all $n \geq N$. Then the series converges.*

Proof. The hypothesis implies that $a_n \leq r^n$ for all $n \geq N$. Hence the series $\sum_{j=0}^{\infty} a_j$ converges by comparison with the convergent geometric series $\sum_{j=0}^{\infty} r^j$. $\square$

Corollary 3.8.2 (Root Test). *Let $\sum_{j=0}^{\infty} a_j$ be a series of nonnegative terms and let*

$$\lim_{n \to \infty} a_n^{1/n} = L$$

for some number L. Then the series converges if $L < 1$ and diverges if $L > 1$.

Proof. If $L < 1$, then there exist numbers $N_1 > 0$ and $r < 1$ such that $a_n^{1/n} < r$ for all $n \geq N_1$. In this case the result follows immediately from Theorem 3.8.1.

On the other hand, suppose $L > 1$. There exists N_2 such that $a_n^{1/n} > 1$ for all $n \geq N_2$. For each such n it follows that $a_n > 1$. Hence $\{a_n\}$ cannot converge to 0 and the given series diverges. $\square$

Remark. As in the ratio test, no inference may be drawn if $L = 1$. Consider the series $\sum_{j=1}^{\infty} 1/j$ and $\sum_{j=1}^{\infty} 1/j^2$ as in the remark following the introduction of the ratio test. Using the fact that $n^{1/n} \to 1$ as $n \to \infty$ (Example 2.5.1), we see that $L = 1$ for both series, but one series diverges and the other converges.

Example 3.8.1. Test the series

$$\sum_{j=1}^{\infty} \left(\frac{j}{j+1} \right)^{j^2}.$$

Solution. Setting

$$a_n = \left(\frac{n}{n+1} \right)^{n^2}$$

for all $n > 0$, we obtain

$$a_n^{1/n} = \left(\frac{n}{n+1}\right)^n = \frac{1}{\left(1+\frac{1}{n}\right)^n} \to \frac{1}{e} < 1.$$

Hence the series converges by the root test.   △

Remark. If $a_n^{1/n} < 1$ for all $n > 0$, it does not necessarily follow that the series $\sum_{j=0}^{\infty} a_j$ converges. Indeed, the harmonic series provides an example of a divergent series where this inequality holds for all $n > 1$.

We now show that if the ratio test is applicable, then so is the root test. In fact, the root test is stronger.

Theorem 3.8.3. *Let $\{a_n\}$ be a sequence of positive terms. If*

$$\lim_{n \to \infty} \frac{a_{n+1}}{a_n} = L,$$

where L may be a number or ∞, then

$$\lim_{n \to \infty} a_n^{1/n} = L.$$

Proof. Since $a_n > 0$ for all n, it follows that if L is a number then it must be nonnegative.

Case 1: Suppose $L > 0$, and choose $\varepsilon \in (0, L)$. There exists N_1 such that

$$L - \varepsilon < \frac{a_{n+1}}{a_n} < L + \varepsilon$$

for all $n \geq N_1$. Thus

$$(L - \varepsilon)a_n < a_{n+1} < (L + \varepsilon)a_n \tag{3.5}$$

for all such n. In particular,

$$(L - \varepsilon)a_{N_1} < a_{N_1+1} < (L + \varepsilon)a_{N_1}.$$

Suppose that

$$(L - \varepsilon)^m a_{N_1} < a_{N_1+m} < (L + \varepsilon)^m a_{N_1} \tag{3.6}$$

for some $m \in \mathbb{N}$. Then, using inequality (3.5) and the fact that $L - \varepsilon > 0$, we find that

$$(L-\varepsilon)^{m+1} a_{N_1} < (L-\varepsilon)a_{N_1+m} < a_{N_1+m+1} < (L+\varepsilon)a_{N_1+m} < (L+\varepsilon)^{m+1} a_{N_1}.$$

Therefore inequality (3.6) holds for all positive integers m, by induction. Any integer $n > N_1$ can be written as $n = N_1 + m$, where $m = n - N_1 > 0$, and so inequality (3.6) can be rewritten as

$$(L - \varepsilon)^n \frac{a_{N_1}}{(L - \varepsilon)^{N_1}} < a_n < (L + \varepsilon)^n \frac{a_{N_1}}{(L + \varepsilon)^{N_1}}.$$

Thus,

$$(L - \varepsilon)\left(\frac{a_{N_1}}{(L - \varepsilon)^{N_1}}\right)^{1/n} < a_n^{1/n} < (L + \varepsilon)\left(\frac{a_{N_1}}{(L + \varepsilon)^{N_1}}\right)^{1/n} \qquad (3.7)$$

for all $n > N_1$, since $L - \varepsilon > 0$. Denoting the left- and right-hand sides of inequality (3.7) by s_n and t_n, respectively, and using the result of Example 2.5.2, we find that $s_n \to L - \varepsilon$ and $t_n \to L + \varepsilon$ as $n \to \infty$. Therefore there exist N_2 and N_3 such that

$$-\varepsilon < s_n - L + \varepsilon < \varepsilon$$

for all $n \geq N_2$ and

$$-\varepsilon < t_n - L - \varepsilon < \varepsilon$$

for all $n \geq N_3$. For all $n > \max\{N_1, N_2, N_3\}$ it follows that

$$L - 2\varepsilon < s_n < a_n^{1/n} < t_n < L + 2\varepsilon,$$

and so

$$\lim_{n \to \infty} a_n^{1/n} = L,$$

as required.

Case 2: Suppose $L = 0$. For every $\varepsilon > 0$ there exists N such that

$$\frac{a_{n+1}}{a_n} < \varepsilon$$

whenever $n \geq N$. For all such n we therefore have

$$a_{n+1} < \varepsilon a_n.$$

Arguing as in case 1, we see by induction that

$$a_{N+m} < \varepsilon^m a_N$$

for all positive integers m, and therefore that

$$a_n < \varepsilon^{n-N} a_N = \varepsilon^n \cdot \frac{a_N}{\varepsilon^N}$$

for all $n > N$. Hence

$$a_n^{1/n} < \varepsilon \left(\frac{a_N}{\varepsilon^N}\right)^{1/n}$$

for all such n. Denoting the right-hand side of this inequality by s_n, we see that $s_n \to \varepsilon$ as $n \to \infty$. We then argue as in case 1 to show that

$$0 < a_n^{1/n} < s_n < 2\varepsilon$$

for large enough n, and the result follows in this case also.

Case 3: Suppose finally that

$$\lim_{n \to \infty} \frac{a_{n+1}}{a_n} = \infty,$$

and choose a number M. There exists N such that

$$\frac{a_{n+1}}{a_n} > M + 1$$

for all $n \geq N$. Thus

$$a_{n+1} > (M + 1)a_n$$

for all such n. Arguing as in the previous cases, we see by induction that

$$a_n > (M + 1)^n \frac{a_N}{(M + 1)^N}$$

for all $n > N$ and hence that

$$a_n^{1/n} > (M + 1) \left(\frac{a_N}{(M + 1)^N}\right)^{1/n}.$$

If we denote the right-hand side of this inequality by s_n, then $s_n \to M + 1$ as $n \to \infty$. For large enough n we therefore have

$$a_n^{1/n} > s_n > M,$$

and the proof is complete. $\qquad\square$

Example 3.8.2. Consider the series

$$\sum_{j=0}^{\infty} \frac{1}{2^{j+(-1)^j}}.$$

Letting

$$a_n = \frac{1}{2^{n+(-1)^n}}$$

for all n, we find that

$$a_n = \begin{cases} \frac{1}{2^{n+1}} & \text{if } n \text{ is even,} \\ \frac{1}{2^{n-1}} & \text{if } n \text{ is odd.} \end{cases}$$

Thus, if n is even so that $n+1$ is odd, then

$$\frac{a_{n+1}}{a_n} = \frac{2^{n+1}}{2^n} = 2,$$

but if n is odd, then

$$\frac{a_{n+1}}{a_n} = \frac{2^{n-1}}{2^{n+2}} = \frac{1}{8}.$$

Therefore $\lim_{n\to\infty} a_{n+1}/a_n$ does not exist. However,

$$a_n^{1/n} = \begin{cases} \frac{1}{2 \cdot 2^{1/n}} & \text{if } n \text{ is even,} \\ \frac{2^{1/n}}{2} & \text{if } n \text{ is odd.} \end{cases}$$

Hence

$$\lim_{n\to\infty} a_n^{1/n} = \frac{1}{2} < 1,$$

so that the series converges by the root test.

Note that the comparison test could also have been used, since $a_n \leq 1/2^{n-1}$ for every n. △

Exercises 3.7. 1. Prove that

$$\lim_{n\to\infty} \frac{n}{(n!)^{1/n}} = e.$$

(Hint: Apply Theorem 3.8.3.)

2. Apply the root test to investigate the convergence of the series

$$\sum_{j=0}^{\infty} \frac{e^j}{j!}.$$

3. Test the convergence of the following series:

(a) $\sum_{j=1}^{\infty} \left(\frac{j}{2j-1}\right)^j$;

(b) $\sum_{j=2}^{\infty} (j^{1/j} - 1)^j$;

(c) $\sum_{j=0}^{\infty} \frac{3^{j/2}}{2^j}$;

(d) $\sum_{j=1}^{\infty} j^2 e^{\alpha j}$ for every real α;

(e) $\sum_{j=1}^{\infty} \frac{x^j}{\left(\frac{j+1}{j}\right)^{j^2}}$ for every $x > 0$;

(f) $\sum_{j=0}^{\infty} \left(\frac{j^2+1}{j^2+j+1}\right)^{-j^2}$;

(g) $\sum_{j=1}^{\infty} \left(1 - \frac{1}{j}\right)^{j^2}$;

(h) $\sum_{j=1}^{\infty} \frac{j^j}{2^{j^2}}$.

4. Consider the series

$$\sum_{j=2}^{\infty} \frac{1}{j} \left(1 - \frac{1}{j}\right)^j.$$

   (a) Show that the root test is not applicable.
   (b) Test the convergence of the series. (Hint: You may find Proposition 2.7.3 useful.)

5. Let $0 < \alpha < \beta < 1$, and for each nonnegative integer n define

$$a_n = \begin{cases} \alpha^n & \text{if } n \text{ is odd}, \\ \beta^n & \text{if } n \text{ is even}. \end{cases}$$

Determine the convergence of $\sum_{j=0}^{\infty} a_j$. (Note that the ratio and root tests both fail.)

3.9 The Kummer–Jensen Test

The ratio test is an application of the comparison test. In fact, it is a special case of a more general application due to Kummer and Jensen.

Theorem 3.9.1 (Kummer–Jensen Test). *Let $\sum_{j=0}^{\infty} a_j$ be a series of positive terms and $\{b_n\}$ a sequence of positive terms. Let*

$$c_n = b_n - \frac{a_{n+1}}{a_n} b_{n+1}$$

for all n.

1. If $\lim_{n \to \infty} c_n > 0$, then $\sum_{j=0}^{\infty} a_j$ converges.
2. If $\sum_{j=0}^{\infty} 1/b_j$ diverges and there exists N such that $c_n \leq 0$ for all $n \geq N$, then $\sum_{j=0}^{\infty} a_j$ diverges.

Proof. 1. Suppose that

$$\lim_{n \to \infty} c_n = L > 0$$

and choose r such that $0 < r < L$. Then there exists $N > 0$ such that

$$c_n = b_n - \frac{a_{n+1}}{a_n} b_{n+1} > r$$

for all $n \geq N$. Hence

$$a_n b_n - a_{n+1} b_{n+1} > r a_n \qquad (3.8)$$

for each such n.

Define

$$S_n = \sum_{j=0}^{n} a_j$$

for all $n \geq 0$. Since $a_j > 0$ for all j, the sequence $\{S_n\}$ of partial sums is increasing. It therefore suffices to show that it is bounded above. Using (3.8) and the telescoping property, for all integers $m \geq N$ we have

$$
\begin{aligned}
r(S_m - S_{N-1}) &= r \sum_{n=N}^{m} a_n \\
&< \sum_{n=N}^{m} (a_n b_n - a_{n+1} b_{n+1}) \\
&= a_N b_N - a_{m+1} b_{m+1} \\
&< a_N b_N.
\end{aligned}
$$

Hence

$$S_m < \frac{a_N b_N}{r} + S_{N-1}$$

for all such m, and the sequence $\{S_n\}$ is indeed bounded above, as desired.

2. By hypothesis, we have

$$a_n b_n - a_{n+1} b_{n+1} \leq 0$$

for all $n \geq N$. Hence the sequence $\{a_n b_n\}$ is nondecreasing when $n \geq N$, and so $a_n b_n \geq a_N b_N$ for all such n. Thus

$$\frac{a_N b_N}{b_n} \leq a_n,$$

and the divergence of $\sum_{j=0}^{\infty} 1/b_j$ implies that of $\sum_{j=0}^{\infty} a_j$ by the comparison test. □

The ratio test is obtained immediately by putting $b_n = 1$ for all n. In view of the limit comparison test, the divergence of the harmonic series shows that another possibility is to have $b_n = n - 1$ for all n. We then obtain

$$c_n = n - 1 - \frac{a_{n+1}}{a_n} n$$

$$= n\left(1 - \frac{a_{n+1}}{a_n}\right) - 1$$

for all n, and so we deduce the following result, due to Raabe.

Corollary 3.9.2 (Raabe's Test). *Let $\sum_{j=0}^{\infty} a_j$ be a series of positive terms and suppose that*

$$\lim_{n \to \infty} n\left(1 - \frac{a_{n+1}}{a_n}\right) = L$$

for some number L. Then the series converges if $L > 1$ and diverges if $L < 1$.

In fact, the Kummer–Jensen test shows that in order to establish divergence, it suffices to find an N such that

$$n\left(1 - \frac{a_{n+1}}{a_n}\right) \leq 1$$

whenever $n \geq N$. Such an N certainly exists if $L < 1$.

Example 3.9.1. Test the series

$$\sum_{j=1}^{\infty} \left(\frac{(2j-1)!}{2^{2j-1}(j-1)!j!}\right)^m$$

for convergence, where $m \in \mathbb{N}$.

Solution. Writing

$$a_n = \left(\frac{(2n-1)!}{2^{2n-1}(n-1)!n!} \right)^m$$

for all $n \in \mathbb{N}$, we compute

$$\frac{a_{n+1}}{a_n} = \left(\frac{(2n+1)!}{2^{2n+1}n!(n+1)!} \cdot \frac{2^{2n-1}(n-1)!n!}{(2n-1)!} \right)^m$$

$$= \left(\frac{2n+1}{2n+2} \right)^m.$$

Thus

$$\lim_{n \to \infty} \frac{a_{n+1}}{a_n} = 1,$$

so that the ratio test is inconclusive.

Let us try Raabe's test. We have

$$n\left(1 - \frac{a_{n+1}}{a_n} \right) = n\left(1 - \left(1 - \frac{1}{2n+2} \right)^m \right)$$

$$= n\left(1 - \left(1 - \frac{m}{2n+2} + \sum_{j=2}^{m} \binom{m}{j} \frac{(-1)^j}{(2n+2)^j} \right) \right)$$

$$= \frac{nm}{2n+2} - n\sum_{j=2}^{m} \binom{m}{j} \frac{(-1)^j}{(2n+2)^j}$$

$$\to \frac{m}{2}$$

as $n \to \infty$. Raabe's test therefore shows that the series diverges for $m = 1$ and converges for $m > 2$.

We now apply the Kummer–Jensen test to the case $m = 2$, where Raabe's test is not applicable. Taking $b_n = n$ for all n, we find that

$$b_n - \frac{a_{n+1}}{a_n} b_{n+1} = n - \frac{(2n+1)^2(n+1)}{4(n+1)^2}$$

$$= -\frac{1}{4(n+1)}$$

$$< 0.$$

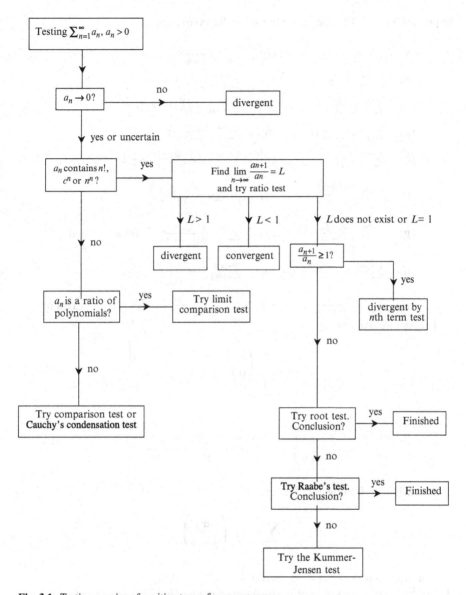

Fig. 3.1 Testing a series of positive terms for convergence

Recalling that the harmonic series diverges, we conclude by the Kummer–Jensen test that the given series diverges. △

Figure 3.1 suggests a procedure for testing a series of positive terms for convergence.

Exercises 3.8. 1. Test the following series for convergence:

(a) $\sum_{j=1}^{\infty} \frac{j^5}{2^j}$;

(b) $\sum_{j=0}^{\infty} \frac{100^j}{j!}$;

(c) $\sum_{j=1}^{\infty} \frac{j+5}{j^2}$;

(d) $\sum_{j=0}^{\infty} \frac{1}{(2j+1)!}$;

(e) $\sum_{j=0}^{\infty} \frac{\sqrt{j+1}}{j^2+1}$;

(f) $\sum_{j=1}^{\infty} \frac{\left(1+\frac{1}{j}\right)^j}{j}$;

(g) $\sum_{j=0}^{\infty} \frac{3j+5\sqrt{j}+7}{j^3+3j^{3/2}+5}$;

(h) $\sum_{j=1}^{\infty} \frac{1+2+\ldots+j}{j^3}$;

(i) $\sum_{j=1}^{\infty} \frac{1}{j^{3+(-1)^j}}$;

(j) $\sum_{j=0}^{\infty} \frac{1}{2^{2j}+(-1)^j}$;

(k) $\sum_{j=1}^{\infty} \frac{j}{e^{j^2}}$;

(l) $\sum_{j=0}^{\infty} \frac{(2j)!}{2^{2j}(j!)^2}$;

(m) $\sum_{j=1}^{\infty} \frac{(2j-1)!}{2^{2j}(j!)^2}$;

(n) $\sum_{j=1}^{\infty} \frac{(2j-1)!}{2^{2j-1}(j-1)!(j+1)!}$;

(o) $\sum_{j=0}^{\infty} \frac{2^{3j-1}(j!)^3}{(j+1)(2j+1)!}$;

(p) $\sum_{j=1}^{\infty} \frac{\prod_{k=1}^{j}(k+a-1)}{\prod_{k=1}^{j}(k+b-1)}$ for $a > 0$ and $b > 0$;

(q) $\sum_{j=1}^{\infty} \frac{\prod_{k=1}^{j}(3k-1)^2(3k+1)^2}{\prod_{k=1}^{j}(3k)^4}$;

(r) $\sum_{j=1}^{\infty} \frac{\alpha(\alpha+1)\cdots(\alpha+j-1)}{j!}$ for every real α.

2. For each real α define

$$\binom{\alpha}{0} = 1$$

and

$$\binom{\alpha}{n} = \frac{\alpha(\alpha - 1) \cdots (\alpha - n + 1)}{n!}$$

for each $n \in \mathbb{N}$. Show that the series

$$\sum_{j=1}^{\infty} (-1)^j \binom{\alpha}{j-1}$$

converges if $\alpha \geq 0$ and diverges if $\alpha < 0$.

3. Let $\sum_{j=0}^{\infty} a_j$ be a series of positive terms and let $\{b_n\}$ be a sequence of positive terms. For all n define

$$c_n = \frac{a_n}{a_{n+1}} b_n - b_{n+1}$$

and

$$r_n = n \left(\frac{a_n}{a_{n+1}} - 1 \right).$$

Prove the following variants of the Kummer–Jensen test and Raabe's test:

(a) If $c_n > t > 0$ for all n and some fixed t, then the series converges.
(b) If $\sum_{j=0}^{\infty} 1/b_j$ diverges and $c_n \leq 0$ for all n, then the series diverges.
(c) If $r_n > t > 1$ for all n and some fixed t, then the series converges.
(d) If $r_n \leq 1$ for all n, then the series diverges.

3.10   Alternating Series

In the previous several sections we dealt with series of nonnegative terms. Clearly, the results can be applied also if the terms are all negative. If neither of these conditions obtains, the series may be difficult to handle. However, there is a convenient test, due to Leibniz, that can be applied to what is known as an alternating series.

Definition 3.10.1. The series

$$\sum_{j=0}^{\infty}(-1)^j b_j \tag{3.9}$$

is **alternating** if each b_j is positive.

Theorem 3.10.1 (Leibniz's Test). *The alternating series (3.9) is convergent if $\{b_n\}$ is a nonincreasing sequence of positive terms converging to 0.*

Proof. For each integer $n \geq 0$ let

$$S_n = \sum_{j=0}^{n}(-1)^j b_j.$$

We need to show that the sequence $\{S_n\}$ converges. We achieve this result by proving that the subsequences $\{S_{2n}\}$ and $\{S_{2n+1}\}$ both converge to some number S and then appealing to Theorem 2.4.3.

First we have

$$S_{2n+1} = \sum_{j=0}^{2n+1}(-1)^j b_j$$

$$= b_0 + \sum_{j=1}^{2n}(-1)^j b_j - b_{2n+1}$$

$$= b_0 + \sum_{j=1}^{n}(b_{2j} - b_{2j-1}) - b_{2n+1}.$$

Note that $b_{2n+1} > 0$ and, since $\{b_n\}$ is nonincreasing, $b_{2j} \leq b_{2j-1}$ for all j. Hence, the sequence $\{S_{2n+1}\}$ is bounded above by b_0. It is nondecreasing, because for all $n > 0$ we have

$$S_{2n+1} - S_{2n-1} = b_{2n} - b_{2n+1} \geq 0.$$

We conclude that $\{S_{2n+1}\}$ converges to some number S. It follows that

$$\lim_{n \to \infty} S_{2n} = \lim_{n \to \infty} (S_{2n+1} - b_{2n+1})$$
$$= S - 0$$
$$= S,$$

and the proof is complete.   □

Example 3.10.1. Consider the series

$$\sum_{j=1}^{\infty} (-1)^j \frac{j}{(j+1)^2}.$$

Setting

$$b_n = \frac{n}{(n+1)^2}$$

for all $n > 0$, we observe that the sequence $\{b_n\}$ converges to 0. To show that it is nonincreasing, note first that

$$b_{n+1} - b_n = \frac{n+1}{(n+2)^2} - \frac{n}{(n+1)^2}$$
$$= \frac{(n+1)^3 - n(n+2)^2}{(n+1)^2(n+2)^2}.$$

As the denominator of this expression is positive, in order to determine the sign of $b_{n+1} - b_n$ it suffices to inspect the numerator. Since

$$(n+1)^3 - n(n+2)^2 = -n^2 - n + 1 < 0$$

for all $n > 0$, we confirm that the sequence $\{b_n\}$ is in fact decreasing. Hence the given series converges by Leibniz's test.   △

Example 3.10.2. Let $a_n = i^n/n$ for all $n > 0$. Then $\mathrm{Re}\,(a_n)$ is equal to 0 if n is odd, to $1/n$ if n is divisible by 4, and to $-1/n$ otherwise. Thus

$$\sum_{j=1}^{\infty} \text{Re}\left(\frac{i^j}{j}\right) = \sum_{k=1}^{\infty} \frac{(-1)^k}{2k},$$

which converges by Leibniz's test. Similarly,

$$\sum_{j=1}^{\infty} \text{Im}\left(\frac{i^j}{j}\right) = \sum_{k=1}^{\infty} \frac{(-1)^{k-1}}{2k-1},$$

which also converges. Hence the series $\sum_{j=1}^{\infty} i^j/j$ converges by Theorem 3.3.2.
The conjugate series

$$\sum_{j=1}^{\infty} \frac{(-i)^j}{j}$$

also converges, by the same argument.   △

Our next example shows that the condition that the sequence $\{b_n\}$ of Leibniz's test be nonincreasing cannot be dropped.

Example 3.10.3. Consider the series $\sum_{j=1}^{\infty}(-1)^{j+1}b_j$, where

$$b_{2j-1} = \frac{1}{j+1}$$

and

$$b_{2j} = \frac{1}{(j+1)^2}$$

for all $j > 0$. The sequence $\{b_n\}$ converges to 0, but it is not monotonic.

If the series were to converge, then we could obtain another convergent series by grouping its terms in any way. However, by grouping the terms in pairs, we obtain

$$\sum_{j=1}^{\infty} \left(\frac{1}{j+1} - \frac{1}{(j+1)^2}\right) = \sum_{j=1}^{\infty} \frac{j}{(j+1)^2},$$

and this series is divergent by the limit comparison test applied to the harmonic series.   △

Often it is difficult to determine to what number S a convergent series converges, but in the situation where Leibniz's test is applicable it is possible to approximate S. Indeed, we see in the following theorem that if we attempt to approximate S by taking the sum of the first few terms of the series, then the error is bounded above by the absolute value of the first term omitted from the sum.

Theorem 3.10.2. *Let $\{b_n\}$ be a nonincreasing sequence of positive terms, and let*

$$\sum_{j=0}^{\infty}(-1)^j b_j = S.$$

Then

$$|S - S_n| \le b_{n+1}$$

for each $n \ge 0$, where

$$S_n = \sum_{j=0}^{n}(-1)^j b_j.$$

Proof. We saw in the proof of Theorem 3.10.1 that the sequence $\{S_{2n+1}\}$ is nondecreasing. Similarly, $\{S_{2n}\}$ is nonincreasing, since

$$S_{2n+2} - S_{2n} = b_{2n+2} - b_{2n+1} \le 0$$

for all n. Therefore S is an upper bound for $\{S_{2n+1}\}$ and a lower bound for $\{S_{2n}\}$. Hence

$$S_{2n-1} \le S_{2n+1} \le S \le S_{2n} \tag{3.10}$$

for all $n > 0$, so that

$$\begin{aligned} 0 &\le S - S_{2n-1} \\ &\le S_{2n} - S_{2n-1} \\ &= b_{2n}, \end{aligned}$$

as desired. The last two inequalities of (3.10) hold even if $n = 0$. Therefore

$$\begin{aligned} 0 &\le S_{2n} - S \\ &\le S_{2n} - S_{2n+1} \\ &= -(-b_{2n+1}) \\ &= b_{2n+1} \end{aligned}$$

for all $n \ge 0$, and the proof is complete.   □

Remark. The theorem shows that

$$S_n - b_{n+1} \le S \le S_n + b_{n+1}$$

for all $n \geq 0$. Moreover it is easily seen that the inequalities are strict if the sequence $\{b_n\}$ is decreasing. In particular, since we then have $b_0 > b_1$, it follows that $S > 0$.

Example 3.10.4. Consider

$$\sum_{j=1}^{\infty} \frac{(-1)^j}{j^2}.$$

This series converges to some number S by Leibniz's test, since the sequence $\{1/n^2\}$ is decreasing and converges to 0. We approximate S by the partial sum

$$S_n = \sum_{j=1}^{n} \frac{(-1)^j}{j^2}$$

for some $n > 0$. If we require accuracy to within a given positive number ε, then according to Theorem 3.10.2, we need to choose n so that

$$\frac{1}{(n+1)^2} < \varepsilon.$$

Thus we must have

$$(n+1)^2 > \frac{1}{\varepsilon},$$

and to ensure that this inequality holds, we take

$$n > \frac{1}{\sqrt{\varepsilon}} - 1.$$

The exact value of the sum requires ideas that are beyond the scope of this book. $\triangle$

Exercises 3.9. 1. Determine the convergence of the following series:

(a) $\sum_{j=1}^{\infty}(-1)^{j+1}\frac{j^2}{j^3+1}$;

(b) $\sum_{j=1}^{\infty}(-1)^{j+1}\frac{2j+1}{j(j+1)}$;

(c) $\sum_{j=0}^{\infty}(-1)^{j+1}\frac{j-5}{7j+3}$;

(d) $\sum_{j=2}^{\infty}(-1)^{j+1}(j^{1/j} - 1)$;

(e) $\sum_{j=1}^{\infty}(-1)^{j+1}(a^{1/j} - 1)$, where $a > 0$;

(f) $\sum_{j=1}^{\infty}(-1)^{j+1}\left(\left(1 + \frac{1}{j}\right)^{j+1} - e\right)$;

(g) $\sum_{j=1}^{\infty} \frac{(-1)^j - i^j}{\sqrt{j}}$.

2. Show that the series

$$S = \sum_{j=2}^{\infty} \frac{(-1)^{j+1}}{j - \sqrt{j}}$$

converges. Letting S_n be the nth partial sum of the series, find an upper bound for $|S_{15} - S|$.

3. Let $\{a_n\}$ be a decreasing null sequence. Show that

$$\sum_{j=1}^{\infty} (-1)^{j+1} \frac{a_1 + a_2 + \ldots + a_j}{j}$$

converges.

4. If $\sum_{j=0}^{\infty} a_j$ converges, does it follow that $\sum_{j=0}^{\infty} (-1)^j a_j$ converges? Give a proof or a counterexample.

5. Determine the convergence of

$$\sum_{j=0}^{\infty} \frac{(-1)^{j+1}}{j + 2 + (-1)^j}.$$

[Hint:

$$\frac{1}{j + 2 + (-1)^j} = \frac{j + 2 - (-1)^j}{(j + 2)^2 - 1}.]$$

3.11 Dirichlet's Test

All the tests we have learned so far help us to determine the convergence of series whose terms are all positive or alternate in sign. In this section we provide a useful test that does not require these assumptions. The proof relies on Abel's partial summation identity.

Lemma 3.11.1 (Abel's Partial Summation Identity). *Let $\{u_n\}$ and $\{v_n\}$ be two sequences of complex numbers, and let $\{U_n\}$ be a sequence such that $U_n - U_{n-1} = u_n$ for all $n \in \mathbb{N}$. Then*

$$\sum_{j=m+1}^{n} u_j v_j = U_n v_n - U_m v_{m+1} - \sum_{j=m+1}^{n-1} U_j (v_{j+1} - v_j) \qquad (3.11)$$

whenever $m < n$.

Proof. Since $U_j - U_{j-1} = u_j$, we have

$$\sum_{j=m+1}^{n} u_j v_j = \sum_{j=m+1}^{n} v_j \left(U_j - U_{j-1}\right)$$

$$= \sum_{j=m+1}^{n} v_j U_j - \sum_{j=m+1}^{n} v_j U_{j-1}$$

$$= \sum_{j=m+1}^{n-1} v_j U_j + v_n U_n - v_{m+1} U_m - \sum_{j=m+2}^{n} v_j U_{j-1}$$

$$= v_n U_n - v_{m+1} U_m + \sum_{j=m+1}^{n-1} v_j U_j - \sum_{j=m+1}^{n-1} v_{j+1} U_j$$

$$= U_n v_n - U_m v_{m+1} - \sum_{j=m+1}^{n-1} U_j \left(v_{j+1} - v_j\right).$$

$\square$

The next result is due to Shiu [14].

Theorem 3.11.2. *Let $\{u_n\}$ and $\{v_n\}$ be sequences of complex numbers and suppose that*

1. *$\sum_{j=0}^{\infty} |v_{j+1} - v_j|$ converges,*
2. *$\{v_n\}$ converges to 0, and*
3. *there is a constant K such that*

$$\left| \sum_{j=0}^{n} u_j \right| \le K$$

for all $n \ge 0$.

Then $\sum_{j=0}^{\infty} u_j v_j$ is convergent.

Proof. Choose $\varepsilon > 0$. The convergent sequence

$$\left\{ \sum_{j=0}^{n} |v_{j+1} - v_j| \right\}$$

is Cauchy. Therefore by hypotheses (1) and (2) there exists N such that for all $n \geq N$ we have

$$\sum_{j=N+1}^{n} |v_{j+1} - v_j| < \varepsilon$$

and

$$|v_n| < \varepsilon.$$

For each $n \geq 0$ let

$$S_n = \sum_{j=0}^{n} u_j v_j$$

and

$$U_n = \sum_{j=0}^{n} u_j.$$

Thus $|U_n| \leq K$ for all $n \geq 0$, by hypothesis. For all integers m and n such that $n > m \geq N$, Abel's identity therefore gives

$$
\begin{aligned}
|S_n - S_m| &= \left| \sum_{j=m+1}^{n} u_j v_j \right| \\
&\leq |U_n v_n| + |U_m v_{m+1}| + \sum_{j=m+1}^{n-1} |U_j| |v_{j+1} - v_j| \\
&\leq K|v_n| + K|v_{m+1}| + K \sum_{j=m+1}^{n-1} |v_{j+1} - v_j| \\
&< 3K\varepsilon,
\end{aligned}
$$

as it may be assumed that $K > 0$. The desired conclusion follows from Cauchy's criterion.   □

Corollary 3.11.3 (Dirichlet's Test). *Let $\{u_n\}$ be a complex sequence and let $\{v_n\}$ be a real sequence. Suppose that*

1. $\{v_n\}$ is monotonic and converges to 0, and

2. there is a constant K such that

$$\left| \sum_{j=0}^{n} u_j \right| \leq K$$

for all $n \geq 0$.

Then $\sum_{j=0}^{\infty} u_j v_j$ is convergent.

Proof. According to Theorem 3.11.2, it is enough to show that

$$\sum_{j=0}^{\infty} |v_{j+1} - v_j|$$

converges. Choose $\varepsilon > 0$. By condition (1), there exists N such that

$$|v_n| < \varepsilon$$

for all $n \geq N$. For each such n, define

$$T_n = \sum_{j=0}^{n} |v_{j+1} - v_j|.$$

We must show that $\{T_n\}$ is convergent.

Since $\{v_j\}$ is monotonic, either $\{v_{j+1} - v_j\}$ is a sequence of nonnegative terms or it is a sequence of nonpositive terms. Therefore for all integers m and n such that $n > m \geq N$, it follows that

$$\begin{aligned}
|T_n - T_m| &= \sum_{j=m+1}^{n} |v_{j+1} - v_j| \\
&= \left| \sum_{j=m+1}^{n} (v_{j+1} - v_j) \right| \\
&= |v_{n+1} - v_{m+1}| \\
&\leq |v_{n+1}| + |v_{m+1}| \\
&< 2\varepsilon.
\end{aligned}$$

The desired conclusion thus follows from Cauchy's criterion. □

Remark. We obtain Leibniz's test by taking $u_n = (-1)^n$ for all n in Dirichlet's test.

Example 3.11.1. Let $|z| = 1$ and $z \neq 1$. We show that

$$\sum_{j=1}^{\infty} \frac{z^j}{j^p}$$

is convergent for every rational $p > 0$.

Let us take $v_n = 1/n^p$ and $u_n = z^n$ for all n. Then

$$\left| \sum_{j=0}^{n} z^j \right| = \left| \frac{1 - z^{n+1}}{1 - z} \right| \leq \frac{1 + |z|^{n+1}}{|1 - z|} = \frac{2}{|1 - z|}$$

for all n. Therefore the series converges by Dirichlet's test.   △

Exercises 3.10. 1. (a) Let $\{v_n\}$ be a monotonic sequence, and let $\{u_n\}$ be a sequence for which there is a constant K such that

$$\left| \sum_{j=0}^{n} u_j \right| \leq K$$

for all $n \geq 0$. Prove that

$$\left| \sum_{j=m+1}^{n} u_j v_j \right| \leq 2K(|v_{m+1}| + |v_n|)$$

for all positive integers m and n such that $m < n$.

(b) Use part (a) to prove Dirichlet's test.

2. Suppose that $\{v_n\}$ is a monotonic bounded sequence and that $\sum_{j=0}^{\infty} u_j$ converges. Prove that $\sum_{j=0}^{\infty} u_j v_j$ converges. (This result is known as Abel's test.)

3.12 Absolute and Conditional Convergence

We have seen the advantages of considering series whose terms are all nonnegative. Given an arbitrary series, we may obtain a series of nonnegative terms by replacing each term with its absolute value. If the resulting series converges, then so does the original series. That is the content of our next theorem. We begin its proof with a useful lemma.

Lemma 3.12.1. *Let $\{z_n\}$ be a sequence of complex numbers. Then $\sum_{j=0}^{\infty} |z_j|$ converges if and only if $\sum_{j=0}^{\infty} |\mathrm{Re}\,(z_j)|$ and $\sum_{j=0}^{\infty} |\mathrm{Im}\,(z_j)|$ converge.*

Proof. Recalling that $|\text{Re}\,(z_n)| \leq |z_n|$ and $|\text{Im}\,(z_n)| \leq |z_n|$ for all n, we find that the lemma holds by the comparison test if $\sum_{j=0}^{\infty} |z_j|$ converges. The converse follows from the fact that

$$|z_n| \leq |\text{Re}\,(z_n)| + |\text{Im}\,(z_n)|$$

for all n, by the triangle inequality.   □

Theorem 3.12.2. *Let $\{z_n\}$ be a sequence (not necessarily real). If $\sum_{j=0}^{\infty} |z_j|$ converges, then so does $\sum_{j=0}^{\infty} z_j$.*

Proof. In view of Lemma 3.12.1, we may assume that each z_n is real. Thus we write $x_n = z_n$ for each n. Certainly,

$$0 \leq x_n + |x_n| \leq 2|x_n|$$

for each n. Given that $\sum_{j=0}^{\infty} |x_j|$ converges, so does

$$\sum_{j=0}^{\infty} (x_j + |x_j|),$$

by the comparison test. As

$$\sum_{j=0}^{\infty} x_j = \sum_{j=0}^{\infty} (x_j + |x_j|) - \sum_{j=0}^{\infty} |x_j|$$

by Theorem 3.3.1, the result follows.   □

Remark. The converse of the theorem above is not always true. For example, the alternating series

$$\sum_{j=1}^{\infty} \frac{(-1)^j}{j}$$

converges by Leibniz's test, but the harmonic series diverges.

A series $\sum_{j=0}^{\infty} z_j$ is said to be **absolutely convergent** if $\sum_{j=0}^{\infty} |z_j|$ converges. Thus every absolutely convergent series does in fact converge, according to Theorem 3.12.2. A series that is convergent but not absolutely is said to be **conditionally convergent**.

Example 3.12.1. The series

$$\sum_{j=1}^{\infty} \frac{i^j}{j^2}$$

is absolutely convergent since the series

$$\sum_{j=1}^{\infty} \left| \frac{i^j}{j^2} \right| = \sum_{j=1}^{\infty} \frac{1}{j^2}$$

converges.   △

Example 3.12.2. Test the series

$$\sum_{j=1}^{\infty} (-1)^j \frac{j}{2^j}.$$

Solution. Note that the series

$$\sum_{j=1}^{\infty} \frac{j}{2^j}$$

converges by the ratio test, since

$$\frac{n+1}{2^{n+1}} \cdot \frac{2^n}{n} = \frac{n+1}{2n} \to \frac{1}{2} < 1.$$

Hence the given series converges absolutely.   △

The work in this example suggests that the ratio test can be extended to series whose terms are not necessarily positive or even real. In fact, we can prove the following theorem.

Theorem 3.12.3 (Generalized Ratio Test). *Let* $\sum_{j=0}^{\infty} z_j$ *be a series (not necessarily real) and suppose that*

$$\lim_{n \to \infty} \left| \frac{z_{n+1}}{z_n} \right| = L$$

for some number L. Then the series converges absolutely if $L < 1$ *but diverges if* $L > 1$.

Proof. If $L < 1$, then the series $\sum_{j=0}^{\infty} |z_j|$ converges by the ratio test, and so the given series is absolutely convergent. If $L > 1$, then, arguing as in the proof of the ratio test, we see that the sequence $\{|z_n|\}$ cannot converge to 0. Hence the sequence $\{z_n\}$ cannot converge to 0 either. Thus the given series fails the nth term test and therefore cannot converge.   □

There is also a generalized root test, which has a similar proof.

Theorem 3.12.4 (Generalized Root Test). *Let $\sum_{j=0}^{\infty} z_j$ be a series and let*

$$\lim_{n \to \infty} |z_n|^{1/n} = L$$

for some number L. Then the series converges absolutely if $L < 1$ but diverges if $L > 1$.

Proof. If $L < 1$, then the series is absolutely convergent by the root test. If $L > 1$, then, as in the proof of the root test, the sequence $\{z_n\}$ cannot converge to 0. □

Theorem 3.12.5. *Let $\{x_n\}$ and $\{y_n\}$ be real sequences, and suppose that $\sum_{j=0}^{\infty} x_j^2$ and $\sum_{j=0}^{\infty} y_j^2$ converge. Then the series $\sum_{j=0}^{\infty} x_j y_j$ is absolutely convergent.*

Proof. Note first that

$$(|x_j| - |y_j|)^2 \geq 0$$

for all j. Hence

$$\frac{1}{2}(x_j^2 + y_j^2) \geq |x_j y_j|.$$

The result now follows from the comparison test. □

Exercises 3.11. 1. Use Cauchy's principle to prove that if a complex series is absolutely convergent, then it is convergent.

2. (a) Show that if $\sum_{j=0}^{\infty} x_j$ converges absolutely, then so does $\sum_{j=0}^{\infty} x_j^2$.
   (b) Is the converse true?
   (c) Is it true that if $\sum_{j=0}^{\infty} x_j$ converges, then so does $\sum_{j=0}^{\infty} x_j^2$?

3. Show that if $\sum_{j=0}^{\infty} z_j$ and $\sum_{j=0}^{\infty} w_j$ are absolutely convergent complex series, then so are $\sum_{j=0}^{\infty} z_j w_j$ and $\sum_{j=0}^{\infty} (\alpha z_j + \beta w_j)$ for all complex numbers α and β.

4. Show that if $\sum_{j=0}^{\infty} z_j$ is an absolutely convergent complex series, then

$$\left| \sum_{j=0}^{\infty} z_j \right| \leq \sum_{j=0}^{\infty} |z_j|.$$

5. Test the following series for absolute convergence and conditional convergence:

   (a) $\sum_{j=2}^{\infty} \frac{(-1)^{j+1}}{j - \sqrt{j}}$;

   (b) $\sum_{j=0}^{\infty} (-1)^{j+1} \left(\frac{j+2}{3j-1} \right)^j$;

   (c) $\sum_{j=0}^{\infty} (j+1)^2 \left(\frac{x}{x+2} \right)^j$ for all real $x \neq -2$;

   (d) $\sum_{j=0}^{\infty} (-1)^{j+1} \frac{(j-i)^2}{2^j}$;

   (e) $\sum_{j=1}^{\infty} \frac{i^j}{j^{3/2}}$.

3.13   Rearrangements of Series

The notion of a conditionally convergent series was introduced in the previous section. Part of the reason for this terminology is an unpleasant and perhaps counterintuitive property: The convergence turns out to depend on the order in which the terms of the series are written. In fact, we have the following remarkable theorem, due to Riemann.

Theorem 3.13.1 (Riemann). *For each conditionally convergent real series and given number S, the terms of the series may be rearranged to yield a series that converges to S. There is also a rearrangement of the terms so that the resulting series diverges.*

Before proving Theorem 3.13.1, we make some important observations. Given a real sequence $\{a_n\}$, let us define two new sequences $\{P_n\}$ and $\{Q_n\}$, where

$$P_n = \frac{|a_n| + a_n}{2} \tag{3.12}$$

and

$$Q_n = \frac{|a_n| - a_n}{2} \tag{3.13}$$

for all n. Thus, if $a_n \geq 0$, then $P_n = a_n = |a_n|$ and $Q_n = 0$, but if $a_n < 0$, then $P_n = 0$ and $Q_n = -a_n = |a_n|$. We can therefore think of $\sum_{j=1}^{\infty} P_j$ and $-\sum_{j=1}^{\infty} Q_j$ as the series composed of the nonnegative and negative terms, respectively, of the series $\sum_{j=1}^{\infty} a_j$. It is easy to see that if $\sum_{j=1}^{\infty} a_j$ is conditionally convergent, then both $\sum_{j=1}^{\infty} P_j$ and $\sum_{j=1}^{\infty} Q_j$ diverge. Indeed, we have

$$|a_n| = 2P_n - a_n$$

for all n, and so if $\sum_{j=1}^{\infty} P_j$ were to converge, then it follows from Theorem 3.3.1 that $\sum_{j=1}^{\infty} |a_j|$ would converge. This conclusion would contradict the conditional convergence of $\sum_{j=1}^{\infty} a_j$. The proof that $\sum_{j=1}^{\infty} Q_j$ diverges is similar.

Note also that if $n > 1$ and $x_1, x_2, \ldots, x_n$ are numbers such that

$$\sum_{j=1}^{n-1} x_j \leq S,$$

then

$$\sum_{j=1}^{n} x_j - S = x_n + \sum_{j=1}^{n-1} x_j - S \leq x_n. \tag{3.14}$$

Similarly, if

$$\sum_{j=1}^{n-1} x_j \geq S,$$

then

$$S - \sum_{j=1}^{n} x_j = S - \sum_{j=1}^{n-1} x_j - x_n \leq -x_n. \tag{3.15}$$

The proof of the theorem involves some technical details, but the idea is simple. From the terms of the series $\sum_{j=1}^{\infty} a_j$, we construct a series converging to S as follows. First, we take just enough positive terms to obtain a partial sum that exceeds S. Next, we throw in just enough negative terms to produce a partial sum below S. Then we add some more positive terms, just enough to bring the partial sum above S once more. We continue in this fashion, so that the partial sums oscillate about S. The resulting series converges to S.

Proof of Theorem 3.13.1.   Let us first try to use the terms of a conditionally convergent series $\sum_{j=1}^{\infty} a_j$ to construct a series that converges to S. For all $n > 0$ define P_n and Q_n as in (3.12) and (3.13), respectively. Let M_1 be the smallest positive integer such that

$$\sum_{j=1}^{M_1} P_j > S.$$

Certainly M_1 exists: The sequence $\{\sum_{j=1}^{n} P_j\}$ is nondecreasing and therefore cannot be bounded above since it diverges. Next let N_1 be the smallest positive integer such that

$$\sum_{j=1}^{M_1} P_j - \sum_{j=1}^{N_1} Q_j < S;$$

N_1 also exists since $\sum_{j=1}^{\infty} Q_j$ diverges. In fact, the choice of N_1 shows that

$$\sum_{j=1}^{M_1} P_j - \sum_{j=1}^{N_1-1} Q_j \geq S;$$

hence

$$0 < S - \left(\sum_{j=1}^{M_1} P_j - \sum_{j=1}^{N_1} Q_j \right) \leq -(-Q_{N_1}) = Q_{N_1}$$

by (3.15).

Continuing by induction, suppose that $n > 0$ and that

$$M_1, M_2, \ldots, M_n, N_1, N_2, \ldots, N_n$$

have been defined so that

$$0 < S - T_n \le Q_{N_n}, \tag{3.16}$$

where

$$T_n = \sum_{k=0}^{n-1} \left(\sum_{j=M_k+1}^{M_{k+1}} P_j - \sum_{j=N_k+1}^{N_{k+1}} Q_j \right)$$

and $M_0 = N_0 = 0$. Continue by letting M_{n+1} be the smallest integer greater than M_n such that

$$T_n + \sum_{j=M_n+1}^{M_{n+1}} P_j > S$$

and letting N_{n+1} be the smallest integer greater than N_n such that

$$T_{n+1} = T_n + \sum_{j=M_n+1}^{M_{n+1}} P_j - \sum_{j=N_n+1}^{N_{n+1}} Q_j < S;$$

both M_{n+1} and N_{n+1} exist. The choice of N_{n+1} implies that

$$T_n + \sum_{j=M_n+1}^{M_{n+1}} P_j - \sum_{j=N_n+1}^{N_{n+1}-1} Q_j \ge S,$$

so that

$$0 < S - T_{n+1} \le Q_{N_{n+1}},$$

by (3.15).

We must show that the resulting series converges to S. This goal is achieved by studying the partial sum T_n and those between T_n and T_{n+1}. From the choice of M_{n+1} and inequalities (3.16) it follows that

$$0 \le S - T_n - \sum_{j=M_n+1}^{m} P_j \le Q_{N_n} \tag{3.17}$$

for each m such that $M_n < m < M_{n+1}$, since $P_j \geq 0$ for all j. Moreover the choice of M_{n+1} also shows that

$$0 < T_n + \sum_{j=M_n+1}^{M_{n+1}} P_j - S \leq P_{M_{n+1}}, \tag{3.18}$$

by (3.14), whence

$$0 \leq T_n + \sum_{j=M_n+1}^{M_{n+1}} P_j - \sum_{j=N_n+1}^{m} Q_j - S \leq P_{M_{n+1}} \tag{3.19}$$

for all m such that $N_n < m < N_{n+1}$. From (3.16–3.19) we find that the partial sums in question all differ from S by an amount no greater than

$$\max\{P_{M_{n+1}}, Q_{N_n}\}.$$

But the sequence $\{a_n\}$ converges to 0 by the convergence of $\sum_{j=1}^{\infty} a_j$. Hence $\{|a_n|\}$ also converges to 0 and therefore so do its subsequences $\{P_{M_{n+1}}\}$ and $\{Q_{N_n}\}$, as required.

It's easier to construct a divergent series from the terms of the series $\sum_{j=1}^{\infty} a_j$. Defining P_n and Q_n as before, let K_1 be the smallest positive integer such that

$$\sum_{j=1}^{K_1} P_j - Q_1 > 1.$$

Suppose that $K_1, K_2, \ldots, K_n$ have been defined for some positive integer n so that

$$\sum_{k=0}^{n-1} \left(\sum_{j=K_k+1}^{K_{k+1}} P_j - Q_{k+1} \right) > n,$$

where $K_0 = 0$. Let K_{n+1} be the smallest integer greater than K_n such that

$$\sum_{k=0}^{n} \left(\sum_{j=K_k+1}^{K_{k+1}} P_j - Q_{k+1} \right) > n + 1.$$

The series defined inductively by this procedure diverges. $\qquad\qquad\square$

By contrast, Dirichlet proved that an absolutely convergent series converges to the same number regardless of the order in which the terms are written. In order to make the discussion rigorous, let us first introduce the following definition.

Definition 3.13.1. Let $\{k_n\}$ be a sequence of nonnegative integers in which each nonnegative integer appears exactly once. Then the sequence $\{a_{k_n}\}$ is a **rearrangement** of a sequence $\{a_n\}$ and the series $\sum_{j=0}^{\infty} a_{k_j}$ is a **rearrangement** of the series $\sum_{j=0}^{\infty} a_j$.

Theorem 3.13.2 (Dirichlet). *All rearrangements of an absolutely convergent series are absolutely convergent and converge to the same number.*

Proof. Let $\sum_{j=0}^{\infty} w_j$ be a rearrangement of an absolutely convergent series $\sum_{j=0}^{\infty} z_j$. For all $n \in \mathbb{N}$ write

$$S_n = \sum_{j=0}^{n} z_j,$$

$$S'_n = \sum_{j=0}^{n} |z_j|,$$

$$T_n = \sum_{j=0}^{n} w_j,$$

and

$$T'_n = \sum_{j=0}^{n} |w_j|.$$

Every convergent sequence is bounded, and so there exists M such that $S'_n \leq M$ for all n. As $\sum_{j=0}^{\infty} |w_j|$ is a rearrangement of $\sum_{j=0}^{\infty} |z_j|$, we also have $T'_n \leq M$ for all n. Hence $\sum_{j=0}^{\infty} w_j$ is absolutely convergent.

Let

$$S = \sum_{j=0}^{\infty} z_j$$

and

$$T = \sum_{j=0}^{\infty} w_j.$$

It remains only to show that $S = T$. Choose $\varepsilon > 0$. We can find N such that

$$|S - S_n| < \varepsilon$$

and

$$S'_m - S'_n < \varepsilon$$

whenever $m > n \geq N$. Fix $n \geq N$ and choose p large enough so that

$$|T_p - T| < \varepsilon$$

and the sum T_p includes $z_0, z_1, \ldots, z_n$ among its terms. Now choose $m > n$ so large that every term of T_p is also a term of S_m. Thus $S_m - T_p$ is a summation whose terms form a subset of $\{z_{n+1}, z_{n+2}, \ldots, z_m\}$, so that

$$|S_m - T_p| \leq S'_m - S'_n$$

$$< \varepsilon.$$

Hence

$$|S - T| \leq |S - S_m| + |S_m - T_p| + |T_p - T|$$

$$< 3\varepsilon.$$

Since ε is arbitrary, we conclude that $S = T$, as required. $\qquad\square$

3.14 Products of Series

In order to understand products of series, we must first carefully consider the process of multiplying finite sums. Suppose that $a_m, a_{m+1}, \ldots, a_n, b_p, b_{p+1}, \ldots, b_q$ are numbers and we wish to form the product

$$\left(\sum_{j=m}^{n} a_j \right) \left(\sum_{k=p}^{q} b_k \right). \tag{3.20}$$

In fact, this product may be written without the parentheses, as we now show. Using the distributive law (1.17) twice, first with $s = \sum_{j=m}^{n} a_j$ and then with $s = b_k$, gives

$$\left(\sum_{j=m}^{n} a_j \right) \left(\sum_{k=p}^{q} b_k \right) = \sum_{k=p}^{q} \left(\sum_{j=m}^{n} a_j \right) b_k$$

$$= \sum_{k=p}^{q} \left(\sum_{j=m}^{n} a_j b_k \right)$$

$$= \sum_{k=p}^{q} \sum_{j=m}^{n} a_j b_k.$$

$$
\begin{array}{ccccccc}
a_0 b_0 & & a_0 b_1 & & a_0 b_2 & \cdots & a_0 b_{q-1} & & a_0 b_q \\
& \swarrow & & \swarrow & & & & \swarrow & \\
a_1 b_0 & & a_1 b_1 & & a_1 b_2 & \cdots & a_1 b_{q-1} & & a_1 b_q \\
& \swarrow & & \swarrow & & & & \swarrow & \\
a_2 b_0 & & a_2 b_1 & & a_2 b_2 & \cdots & a_2 b_{q-1} & & a_2 b_q \\
\vdots & & \vdots & & \vdots & \ddots & \vdots & & \vdots \\
a_{n-1} b_0 & & a_{n-1} b_1 & & a_{n-1} b_2 & \cdots & a_{n-1} b_{q-1} & & a_{n-1} b_q \\
& \swarrow & & \swarrow & & & & \swarrow & \\
a_n b_0 & & a_n b_1 & & a_n b_2 & \cdots & a_n b_{q-1} & & a_n b_q
\end{array}
$$

Fig. 3.2 The terms of $(\sum_{j=m}^{n} a_j)(\sum_{k=p}^{q} b_k)$

As multiplication is commutative, we conclude that

$$
\sum_{k=p}^{q} \sum_{j=m}^{n} a_j b_k = \left(\sum_{k=p}^{q} b_k \right) \left(\sum_{j=m}^{n} a_j \right)
$$

$$
= \sum_{j=m}^{n} \sum_{k=p}^{q} a_j b_k .
$$

It now follows that

$$
\left(\sum_{j=m}^{n} a_j \right) \left(\sum_{k=p}^{q} b_k \right) = \sum_{j=m}^{n} \sum_{k=p}^{q} a_j b_k
$$

$$
= \sum_{j=m}^{n} \left(a_j \sum_{k=p}^{q} b_k \right)
$$

$$
= \sum_{j=m}^{n} a_j \sum_{k=p}^{q} b_k ,
$$

where the penultimate line was obtained by another application of distributivity.

In order to understand better how to evaluate the expression (3.20), let us begin by considering the case where $m = p = 0$. Basically, the product (3.20) is calculated by adding up the products of the form $a_j b_k$ for all relevant j and k. In other words, we simply add all the entries in Fig. 3.2. In the calculation above, the terms $a_j b_k$ are ordered so that all those containing a_0 are listed first, then all those containing a_1, and so forth. In other words, the entries of the figure are added row by row. The sum of the entries in the $(j + 1)$th row is $a_j \sum_{k=0}^{q} b_k$.

However, there is another way of performing this calculation which is of some importance. Let us assume first that $n = q$, so that Fig. 3.2 gives a square array. Instead of ordering the terms in Fig. 3.2 row by row, we may order them as indicated by the arrows. In other words, we begin with the term $a_0 b_0$ in the top left corner, followed by the terms in the diagonal from $a_0 b_1$ to $a_1 b_0$, then those in the diagonal from $a_0 b_2$ to $a_2 b_0$, and so on, up to and including the diagonal that contains $a_0 b_q = a_0 b_n$, the entry in the top right corner. As the array is assumed to be square, this diagonal also contains $a_n b_0$. We then continue with the diagonal from $a_1 b_n$ to $a_n b_1$, then that from $a_2 b_n$ to $a_n b_2$, and so on. The last diagonal consists only of the term $a_n b_n$. In each diagonal the sum of the subscripts is a constant between 0 and $2n$ (inclusive), and this constant is different from the corresponding constant for each other diagonal. For the diagonal where this constant is t, the sum of the terms in the diagonal is

$$\sum_{s=0}^{t} a_s b_{t-s},$$

provided we agree that $a_s = 0$ whenever $s > n$ and $b_{t-s} = 0$ whenever $t - s > n$. With this understanding, we conclude that

$$\sum_{j=0}^{n} a_j \sum_{k=0}^{n} b_k = \sum_{t=0}^{2n} \sum_{s=0}^{t} a_s b_{t-s}.$$

If $n > q$, then we may set $b_k = 0$ for all $k > q$ and apply the result of the calculation above. Similarly, if $n < q$, then we set $a_j = 0$ for all $j > n$. In both cases we reach the conclusion that

$$\sum_{j=0}^{n} a_j \sum_{k=0}^{q} b_k = \sum_{t=0}^{n+q} \sum_{s=0}^{t} a_s b_{t-s},$$

since $a_s b_{t-s} = 0$ whenever $t = s + (t - s) > n + q$. In general,

$$\sum_{j=m}^{n} a_j \sum_{k=p}^{q} b_k = \sum_{j=0}^{n-m} a_{j+m} \sum_{k=0}^{q-p} b_{k+p}$$

$$= \sum_{t=0}^{n-m+q-p} \sum_{s=0}^{t} a_{s+m} b_{t-s+p}.$$

Setting $v = s + m$ and $u = t + m + p$, we therefore obtain

$$\sum_{j=m}^{n} a_j \sum_{k=p}^{q} b_k = \sum_{u=m+p}^{n+q} \sum_{v=m}^{u-p} a_v b_{u-v},$$

since $t + p - s = u - m - s = u - v$.

Using Eq. (1.15), we may summarize the result of this calculation in the following theorem.

Theorem 3.14.1. *For integers m, n, p, q such that $m \leq n$ and $p \leq q$, we have*

$$\sum_{j=m}^{n} a_j \sum_{k=p}^{q} b_k = \sum_{j=m+p}^{n+q} \sum_{k=m}^{j-p} a_k b_{j-k},$$

where $a_l = 0$ for all $l > n$ and $b_l = 0$ for all $l > q$. In particular,

$$\sum_{j=0}^{n} a_j \sum_{k=0}^{q} b_k = \sum_{j=0}^{n+q} \sum_{k=0}^{j} a_k b_{j-k},$$

for all nonnegative integers n and q.

We now turn our attention to the multiplication of absolutely convergent series

$$A = \sum_{j=0}^{\infty} a_j$$

and

$$B = \sum_{j=0}^{\infty} b_j.$$

Thus we need to find the sum of all products of the form $a_j b_k$ for nonnegative integers j and k. This sum may be written as the series

$$\sum_{j=0}^{\infty} \left(\sum_{k=0}^{j} a_j b_k + \sum_{l=0}^{j-1} a_l b_j \right) \tag{3.21}$$

$$= a_0 b_0 + a_1 b_0 + a_1 b_1 + a_0 b_1 + a_2 b_0 + a_2 b_1 + a_2 b_2 + a_0 b_2 + a_1 b_2 + \cdots .$$

For each nonnegative integer n let A_n, B_n, S_n be the nth partial sums of A, B, and the series (3.21), respectively. Thus,

$$A_n = \sum_{j=0}^{n-1} a_j$$

and

$$B_n = \sum_{j=0}^{n-1} b_j.$$

The product $A_n B_n$ has n^2 terms, and it follows that

$$S_{n^2} = A_n B_n.$$

For each nonnegative integer n let T_n be the sum of the absolute values of the first n terms of series (3.21). Since A and B are absolutely convergent, we find that

$$T_n \le T_{n^2}$$

$$= \sum_{j=0}^{n-1} |a_j| \sum_{j=0}^{n-1} |b_j|$$

$$\le \sum_{j=0}^{\infty} |a_j| \sum_{j=0}^{\infty} |b_j|.$$

Thus the series (3.21) converges absolutely. Since $\{S_{n^2}\}$ is a subsequence of $\{S_n\}$, we find that

$$\sum_{j=0}^{\infty} \left(\sum_{k=0}^{j} a_j b_k + \sum_{l=0}^{j-1} a_l b_j \right) = \lim_{n\to\infty} S_n$$

$$= \lim_{n\to\infty} S_{n^2}$$

$$= \lim_{n\to\infty} A_n B_n$$

$$= AB.$$

Because of the absolute convergence of this series, Dirichlet's theorem shows that every rearrangement must also converge absolutely to AB. We have now proved the following theorem.

Theorem 3.14.2. *If $\sum_{j=0}^{\infty} a_j$ and $\sum_{j=0}^{\infty} b_j$ converge absolutely to A and B, respectively, then*

$$\sum_{j=0}^{\infty} \sum_{k=0}^{j} a_k b_{j-k} = AB$$

and the convergence is absolute.

3.15   Introduction to Power Series

A **power series** in a variable z is defined as a series of the form

$$\sum_{j=0}^{\infty} c_j (z - c)^j, \qquad (3.22)$$

where $c, c_0, c_1, \ldots$ are constants, not necessarily real. Throughout this discussion we shall take 0^0 to be 1. Thus the first term of the power series is c_0. We refer to c as the **center** of the power series.

Each term of a power series is therefore a function of z. Series of functions played a major role in the development of analysis in the nineteenth century. Their study was a driving force behind the development of a satisfactory standard of rigor. The general theory of series of functions is by no means simple.

Our first task is to find all z for which the power series converges. Often the ratio test is very useful. Let us try some examples.

Example 3.15.1. Discuss the convergence of the series

$$\sum_{j=1}^{\infty} \frac{jx^j}{(j+1)^2},$$

where x is a real variable.

Solution. Certainly the series converges if $x = 0$. Suppose $x \neq 0$. Setting

$$a_n = \frac{nx^n}{(n+1)^2}$$

for all $n > 0$, we find that

$$\left| \frac{a_{n+1}}{a_n} \right| = \frac{(n+1)|x|^{n+1}}{(n+2)^2} \cdot \frac{(n+1)^2}{n|x|^n}$$

$$= \frac{(n+1)^3}{n(n+2)^2} |x|$$

$$\to |x|$$

as $n \to \infty$. Hence the series converges absolutely if $|x| < 1$. If $|x| > 1$, then there exists N such that $|a_{n+1}| > |a_n|$ for all $n \geq N$. Thus the series fails the nth-term test and therefore diverges. (If x is not real, it is also immediate from these arguments that the series converges absolutely if $|x| < 1$ and diverges if $|x| > 1$.)

It remains to check the cases where $x = \pm 1$. With $x = -1$, we obtain the alternating series

$$\sum_{j=1}^{\infty} \frac{(-1)^j j}{(j+1)^2},$$

whose convergence has already been confirmed in Example 3.10.1.

Suppose therefore that $x = 1$. Then the series becomes

$$\sum_{j=1}^{\infty} \frac{j}{(j+1)^2}.$$

We use the limit comparison test, comparing this series with the harmonic series. Putting $a_n = n/(n+1)^2$ and $b_n = 1/n$ for all $n > 0$, we find that

$$\frac{a_n}{b_n} = \frac{n^2}{(n+1)^2} \to 1$$

as $n \to \infty$. Hence $a_n \sim b_n$ and our series diverges by the limit comparison test.

We conclude that the given series converges if and only if $-1 \le x < 1$. The convergence is absolute if $|x| < 1$ but not if $x = -1$. $\triangle$

Example 3.15.2. Test the series

$$\sum_{j=0}^{\infty} j^j (z-1)^j.$$

Solution. The series converges if $z = 1$. Suppose $z \ne 1$. Then for all $n > 0$ we have

$$\left| \frac{(n+1)^{n+1}(z-1)^{n+1}}{n^n(z-1)^n} \right| = (n+1)\left(\frac{n+1}{n}\right)^n |z-1|$$

$$> (n+1)|z-1|$$

$$\to \infty$$

as $n \to \infty$. Hence the series converges if and only if $z = 1$. $\triangle$

Example 3.15.3. Consider the series

$$\sum_{j=1}^{\infty} \frac{z^j}{j}.$$

The series converges if $z = 0$. If $z \ne 0$, then

$$\left| \frac{z^{n+1}}{n+1} \cdot \frac{n}{z^n} \right| = \frac{n|z|}{n+1} \to |z|$$

as $n \to \infty$. Hence the series converges absolutely by the ratio test if $|z| < 1$ and diverges by the nth-term test if $|z| > 1$. We have already seen in Example 3.11.1 that the series converges if $|z| = 1$ and $z \ne 1$. If $z = 1$, then the series is harmonic and therefore diverges. $\triangle$

Example 3.15.4. We have already seen in Theorem 2.7.6 that the series

$$\sum_{j=0}^{\infty} \frac{x^j}{j!}$$

converges to e^x for all real x. $\triangle$

Our next theorem shows that there are three possibilities: The power series (3.22) converges absolutely for all z; it converges only when $z = c$; or there exists a positive number r such that the power series converges absolutely if $|z - c| < r$ and diverges if $|z - c| > r$. The number r in the third case is called the **radius of convergence** of the power series. We write $r = 0$ in the second case, and in the first case we say that the power series has an infinite radius of convergence. Thus every power series in a real variable has associated with it an interval on which it converges (provided we consider each real number to constitute an interval and the set $\mathbb{R}$ of all real numbers also to be an interval). This interval is called the **interval of convergence** of the power series. In our four examples above, the respective intervals of convergence are $[-1, 1)$, $\{1\}$, $[-1, 1)$, and $\mathbb{R}$. Similarly, a power series in a complex variable has associated with it a **circle of convergence** in the interior of which the power series is absolutely convergent and in the exterior of which it diverges.

But we are getting ahead of ourselves. The result referred to in the previous paragraph still needs to be proved. First, though, observe that we can simplify the expression (3.22) by substituting $w = z - c$. This change of variable shows that there is no loss of generality in assuming that $c = 0$.

Theorem 3.15.1. *For the power series $\sum_{j=0}^{\infty} a_j z^j$ one of the following possibilities must hold:*

1. *the series converges only when $z = 0$;*
2. *the series is absolutely convergent for all z;*
3. *there exists $r > 0$ such that the series converges absolutely whenever $|z| < r$ and diverges whenever $|z| > r$.*

Proof. Certainly the power series converges when $z = 0$. We may also choose a particular $w \neq 0$ for which it converges, for if no such w exists, then possibility (1) obtains.

We show first that

$$\sum_{j=0}^{\infty} a_j z^j \tag{3.23}$$

converges absolutely for each z such that $|z| < |w|$. Since

$$\sum_{j=0}^{\infty} a_j w^j$$

converges, it follows that

$$\lim_{n \to \infty} |a_n w^n| = 0.$$

Thus the sequence $\{|a_n w^n|\}$ is bounded above by some number M. Therefore

$$|a_n z^n| = |a_n w^n| \left| \frac{z^n}{w^n} \right| \le M \left| \frac{z}{w} \right|^n .$$

But $|z/w| < 1$, and so the series

$$\sum_{j=0}^{\infty} M \left| \frac{z}{w} \right|^j$$

converges. Hence

$$\sum_{j=0}^{\infty} |a_j z^j|$$

also converges, by the comparison test.

One conclusion to be drawn is that if the series (3.23) converges for all z, then it converges absolutely for all z, because for every z there exists w, of modulus greater than $|z|$, for which the series converges. In this case possibility (2) holds. We may therefore suppose that there exists z_0 for which (3.23) diverges. The series must then diverge for every w_0 such that $|w_0| > |z_0|$. Therefore the set of moduli of all z such that (3.23) converges is bounded above by $|z_0|$, and since it contains 0, it must have a supremum, r. The series therefore converges absolutely for all z such that $|z| < r$ (because there exists w for which $|z| < |w| \le r$ and for which the series converges) and diverges for all z such that $|z| > r$ by the definition of r. $\qquad\square$

Example 3.15.5. Find the radius of convergence for the power series

$$\sum_{j=1}^{\infty} \frac{2 + (-i)^j}{j^2} z^j .$$

Solution. Set

$$a_n = \frac{2 + (-i)^n}{n^2} z^n$$

for each $n > 0$. If $|z| \le 1$, then $|a_n| \le 3/n^2$. Since

$$\sum_{j=1}^{\infty} \frac{3}{j^2}$$

converges, the given series is absolutely convergent by the comparison test. Suppose therefore that $|z| > 1$. Since

$$|a_n| \geq \frac{|z|^n}{n^2}$$

for all $n > 0$, it follows from Example 2.5.5 that the sequence $\{a_n\}$ does not converge to 0 and the given series therefore diverges. Hence the radius of convergence is 1.

Note that the ratio test could not have been applied in this example, since

$$\lim_{n \to \infty} \left| \frac{a_{n+1}}{a_n} \right|$$

does not exist. $\triangle$

We conclude this section with some simple observations about the addition and multiplication of power series. Given numbers s and t and convergent power series $\sum_{j=0}^{\infty} a_j z^j$ and $\sum_{j=0}^{\infty} b_j z^j$, Theorem 3.3.1 shows that

$$s \sum_{j=0}^{\infty} a_j z^j + t \sum_{j=0}^{\infty} b_j z^j = \sum_{j=0}^{\infty} (s a_j z^j + t b_j z^j)$$

$$= \sum_{j=0}^{\infty} (s a_j + t b_j) z^j.$$

Similarly, Theorem 3.14.2 gives

$$\sum_{j=0}^{\infty} a_j z^j \sum_{j=0}^{\infty} b_j z^j = \sum_{j=0}^{\infty} \sum_{k=0}^{j} a_k z^k b_{j-k} z^{j-k}$$

$$= \sum_{j=0}^{\infty} \sum_{k=0}^{j} a_k b_{j-k} z^j.$$

Exercises 3.12. 1. Consider the power series $\sum_{j=0}^{\infty} c_j (z - c)^j$, where $c_j \neq 0$ for all j.

(a) If

$$\lim_{n \to \infty} \left| \frac{c_{n+1}}{c_n} \right| = L,$$

prove that the radius of convergence of the series is $1/L$ (take $1/L = \infty$ if $L = 0$ and $1/L = 0$ if $L = \infty$).

(b) If

$$\lim_{n \to \infty} |c_n|^{1/n} = L,$$

prove that the radius of convergence of the series is $1/L$.

2. Let the radii of convergence of

$$\sum_{j=0}^{\infty} a_j z^j, \sum_{j=0}^{\infty} b_j z^j, \sum_{j=0}^{\infty} (a_j + b_j) z^j$$

be R_1, R_2, R, respectively.

(a) Show that if $R_1 \neq R_2$, then $R = \min\{R_1, R_2\}$.
(b) Give an example with the property that $R > R_1 = R_2$.

3. Can the power series $\sum_{j=0}^{\infty} c_j (z - 2)^j$ converge at 0 but diverge at 3?
4. For each of the following series find all real x for which the series converges:

(a) $\sum_{j=1}^{\infty} (-1)^j \frac{x^j}{j}$;   (f) $\sum_{j=0}^{\infty} \left(\frac{1}{2^{2j-1}} + \frac{1}{3^{2j}} \right) x^j$;

(b) $\sum_{j=1}^{\infty} j 2^j x^j$;   (g) $\sum_{j=0}^{\infty} \frac{j!}{j^j} x^j$;

(c) $\sum_{j=0}^{\infty} (-1)^j \frac{x^{2j}}{(2j)!}$;   (h) $\sum_{j=1}^{\infty} (-1)^j \frac{(x+4)^j}{j 2^j}$;

(d) $\sum_{j=0}^{\infty} \frac{(j!)^2}{(2j)!} x^j$;   (i) $\sum_{j=1}^{\infty} \frac{j}{(j+1)(j+2)} x^j$;

(e) $\sum_{j=1}^{\infty} \frac{x^j}{\sqrt{j}}$;   (j) $\sum_{j=1}^{\infty} \frac{x^j}{j\sqrt{j}}$.

5. For each of the following series find all complex z for which the series converges:

(a) $\sum_{j=0}^{\infty} z^{j^2}$;   (e) $\sum_{j=0}^{\infty} \frac{(z+i)^j}{e^j}$;

(b) $\sum_{j=1}^{\infty} \frac{(i-1)^j}{j} z^j$;   (f) $\sum_{j=1}^{\infty} (-1)^j \frac{2^j}{j} (z - i)^j$;

(c) $\sum_{j=0}^{\infty} \frac{j!}{2^j} (z - i)^j$;   (g) $\sum_{j=0}^{\infty} \frac{z^{2j}}{(-3)^j}$;

(d) $\sum_{j=0}^{\infty} (3 + (-1)^j) z^j$.

6. Find the radius of convergence of the series $\sum_{j=0}^{\infty} a_j z^j$, where $a_{2j} = 1$ and $a_{2j+1} = 2$ for all $j \geq 0$.

3.16 The Exponential, Sine, and Cosine Functions

We now use the work of the previous section to generalize our earlier results concerning the exponential function. Recall that, for real x,

$$e^x = \sum_{j=0}^{\infty} \frac{x^j}{j!},$$

by Theorem 2.7.6. More generally, for every complex z let us define

$$e^z = \sum_{j=0}^{\infty} \frac{z^j}{j!},$$

noting that the series converges absolutely by the ratio test, since

$$\left| \frac{z^{n+1}}{(n+1)!} \cdot \frac{n!}{z^n} \right| = \left| \frac{z}{n+1} \right| \to 0.$$

With this definition it is easily seen that parts (2) and (3) of Theorem 2.7.10 are valid also for complex sequences. In addition, we define $\exp(z) = e^z$ for all $z \in \mathbb{C}$, and refer to exp as the **exponential** function.

Because of the absolute convergence of the series for e^z, we may apply Theorem 3.14.2 to prove that

$$e^w e^z = \left(\sum_{j=0}^{\infty} \frac{w^j}{j!} \right) \left(\sum_{j=0}^{\infty} \frac{z^j}{j!} \right)$$

$$= \sum_{j=0}^{\infty} \sum_{k=0}^{j} \frac{w^k z^{j-k}}{k!(j-k)!}$$

$$= \sum_{j=0}^{\infty} \frac{1}{j!} \sum_{k=0}^{j} \binom{j}{k} w^k z^{j-k}$$

$$= \sum_{j=0}^{\infty} \frac{(w+z)^j}{j!}$$

$$= e^{w+z}$$

for all complex numbers w and z. This result generalizes Theorem 2.7.9. It follows by an easy induction that

$$(e^z)^n = e^{nz}$$

for every positive integer n.

Now suppose that $z = ix$ for some real x. Then

$$e^{ix} = \sum_{j=0}^{\infty} \frac{i^j x^j}{j!}$$

$$= \sum_{j=0}^{\infty} \left((-1)^j \frac{x^{2j}}{(2j)!} + i(-1)^j \frac{x^{2j+1}}{(2j+1)!} \right). \qquad (3.24)$$

Let $\cos(x) = \operatorname{Re}(e^{ix})$ and $\sin(x) = \operatorname{Im}(e^{ix})$. It follows from Theorem 2.3.11 that

$$\cos(x) = \sum_{j=0}^{\infty} (-1)^j \frac{x^{2j}}{(2j)!}$$

and

$$\sin(x) = \sum_{j=0}^{\infty} (-1)^j \frac{x^{2j+1}}{(2j+1)!}$$

for all real x, and the convergence of these power series is absolute. (See also Example 3.7.2.) Thus

$$e^{ix} = \cos(x) + i\,\sin(x)$$

for all real x. We refer to cos and sin as the **cosine** and **sine** functions, respectively. They will be given a geometric interpretation later.

We may generalize the definitions of the sine and cosine functions to complex numbers. Thus

$$e^{iz} = \sum_{j=0}^{\infty} \left((-1)^j \frac{z^{2j}}{(2j)!} + i(-1)^j \frac{z^{2j+1}}{(2j+1)!} \right)$$

$$= \cos(z) + i\,\sin(z),$$

where

$$\cos(z) = \sum_{j=0}^{\infty} (-1)^j \frac{z^{2j}}{(2j)!}$$

and

$$\sin(z) = \sum_{j=0}^{\infty} (-1)^j \frac{z^{2j+1}}{(2j+1)!}$$

for all complex numbers z. Both series are absolutely convergent.

It is clear from the definition that $\sin(0) = 0$, but $\cos(0) = 1$ by the convention that $0^0 = 1$. It is also immediate that

$$\sin(-z) = -\sin(z)$$

and

$$\cos(-z) = \cos(z)$$

for all z. We define

$$\tan(z) = \frac{\sin(z)}{\cos(z)}$$

for all z for which $\cos(z) \neq 0$. The function so defined is called the **tangent** function.

Given a real- or complex-valued function f, let g be the function defined by

$$g(z) = \frac{1}{f(z)}$$

for all z such that $f(z)$ is defined and nonzero. The function g is called the **reciprocal** of f. The sine, cosine, and tangent functions and their reciprocals are said to be **trigonometric**. The reciprocals of the sine, cosine, and tangent functions are called the **cosecant**, **secant**, and **cotangent** functions, respectively. They are denoted, respectively, by csc, sec, and cot. The parentheses around the arguments of the trigonometric functions are usually omitted.

Since

$$e^{iz} = \cos z + i \sin z$$

for all z, we also have

$$e^{-iz} = \cos z - i \sin z.$$

Thus $e^{iz} + e^{-iz} = 2 \cos z$, and so

$$\cos z = \frac{e^{iz} + e^{-iz}}{2} = \frac{e^{2iz} + 1}{2e^{iz}}.$$

Similarly,

$$\sin z = \frac{e^{iz} - e^{-iz}}{2i} = \frac{e^{2iz} - 1}{2ie^{iz}}.$$

For instance, if x is real, then

$$\cos ix = \frac{e^x + e^{-x}}{2}$$

and

$$\sin ix = \frac{e^{-x} - e^x}{2i} = \frac{i(e^x - e^{-x})}{2},$$

since $-i^2 = 1$. Moreover

$$\sin w \cos z + \cos w \sin z = \frac{(e^{2iw} - 1)(e^{2iz} + 1)}{4ie^{iw}e^{iz}} + \frac{(e^{2iw} + 1)(e^{2iz} - 1)}{4ie^{iw}e^{iz}}$$

$$= \frac{e^{2i(w+z)} + e^{2iw} - e^{2iz} - 1 + e^{2i(w+z)} - e^{2iw} + e^{2iz} - 1}{4ie^{i(w+z)}}$$

$$= \frac{e^{2i(w+z)} - 1}{2ie^{i(w+z)}}$$

$$= \sin(w + z) \tag{3.25}$$

and

$$\cos w \cos z - \sin w \sin z = \frac{(e^{2iw} + 1)(e^{2iz} + 1)}{4e^{iw}e^{iz}} - \frac{(e^{2iw} - 1)(e^{2iz} - 1)}{-4e^{iw}e^{iz}}$$

$$= \frac{e^{2i(w+z)} + e^{2iw} + e^{2iz} + 1 + e^{2i(w+z)} - e^{2iw} - e^{2iz} + 1}{4e^{i(w+z)}}$$

$$= \frac{e^{2i(w+z)} + 1}{2e^{i(w+z)}}$$

$$= \cos(w + z) \tag{3.26}$$

for all complex numbers w and z. Thus

$$\sin 2z = \sin z \cos z + \cos z \sin z = 2 \sin z \cos z$$

and, similarly,

$$\cos 2z = \cos^2 z - \sin^2 z.$$

It also follows by induction that if $\sin z = 0$, then $\sin nz = 0$ for every $n \in \mathbb{N}$, for if this equation holds for a particular $n \in \mathbb{N}$, then

$$\sin((n + 1)z) = \sin(nz + z)$$
$$= \sin nz \cos z + \cos nz \sin z$$
$$= 0.$$

Furthermore

$$1 = \cos 0 = \cos(z - z) = \cos z \cos(-z) - \sin z \sin(-z) = \cos^2 z + \sin^2 z$$

for all z. If x is a real number, we infer that

$$|\sin x| \le 1$$

and

$$|\cos x| \le 1.$$

In any case we also have

$$\cos 2z = \cos^2 z - (1 - \cos^2 z) = 2\cos^2 z - 1;$$

hence

$$\cos^2 z = \frac{1 + \cos 2z}{2}.$$

In addition,

$$|e^{iz}| = \sqrt{\cos^2 z + \sin^2 z} = 1.$$

Thus $e^{iz} \ne 0$, and it follows that $e^z \ne 0$ for each $z \in \mathbb{C}$. Notice, however, that

$$|\sin ix| = \frac{|e^x - e^{-x}|}{2}.$$

We therefore infer from Theorem 2.7.10(1) that the function $|\sin z|$ is unbounded. Similarly, $|\cos z|$ is unbounded.

Note that if $z = x + iy$, where x and y are real, then

$$\begin{aligned}
\sin z &= \sin(x + iy) \\
&= \sin x \cos iy + \cos x \sin iy \\
&= \frac{1}{2}(e^y + e^{-y}) \sin x + \frac{i}{2}(e^y - e^{-y}) \cos x.
\end{aligned}$$

Similarly,

$$\cos z = \frac{1}{2}(e^y + e^{-y}) \cos x - \frac{i}{2}(e^y - e^{-y}) \sin x.$$

If w and z are numbers such that $\tan w \tan z = 1$, then $\cos w \cos z \ne 0$ and

$$\frac{\sin w}{\cos w} \cdot \frac{\sin z}{\cos z} = 1,$$

so that

$$\sin w \sin z = \cos w \cos z,$$

whence

$$\cos(w + z) = \cos w \cos z - \sin w \sin z = 0.$$

Conversely, if $\cos(w + z) = 0 \neq \cos w \cos z$, then $\tan w \tan z = 1$. If neither $\cos w \cos z$ nor $\cos(w + z)$ is equal to 0, then

$$
\begin{aligned}
\tan(w + z) &= \frac{\sin(w + z)}{\cos(w + z)} \\
&= \frac{\sin w \cos z + \cos w \sin z}{\cos w \cos z - \sin w \sin z} \\
&= \frac{\frac{\sin w \cos z + \cos w \sin z}{\cos w \cos z}}{\frac{\cos w \cos z - \sin w \sin z}{\cos w \cos z}} \\
&= \frac{\tan w + \tan z}{1 - \tan w \tan z}.
\end{aligned}
$$

In particular, if $\tan z$ is defined and $|\tan z| \neq 1$, then

$$\tan 2z = \frac{2 \tan z}{1 - \tan^2 z}.$$

Note also that

$$\tan(-z) = \frac{\sin(-z)}{\cos(-z)} = -\tan z.$$

Proposition 3.16.1. *1. If $x \geq 0$, then*

$$\sin x \leq x. \tag{3.27}$$

2. If $x \in \mathbb{R}$, then

$$1 - \frac{x^2}{2} \leq \cos x \leq 1 - \frac{x^2}{2} + \frac{x^4}{24}. \tag{3.28}$$

Proof. 1. We may assume that $x < 1$ since $\sin x \leq 1$ for all real x. Then the sequence $\{x^{2n+1}/(2n + 1)!\}$ is decreasing, and we may invoke inequality (3.10) in the proof of Theorem 3.10.2, with $n = 0$, to conclude that $\sin x \leq x$.
2. We may assume that $x > 0$, since $\cos(-x) = \cos x$ for all x and $\cos 0 = 1$.

In regard to the first inequality, we may assume also that $x < 2$, as $\cos x \geq -1$ for all real x. Then the sequence $\{x^{2n}/(2n)!\}_{n\geq 1}$ is decreasing, since

$$\frac{x^{2n+2}}{(2n+1)!} \cdot \frac{(2n)!}{x^{2n}} = \frac{x^2}{(2n+2)(2n+1)} < \frac{1}{3} < 1$$

when $n \geq 1$ and $0 < x < 2$. Consequently,

$$\cos x - 1 = \sum_{j=1}^{\infty} (-1)^j \frac{x^{2j}}{(2j)!}$$

$$= -\sum_{j=1}^{\infty} (-1)^{j-1} \frac{x^{2j}}{(2j)!}$$

$$= -\sum_{j=0}^{\infty} (-1)^j \frac{x^{2j+2}}{(2j+2)!}$$

$$\geq -\frac{x^2}{2},$$

again using inequality (3.10) with $n = 0$. The first of the required inequalities follows.

For the second inequality we may assume that $x < 2\sqrt{3}$, since $\cos x \leq 1$. The sequence $\{x^{2n}/(2n)!\}_{n\geq 1}$ is still decreasing, and we may appeal to Theorem 3.10.2 to conclude that

$$0 \leq \cos x - 1 + \frac{x^2}{2} \leq \frac{x^4}{24}.$$

$\square$

For ease of reference, in the following theorem we summarize some of the more important results we have proved.

Theorem 3.16.2. *Let w and z be complex numbers. Then*

1. $\cos z = \frac{e^{iz}+e^{-iz}}{2}$,
2. $\sin z = \frac{e^{iz}-e^{-iz}}{2i}$,
3. $\sin(w + z) = \sin w \cos z + \cos w \sin z$,
4. $\cos(w + z) = \cos w \cos z - \sin w \sin z$,
5. $\tan(w + z) = \frac{\tan w + \tan z}{1 - \tan w \tan z}$ *if* $\cos w \cos z \cos(w + z) \neq 0$,
6. $\sin 2z = 2 \sin z \cos z$,
7. $\cos 2z = \cos^2 z - \sin^2 z$,
8. $\tan 2z = \frac{2 \tan z}{1 - \tan^2 z}$ *if* $\cos z \cos 2z \neq 0$,

9. $\cos^2 z + \sin^2 z = 1,$

10. $\cos^2 z = \frac{1 + \cos 2z}{2},$

11. $|e^{iz}| = 1.$

In addition, if x is real, then $|\sin x| \le 1$ *and* $|\cos x| \le 1.$

Example 3.16.1. Since

$$\left|\frac{\sin n}{n^p}\right| \le \frac{1}{n^p}$$

for all $n \in \mathbb{N}$ and $p \in \mathbb{Q}$, the comparison test shows that the series

$$\sum_{j=1}^{\infty} \frac{\sin j}{j^p} \tag{3.29}$$

is absolutely convergent if $p > 1$. We now prove that it is conditionally convergent when $0 < p \le 1$.

We showed in Example 3.11.1 that

$$\sum_{j=1}^{\infty} \frac{z^j}{j^p} \tag{3.30}$$

converges for each $p > 0$ and each $z \ne 1$ such that $|z| = 1$. In particular, if

$$z = e^i = \cos 1 + i \sin 1,$$

then $|z| = 1$ and $z \ne 1$ since $\cos 1 \ne 1$. [If $\cos 1 = 1$, then

$$|\sin 1| = \sqrt{1 - \cos^2 1} = 0.$$

However, $\sin 1 > 0$ by the remark following Theorem 3.10.2, as the sequence

$$\left\{\frac{1}{(2n + 1)!}\right\}$$

is decreasing.] Therefore the series (3.30) converges for this z, and since

$$z^n = e^{in} = \cos n + i \sin n,$$

it follows by Theorem 3.3.2 that the series (3.29) converges.

It remains to show that this series is not absolutely convergent if $0 < p \le 1$. As

$$\left| \frac{\sin n}{n} \right| \le \left| \frac{\sin n}{n^p} \right|$$

whenever $0 < p \le 1$, the comparison test shows that it is enough to establish the divergence of

$$\sum_{j=1}^{\infty} \frac{|\sin j|}{j}.$$

For each $n > 0$ let us write

$$\frac{|\sin n|}{n} = \frac{a_n + b_n}{2},$$

where

$$a_n = \frac{|\sin n| - |\sin(n-1)|}{n}$$

and

$$b_n = \frac{|\sin n| + |\sin(n-1)|}{n}.$$

It suffices to show that $\sum_{j=1}^{\infty} a_j$ converges and $\sum_{j=1}^{\infty} b_j$ diverges.
In regard to the former series, define

$$u_n = |\sin n| - |\sin(n-1)|$$

and

$$v_n = \frac{1}{n}$$

for each $n > 0$. Then $\{v_n\}$ is a decreasing null sequence. Moreover

$$\left| \sum_{j=1}^{n} u_j \right| = \left| \sum_{j=1}^{n} (|\sin j| - |\sin(j-1)|) \right|$$

$$= |\sin n|$$

$$\le 1$$

for all $n > 0$, by the telescoping property. We conclude from Dirichlet's test that $\sum_{j=1}^{\infty} a_j$ converges.

In order to verify the divergence of $\sum_{j=1}^{\infty} b_j$, note first that for all real numbers α and β we have

$$
\begin{aligned}
|\sin(\alpha - \beta)| &= |\sin\alpha\cos\beta - \cos\alpha\sin\beta| \\
&\leq |\sin\alpha||\cos\beta| + |\cos\alpha||\sin\beta| \\
&\leq |\sin\alpha| + |\sin\beta|.
\end{aligned}
$$

Therefore

$$
\sin 1 = |\sin 1| = |\sin(n - (n-1))| \leq |\sin n| + |\sin(n-1)|.
$$

Consequently,

$$
b_n \geq \frac{\sin 1}{n} > 0
$$

for all $n > 0$, and the series $\sum_{j=1}^{\infty} b_j$ diverges by comparison with the harmonic series. $\triangle$

Example 3.16.2. Consider the series

$$
\sum_{j=1}^{\infty} \frac{\sin jx}{j}
$$

for any $x \in \mathbb{R}$. It converges to 0 if $\sin(x/2) = 0$. Suppose therefore that $\sin(x/2) \neq 0$. The trigonometric identity (3.26) implies that

$$
\cos\left(\frac{x}{2} + jx\right) = \cos\frac{x}{2}\cos jx - \sin\frac{x}{2}\sin jx
$$

and

$$
\cos\left(\frac{x}{2} - jx\right) = \cos\frac{x}{2}\cos jx + \sin\frac{x}{2}\sin jx;
$$

hence

$$
2\sin\frac{x}{2}\sin jx = \cos\left(\frac{x}{2} - jx\right) - \cos\left(\frac{x}{2} + jx\right).
$$

Since

$$\cos\left(\frac{x}{2} - jx\right) = \cos\left(\left(\frac{1}{2} - j\right)x\right)$$

$$= \cos\left(\left(j - \frac{1}{2}\right)x\right)$$

$$= \cos\left(\left((j - 1) + \frac{1}{2}\right)x\right),$$

it follows that

$$\sum_{j=1}^{n}\left(\cos\left(\frac{x}{2} - jx\right) - \cos\left(\frac{x}{2} + jx\right)\right)$$

$$= \sum_{j=1}^{n}\left(\cos\left(\left((j - 1) + \frac{1}{2}\right)x\right) - \cos\left(\left(j + \frac{1}{2}\right)x\right)\right)$$

$$= \cos\frac{x}{2} - \cos\left(\left(n + \frac{1}{2}\right)x\right);$$

hence

$$\sum_{j=1}^{n}\sin jx = \frac{\cos\frac{x}{2} - \cos((n + \frac{1}{2})x)}{2\sin\frac{x}{2}}$$

for all $n \in \mathbb{N}$, since $\sin(x/2) \neq 0$. Consequently,

$$\left|\sum_{j=1}^{n}\sin jx\right| \leq \frac{\left|\cos\frac{x}{2}\right| + \left|\cos\left((n + \frac{1}{2})x\right)\right|}{2\left|\sin\frac{x}{2}\right|}$$

$$\leq \frac{1}{\left|\sin\frac{x}{2}\right|}.$$

Dirichlet's test therefore implies that the series converges. (In fact, the previous example shows that it is conditionally convergent if $x = 1$.)   △

Example 3.16.3. Let $\{a_n\}$ be a null sequence of real numbers. We show that

$$\lim_{n\to\infty}\frac{\sin a_n}{a_n} = 1.$$

From the definition we have

$$\frac{\sin a_n}{a_n} = \sum_{j=0}^{\infty} (-1)^j \frac{a_n^{2j}}{(2j+1)!} = 1 + a_n R_n$$

for all n, where

$$R_n = \sum_{j=1}^{\infty} (-1)^j \frac{a_n^{2j-1}}{(2j+1)!}.$$

Since $\{a_n\}$ is convergent, there is an M such that $|a_n| \le M$ for all n. Hence

$$|R_n| \le \sum_{j=1}^{\infty} \left| \frac{a_n^{2j-1}}{(2j+1)!} \right|$$

$$\le \sum_{j=1}^{\infty} \frac{M^{2j-1}}{(2j-1)!}$$

$$= \sum_{j=0}^{\infty} \frac{M^{2j+1}}{(2j+1)!}$$

$$< e^M.$$

Thus

$$\lim_{n \to \infty} \frac{\sin a_n}{a_n} = 1 + \lim_{n \to \infty} a_n R_n = 1.$$

$\triangle$

Exercises 3.13. 1. If u and v are any complex numbers, prove that

$$\sin u - \sin v = 2 \cos \frac{u+v}{2} \sin \frac{u-v}{2}$$

and

$$\cos u - \cos v = -2 \sin \frac{u+v}{2} \sin \frac{u-v}{2}.$$

Hence show that

$$2 \sin \alpha \sin \beta = \cos(\alpha + \beta) - \cos(\alpha - \beta)$$

and

$$2 \cos \alpha \sin \beta = \sin(\alpha + \beta) - \sin(\alpha - \beta)$$

for all real numbers α and β.

2. Use the inequality

$$\left| \sum_{j=1}^{n} \sin j \right| \leq \frac{1}{\sin \frac{1}{2}},$$

for all $n \in \mathbb{N}$, to verify the convergence of series (3.29) for all rational $p > 0$.

3. Prove that if $\{a_n\}$ is a decreasing null sequence, then

$$\sum_{j=1}^{\infty} a_j \sin j$$

and

$$\sum_{j=1}^{\infty} a_j \cos j$$

converge.

4. Let $\{a_n\}$ be a null sequence of real numbers. Show that

$$\lim_{n \to \infty} \cos a_n = 1$$

and

$$\lim_{n \to \infty} \frac{\tan a_n}{a_n} = 1.$$

5. The functions given by

$$\sinh x = \frac{e^x - e^{-x}}{2}$$

and

$$\cosh x = \frac{e^x + e^{-x}}{2},$$

for each x, are called the hyperbolic sine and hyperbolic cosine functions, respectively.

(a) Show that

$$\cosh^2 x - \sinh^2 x = 1$$

for all x.

(b) Let $z = x + iy$, where x and y are real.

   i. Prove the following:

      A. $\sin z = \sin x \cosh y + i \sinh y \cos x$.
      B. $|\sinh y| \le |\sin z| \le |\cosh y|$.
      C. $|\sin z|^2 = \sin^2 x + \sinh^2 y$.

   ii. For every $c \in \mathbb{R}$, show that the function $f(z) = \sin z$ maps the line $y = c$
      to an ellipse and the line $x = c$ to a hyperbola.

6. Express $\sin i$ and $\cos i$ in the form $x + iy$ where x and y are real.
7. Find the sum

$$\sum_{j=0}^{\infty} \sin \frac{1}{2^j} \cos \frac{3}{2^j}.$$

(Hint: Rewrite it as a telescoping series.)
8. Determine the convergence of

$$\sum_{j=1}^{\infty} \left(1 - \cos \frac{1}{j} \right).$$

(Hint: Write each term in terms of the sine function.)
9. Show that $|\cos z|$ is unbounded, where $z \in \mathbb{C}$.

Chapter 4
Limits of Functions

The concept of a limit of a function is central to the study of mathematical analysis. It generalizes the notion of the limit of a sequence (a function whose domain is a set of integers). Indeed, the former can be defined in terms of the latter. Unless an indication to the contrary is given or it is evident that a restriction to real numbers is required, the domain and range of a given function are assumed to be sets of complex numbers.

4.1 Introduction

We start by looking at an example. Consider the real functions f, g, h given by

$$f(x) = g(x) = h(x) = \frac{2x^2 - 2x}{x - 1}$$

for all $x \neq 1$, $g(1) = 1$, and $h(1) = 2$ (see Fig. 4.1). Thus $\mathcal{D}_f = \mathbb{R} - \{1\}$ and $\mathcal{D}_g = \mathcal{D}_h = \mathbb{R}$. For each $x \neq 1$,

$$f(x) = g(x) = h(x) = \frac{2x(x - 1)}{x - 1} = 2x.$$

Let us first contemplate the function f. As x becomes close to 1, $f(x)$ becomes close to 2 provided that $x \neq 1$. More formally, we say that the limit of f, as x approaches 1, is 2. In this case we write

$$\lim_{x \to 1} f(x) = 2.$$

Likewise $\lim_{x \to 1} g(x) = 2$ and $\lim_{x \to 1} h(x) = 2$; the values of these functions at 1 are immaterial.

© Springer Science+Business Media New York 2015   191
C.H.C. Little et al., *Real Analysis via Sequences and Series*, Undergraduate
Texts in Mathematics, DOI 10.1007/978-1-4939-2651-0_4

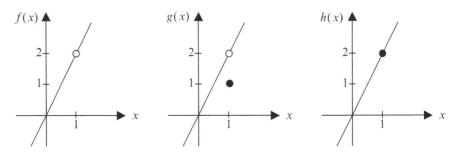

Fig. 4.1 Graphs of f, g and h

Roughly speaking, we can ensure that the distance between $f(x)$ and 2 is as small as we please by choosing x close enough to 1 but distinct from 1. In other words, given any sequence $\{s_n\}$ in $\mathbb{R} - \{1\}$ that converges to 1, the sequence $\{f(s_n)\}$ converges to 2.

4.2 Definition and Examples

We now define the concept formally.

Definition 4.2.1. Let f be a function, c an accumulation point of $\mathcal{D}_f$, and L a number. We write

$$\lim_{z \to c} f(z) = L \tag{4.1}$$

if the sequence $\{f(s_n)\}$ converges to L for every sequence $\{s_n\}$ in $\mathcal{D}_f - \{c\}$ converging to c. We call L a **limit** of f at c. We also say that $f(z)$ **approaches** L **as** z **approaches** c, and we write $f(z) \to L$ as $z \to c$.

Remark 1. The number c may or may not be a member of $\mathcal{D}_f$.

Remark 2. The symbol z in Definition 4.2.1 represents a dummy variable. Thus if $\lim_{z \to c} f(z) = L$, then we also have $\lim_{w \to c} f(w) = L$ for any other symbol w.

Remark 3. The assumption that c is an accumulation point of $\mathcal{D}_f$ guarantees the existence of a sequence in $\mathcal{D}_f - \{c\}$ that converges to c.

We have seen that a sequence has at most one limit. This observation leads immediately to the following theorem.

Theorem 4.2.1. *A function has at most one limit at a given accumulation point of its domain.*

Example 4.2.1. It follows from Theorem 2.7.10(2) that

$$\lim_{x \to a} e^x = e^a$$

for all $a \in \mathbb{R}$. Similarly, Theorem 2.7.10(3) shows that

$$\lim_{h \to 0} \frac{e^h - 1}{h} = 1.$$ $\triangle$

Example 4.2.2. Let $f(x) = i^{1/x}$ for each x in the range of the sequence $\{1/n\}$. Because this sequence is null, we see that 0 is an accumulation point of $\mathcal{D}_f$. We now demonstrate that

$$\lim_{x \to 0} f(x)$$

does not exist. For each positive integer n take

$$s_n = \frac{1}{4n}$$

and

$$t_n = \frac{1}{4n + 2}.$$

Thus

$$\lim_{n \to \infty} s_n = \lim_{n \to \infty} t_n = 0.$$

However,

$$i^{1/s_n} = i^{4n} = 1^n = 1$$

and

$$i^{1/t_n} = i^{2(2n+1)} = (-1)^{2n+1} = -1 \neq 1.$$

Therefore the desired limit does not exist. $\triangle$

Proposition 4.2.2. *Suppose that* $f(z) = k$ *for all* $z \in \mathcal{D}_f$. *Then for every accumulation point* c *of* $\mathcal{D}_f$ *we have*

$$\lim_{z \to c} f(z) = k.$$

Proof. Let $\{s_n\}$ be any sequence in $\mathcal{D}_f - \{c\}$ converging to c. Since $\{f(s_n)\}$ is the constant sequence $\{k\}$, it converges to k.   □

The next two propositions can be proved in a similar way. In both propositions we take c to be an accumulation point of $\mathcal{D}_f$.

Proposition 4.2.3. *Suppose that $f(z) = z$ for all $z \in \mathcal{D}_f$. Then*

$$\lim_{z \to c} f(z) = c.$$

Proposition 4.2.4. *If $f(z) = |z|$ for all $z \in \mathcal{D}_f$, then*

$$\lim_{z \to c} f(z) = |c|.$$

From Example 3.16.3 we obtain the following result.

Proposition 4.2.5.

$$\lim_{x \to 0} \frac{\sin x}{x} = 1.$$

4.3   Basic Properties of Limits

We now present an equivalent formulation of the definition of a limit. Recall that if c is an accumulation point of $\mathcal{D}_f$, then there is an injective sequence $\{s_n\}$ in $\mathcal{D}_f$ that converges to c.

Theorem 4.3.1. *Let f be a function, c an accumulation point of $\mathcal{D}_f$, and L a number. The following statements are equivalent:*

1.

$$\lim_{z \to c} f(z) = L;$$

2. for every $\varepsilon > 0$ there exists $\delta > 0$ such that

$$|f(z) - L| < \varepsilon$$

for each $z \in \mathcal{D}_f$ satisfying $0 < |z - c| < \delta$.

Proof. Suppose that condition (2) holds. Let $\{s_n\}$ be a sequence in $\mathcal{D}_f - \{c\}$ converging to c. Choose $\varepsilon > 0$. There exists $\delta > 0$ such that

$$|f(z) - L| < \varepsilon$$

for each $z \in \mathcal{D}_f$ satisfying $0 < |z - c| < \delta$. Moreover there exists N such that

$$|s_n - c| < \delta$$

for all $n \geq N$. For all such n we have $s_n \in \mathcal{D}_f$ and $s_n \neq c$ by hypothesis, and so

$$|f(s_n) - L| < \varepsilon.$$

Hence $\{f(s_n)\}$ converges to L, and we conclude that condition (1) holds.

Conversely, suppose that condition (2) does not hold. Then, for some $\varepsilon > 0$, every $\delta > 0$ has the property that there exists $z \in \mathcal{D}_f$ satisfying $0 < |z - c| < \delta$ and $|f(z) - L| \geq \varepsilon$. In particular, for each positive integer n we can choose $s_n \in \mathcal{D}_f$ such that

$$0 < |s_n - c| < \frac{1}{n}$$

and $|f(s_n) - L| \geq \varepsilon$. The sequence $\{s_n\}$ evidently converges to c, by Lemma 2.5.1. However,

$$\lim_{n \to \infty} f(s_n) \neq L;$$

otherwise there would exist N such that $|f(s_n) - L| < \varepsilon$ for all $n \geq N$. We conclude that condition (1) does not hold. □

Remark 1. If we use the notation

$$N_r(a) = \{z \mid |z - a| < r\}$$

and

$$N_r^*(a) = N_r(a) - \{a\},$$

where a is any number and r is a positive (real) number, then Theorem 4.3.1 shows that

$$\lim_{z \to c} f(z) = L$$

if and only if for every $\varepsilon > 0$ there exists $\delta > 0$ such that

$$f(N_\delta^*(c) \cap \mathcal{D}_f) \subseteq N_\varepsilon(L).$$

The two parts of Fig. 4.2 depict $N_r(a)$ in $\mathbb{R}$ and $\mathbb{C}$, respectively.

Fig. 4.2 $N_r(a)$ in $\mathbb{R}$ and $\mathbb{C}$

Remark 2. We can clearly assume that $\varepsilon < M$ and $\delta < M$ for some constant M. And, as in Proposition 2.2.3, we can replace ε by $k\varepsilon$, where k is a positive constant.

Remark 3. If x and c are real numbers, we often write the inequality $|x - c| < \delta$ as

$$c - \delta < x < c + \delta.$$

Example 4.3.1. Let us prove from Theorem 4.3.1 that

$$\lim_{z \to 1}(z^2 + 1) = 2.$$

Proof. Choose $\varepsilon > 0$. We need to find a corresponding $\delta > 0$ such that

$$|z^2 + 1 - 2| = |z^2 - 1| < k\varepsilon,$$

for some constant $k > 0$, whenever $0 < |z - 1| < \delta$.
    Now

$$
\begin{aligned}
|z^2 - 1| &= |(z + 1)(z - 1)| \\
&= |z + 1||z - 1| \\
&= |z - 1 + 2||z - 1| \\
&\leq (|z - 1| + 2)|z - 1| \\
&< (\delta + 2)\delta
\end{aligned}
$$

whenever $|z - 1| < \delta$. Thus, if $\delta \leq 1$, then $|z^2 - 1| < 3\delta$, and if in addition $\delta \leq \varepsilon$, then it follows that $|z^2 - 1| < 3\varepsilon$. Hence we can guarantee that $|z^2 - 1| < 3\varepsilon$ if we take

$$\delta = \min\{1, \varepsilon\}.$$

We conclude that

$$\lim_{z \to 1}(z^2 + 1) = 2.$$

△

Example 4.3.2. Now let us prove that

$$\lim_{z \to 1} \frac{1}{z} = 1.$$

Proof. Choose $\varepsilon > 0$. We need to find $\delta > 0$ such that

$$\left| \frac{1}{z} - 1 \right| < k\varepsilon, \tag{4.2}$$

for some constant $k > 0$, whenever $z \neq 0$ and $0 < |z - 1| < \delta$. If $z \neq 0$, then

$$\left| \frac{1}{z} - 1 \right| = \frac{|1 - z|}{|z|} < \frac{\delta}{|z|}$$

whenever

$$\delta > |z - 1| \geq ||z| - 1| \geq 1 - |z|.$$

Taking $\delta \leq 1/2$, we therefore have

$$|z| > 1 - \delta \geq \frac{1}{2},$$

so that

$$\frac{1}{|z|} < 2.$$

If in addition $\delta \leq \varepsilon$, then

$$\left| \frac{1}{z} - 1 \right| < 2\delta \leq 2\varepsilon.$$

Thus we can guarantee that inequality (4.2) holds with $k = 2$ if we take

$$\delta = \min \left\{ \frac{1}{2}, \varepsilon \right\}.$$

We conclude that

$$\lim_{z \to 1} \frac{1}{z} = 1.$$

△

Motivated by our study of sequences, we say that a function f satisfies the **Cauchy condition** at an accumulation point c of $\mathcal{D}_f$ if, for every $\varepsilon > 0$, there exists $\delta > 0$ such that

$$|f(z) - f(w)| < \varepsilon$$

for all z and w in $N_\delta^*(c) \cap \mathcal{D}_f$.

Theorem 4.3.2. *Let c be an accumulation point of the domain of a function f. Then $\lim_{z \to c} f(z) = L$ for some number L if and only if f satisfies the Cauchy condition at c.*

Proof. Suppose first that

$$\lim_{z \to c} f(z) = L.$$

Choose $\varepsilon > 0$. There exists $\delta > 0$ such that $|f(z) - L| < \varepsilon$ for all $z \in \mathcal{D}_f$ satisfying $0 < |z - c| < \delta$. Choose z and w in $N_\delta^*(c) \cap \mathcal{D}_f$. Then $|f(z) - L| < \varepsilon$ and $|f(w) - L| < \varepsilon$. Hence

$$|f(z) - f(w)| \le |f(z) - L| + |L - f(w)| < 2\varepsilon.$$

We conclude that f satisfies the Cauchy condition at c.

Conversely, suppose that f satisfies the Cauchy condition at c. Choose $\varepsilon > 0$. There exists $\delta > 0$ such that

$$|f(z) - f(w)| < \varepsilon \tag{4.3}$$

for all z and w in $N_\delta^*(c) \cap \mathcal{D}_f$.

Since c is an accumulation point of $\mathcal{D}_f$, we can find an injective sequence $\{z_n\}$ in $\mathcal{D}_f$ that converges to c. Choose N such that

$$0 < |z_n - c| < \delta$$

for every $n \ge N$. Thus $z_n \in N_\delta^*(c) \cap \mathcal{D}_f$ for each such n, and so

$$|f(z_n) - f(z_m)| < \varepsilon$$

whenever $m \ge N$ and $n \ge N$. Consequently, the sequence $\{f(z_n)\}$ is Cauchy and therefore converges to some number L.

We conclude that there exists M_1 such that $|f(z_n) - L| < \varepsilon$ for all $n \ge M_1$. Now let $\{w_n\}$ be any sequence in $\mathcal{D}_f - \{c\}$ converging to c. There exists $M_2 \ge \max\{N, M_1\}$ such that $0 < |w_n - c| < \delta$ for all $n \ge M_2$. Choose $n \ge M_2 \ge N$. Then we also have $0 < |z_n - c| < \delta$, and so $|f(w_n) - f(z_n)| < \varepsilon$ by the Cauchy criterion. Therefore

$$|f(w_n) - L| \leq |f(w_n) - f(z_n)| + |f(z_n) - L| < 2\varepsilon,$$

since $n \geq M_1$, and we deduce that $\{f(w_n)\}$ also converges to L.

Hence

$$\lim_{z \to c} f(z) = L,$$

by definition. □

Exercises 4.1. 1. Let f be a function, c an accumulation point of $\mathcal{D}_f$, and L a number. Suppose there exists a real K such that

$$|f(z) - L| \leq K|z - c|$$

for all $z \in \mathbb{C}$. Show that $\lim_{z \to c} f(z) = L$.

2. Let $\alpha > 0$ and $\lim_{x \to 0} f(x) = L$. Show that $\lim_{x \to 0} f(\alpha x) = L$.

3. Suppose that $\lim_{x \to c} f(x) = L$ for some numbers c and L. Show that the function f is bounded on some neighborhood of c.

4. Use Theorem 4.3.1 to prove that $\lim_{x \to 1} \sqrt{x} = 1$.

5. Show that $\lim_{z \to 0}(\overline{z}/z)$ does not exist for any $z \in \mathbb{C}$.

4.4 Algebra of Limits

We shall now write down some theorems that are useful for calculating limits. Most of them follow from the corresponding theorems for sequences. In each theorem we shall assume that the limit is being evaluated at an accumulation point for the domain of the relevant function.

Theorem 4.4.1. *Let f and g be functions, let K and L be numbers, and let c be an accumulation point of $\mathcal{D}_f \cap \mathcal{D}_g$. Suppose that*

$$\lim_{z \to c} f(z) = K$$

and

$$\lim_{z \to c} g(z) = L.$$

Then

1.

$$\lim_{z \to c}(f(z) + g(z)) = K + L,$$

2.

$$\lim_{z \to c} f(z)g(z) = KL,$$

3.

$$\lim_{z \to c} \frac{f(z)}{g(z)} = \frac{K}{L}$$

if $L \neq 0$ and there exists a neighborhood $N_\delta(c)$ such that $g(z) \neq 0$ for all $z \in N_\delta^*(c) \cap \mathcal{D}_g$.

Proof. 1. It follows from the hypothesis that the sequence $\{f(s_n)\}$ converges to K for every sequence $\{s_n\}$ in $\mathcal{D}_f - \{c\}$ converging to c. Similarly, $\{g(t_n)\}$ converges to L for every sequence $\{t_n\}$ in $\mathcal{D}_g - \{c\}$ converging to c.

Choose a sequence $\{s_n\}$ in

$$\mathcal{D}_{f+g} - \{c\} = (\mathcal{D}_f \cap \mathcal{D}_g) - \{c\}$$

converging to c. Then, for all n,

$$(f + g)(s_n) = f(s_n) + g(s_n) \to K + L$$

as $n \to \infty$, and the result follows.
2. The proofs of parts (2) and (3) are similar.   □

Remark 1. Sometimes the limits in the left-hand sides in the conclusion of the theorem exist even if $\lim_{z \to c} f(z)$ or $\lim_{z \to c} g(z)$ do not. For example, let $f(z) = 1/z$ and $g(z) = -1/z$ for all $z \neq 0$. Neither of these functions has a limit at 0, yet their sum and quotient do.

Remark 2. The insistence on c being an accumulation point of $\mathcal{D}_f \cap \mathcal{D}_g$ is inserted in order to guarantee that the limits in question are well defined. (See Theorem 4.2.1.) For instance, if $f(x) = \sqrt{x}$ for all $x \geq 0$ and $g(x) = \sqrt{-x}$ for all $x \leq 0$, then 0 is not an accumulation point of $\mathcal{D}_f \cap \mathcal{D}_g$ since $\mathcal{D}_f \cap \mathcal{D}_g = \{0\}$. In this case the desired limits are undefined.

Corollary 4.4.2. *Let n be a positive integer. Using the notation of the theorem, we have*

1.

$$\lim_{z \to c} f^n(z) = L^n,$$

2.

$$\lim_{z \to c} \frac{1}{f^n(z)} = \frac{1}{L^n}$$

if $L \neq 0$ and there exists a neighborhood $N_\delta(c)$ such that $f(z) \neq 0$ for all $z \in N_\delta^*(c) \cap \mathcal{D}_f$.

Using parts (1) and (2) of the theorem, we also obtain the following corollary.

Corollary 4.4.3. *Let*

$$p(z) = \sum_{k=0}^{n} a_k z^k$$

for all z, where $n, a_0, a_1, \ldots, a_n$ are constants. Then

$$\lim_{z \to c} p(z) = p(c).$$

The following theorem is an analog of the sandwich theorem for sequences.

Theorem 4.4.4. *Let f, g, h be functions and let c be an accumulation point of the set $\mathcal{D}_f \cap \mathcal{D}_g \cap \mathcal{D}_h$. Suppose there exists $\delta > 0$ such that*

$$f(x) \leq g(x) \leq h(x)$$

for each $x \in \mathcal{D}_f \cap \mathcal{D}_g \cap \mathcal{D}_h$ for which $0 < |x - c| < \delta$. If

$$\lim_{x \to c} f(x) = \lim_{x \to c} h(x) = L,$$

then

$$\lim_{x \to c} g(x) = L.$$

Proof. Let $\{s_n\}$ be a sequence in $\mathcal{D}_f - \{c\}$ converging to c. There exists N such that $0 < |s_n - c| < \delta$ for all $n \geq N$. For every such n we have

$$f(s_n) \leq g(s_n) \leq h(s_n).$$

The hypotheses imply that $f(s_n) \to L$ and $h(s_n) \to L$ as $n \to \infty$, and the result now follows from the sandwich theorem. □

Example 4.4.1. Since

$$0 \leq \left| x \sin \frac{1}{x} \right| \leq |x|$$

for all $x \neq 0$, and $\lim_{x \to 0} |x| = 0$, the sandwich theorem for functions (Theorem 4.4.4) shows that

$$\lim_{x \to 0} x \sin \frac{1}{x} = 0.$$

Similarly,

$$\lim_{x \to 0} x \cos \frac{1}{x} = 0.$$

$\triangle$

The next theorem is also analogous to a theorem on sequences from which it follows immediately.

Theorem 4.4.5. *If f is a function such that $m \le f(x) \le M$ and $\lim_{x \to c} f(x) = L$, then $m \le L \le M$.*

Theorem 4.4.6. *Let f be a function and let L and c be numbers such that L is real. Suppose*

$$\lim_{z \to c} f(z) = L \neq 0.$$

1. If $L > 0$, then there exists $\delta > 0$ such that $f(z) > L/2$ for all $z \in N_\delta^*(c) \cap \mathcal{D}_f$.
2. If $L < 0$, then there exists $\delta > 0$ such that $f(z) < L/2$ for all $z \in N_\delta^*(c) \cap \mathcal{D}_f$.

Proof. 1. There exists $\delta > 0$ such that

$$|f(z) - L| < \frac{L}{2}$$

for all $z \in \mathcal{D}_f$ for which $0 < |z - c| < \delta$. For all such z it follows that

$$f(z) > L - \frac{L}{2} = \frac{L}{2}.$$

2. The proof of part (2) is similar. $\square$

Theorem 4.4.7. *Let u and v be real-valued functions. Define*

$$f(z) = u(z) + iv(z)$$

for each $z \in \mathcal{D}_u \cap \mathcal{D}_v$. Let c be an accumulation point of $\mathcal{D}_u \cap \mathcal{D}_v$. Then

$$\lim_{z \to c} f(z) = A + iB \tag{4.4}$$

if and only if $\lim_{z \to c} u(z) = A$ and $\lim_{z \to c} v(z) = B$.

Proof. If $\lim_{z \to c} u(z) = A$ and $\lim_{z \to c} v(z) = B$, then the result follows from Theorem 4.4.1. Conversely, suppose that Eq. (4.4) holds. Considering the real and imaginary parts of $f(z) - A - iB$, we see that

$$|u(z) - A| \le |f(z) - (A + iB)|$$

and

$$|v(z) - B| \le |f(z) - (A + iB)|.$$

The proof now follows by an argument similar to the proof of Theorem 2.3.11.   □

Exercises 4.2. 1. Find the following limits if they exist, giving reasons for any that do not:

(a) $\lim_{x \to 0} \frac{\tan x - x}{\sin x}$;   (f) $\lim_{x \to 1} \frac{\sqrt{x} - 1}{x - 1}$;

(b) $\lim_{x \to 0} \frac{\sin \alpha x}{\sin \beta x}$ for $\beta \ne 0$;   (g) $\lim_{x \to 0} \frac{\sin cx}{x}$ for $c \in \mathbb{R}$;

(c) $\lim_{x \to c} \frac{x^n - c^n}{x - c}$ for $n \in \mathbb{N}$ and $c \in \mathbb{R}$;   (h) $\lim_{x \to 0} \frac{\cos x - 1}{x}$;

(d) $\lim_{z \to 0} \frac{(z-1)^2 - 1}{z}$;   (i) $\lim_{z \to i} \frac{z^2 + 1}{z - i}$;

(e) $\lim_{x \to 0} x \left(\sin \frac{1}{x} + \cos \frac{1}{x} \right)$;   (j) $\lim_{x \to 0} \frac{e^{ix} - 1}{x}$.

2. If $z = x + iy$, where x and y are real, find $\lim_{z \to 0}(x^2/z)$.

4.5   One-Sided Limits

In this section we restrict the domain of our function to a subset of $\mathbb{R}$. For instance, sometimes a function does not have a limit at a point a, but it would if the domain were restricted to a set of numbers greater than a.

Example 4.5.1. Suppose that

$$h(x) = \begin{cases} 1 & \text{if } x > a, \\ 0 & \text{if } x \le a. \end{cases}$$

The function h is known as a **Heaviside** function. If we restrict x to the interval $[a, \infty)$, we obtain a function f given by

$$f(x) = \begin{cases} 1 & \text{if } x > a, \\ 0 & \text{if } x = a. \end{cases}$$

Then

$$\lim_{x \to a} f(x) = 1.$$

We say that the limit of h at a from the right is 1, and write

$$\lim_{x \to a^+} h(x) = 1.$$

Likewise the limit of h at a from the left is 0. We write

$$\lim_{x \to a^-} h(x) = 0.$$

Notice that

$$\lim_{x \to a^+} h(x) \neq \lim_{x \to a^-} h(x).$$

We shall see later that this observation implies that $\lim_{x \to a} h(x)$ does not exist. We can also prove this fact by considering the sequences $\{s_n\}$ and $\{t_n\}$, where

$$s_n = a + \frac{1}{n}$$

and

$$t_n = a - \frac{1}{n}$$

for all $n > 0$. Then

$$\lim_{n \to \infty} s_n = \lim_{n \to \infty} t_n = a,$$

but $h(s_n) = 1$ and $h(t_n) = 0$ for all n. Therefore $\lim_{x \to a} h(x)$ does not exist.   △

Example 4.5.2.   Let $\lfloor x \rfloor$ be the largest integer less than or equal to a real number x. The function f given by $f(x) = \lfloor x \rfloor$ for all x is called the **floor** of x (see Fig. 4.3). For every $n \in \mathbb{N}$ we have

$$\lim_{x \to n^+} \lfloor x \rfloor = n$$

and

$$\lim_{x \to n^-} \lfloor x \rfloor = n - 1.$$ △

Example 4.5.3.   It is easy to show from the definition that

$$\lim_{x \to 0^+} \sqrt{x} = 0.$$

Fig. 4.3 Graph of the floor function

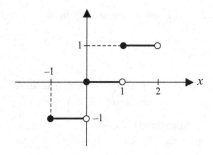

It is also true that

$$\lim_{x \to 0} \sqrt{x} = 0,$$

even though $\sqrt{x}$ is not defined for $x < 0$. △

Let us now define the notion of a one-sided limit formally.

Definition 4.5.1. Let f be a function of a real variable. Let a be an accumulation point of $\{x \in \mathcal{D}_f \mid x < a\}$. Then

$$\lim_{x \to a^-} f(x) = L$$

if the sequence $\{f(s_n)\}$ converges to L for each sequence $\{s_n\}$ in $\{x \in \mathcal{D}_f \mid x < a\}$ that converges to a.

The notation

$$\lim_{x \to a^+} f(x) = L$$

can be defined in a similar way.

The following theorem is analogous to Theorem 4.3.1 and can be proved in a similar manner.

Theorem 4.5.1. *Let f be a function of a real variable and let a and L be numbers.*

1. If a is an accumulation point of $\{x \in \mathcal{D}_f \mid x < a\}$, then

$$\lim_{x \to a^-} f(x) = L$$

if and only if for every $\varepsilon > 0$ there exists $\delta > 0$ such that $|f(x) - L| < \varepsilon$ for all $x \in \mathcal{D}_f$ satisfying $0 < a - x < \delta$.

2. *If a is an accumulation point of* $\{x \in \mathcal{D}_f \mid x > a\}$, *then*

$$\lim_{x \to a^+} f(x) = L$$

if and only if for every $\varepsilon > 0$ there exists $\delta > 0$ such that $|f(x) - L| < \varepsilon$ for all $x \in \mathcal{D}_f$ satisfying $0 < x - a < \delta$.

Note that $0 < a - x < \delta$ if and only if $a - \delta < x < a$ and that $0 < x - a < \delta$ if and only if $a < x < a + \delta$.

Theorem 4.5.2. *Let f be a function and L a number. Let a be an accumulation point of $\{x \in \mathcal{D}_f \mid x > a\}$ and $\{x \in \mathcal{D}_f \mid x < a\}$. Then*

$$\lim_{x \to a} f(x) = L \tag{4.5}$$

if and only if $\lim_{x \to a^+} f(x)$ *and* $\lim_{x \to a^-} f(x)$ *exist and*

$$\lim_{x \to a^+} f(x) = \lim_{x \to a^-} f(x) = L. \tag{4.6}$$

Proof. Suppose first that Eq. (4.5) holds, and choose $\varepsilon > 0$. There exists $\delta > 0$ such that

$$|f(x) - L| < \varepsilon \tag{4.7}$$

for all $x \in \mathcal{D}_f$ such that $0 < |x - a| < \delta$. In particular, inequality (4.7) holds for all $x \in \mathcal{D}_f$ such that $0 < x - a < \delta$. Therefore

$$\lim_{x \to a^+} f(x) = L.$$

A similar argument shows that

$$\lim_{x \to a^-} f(x) = L.$$

Conversely, suppose that Eq. (4.6) holds, and choose $\varepsilon > 0$. There exist $\delta_1 > 0$ and $\delta_2 > 0$ such that inequality (4.7) holds for all $x \in \mathcal{D}_f$ satisfying either $0 < x - a < \delta_1$ or $0 < a - x < \delta_2$. Thus inequality (4.7) holds for all $x \in \mathcal{D}_f$ such that

$$0 < |x - a| < \min\{\delta_1, \delta_2\}.$$

Consequently, Eq. (4.5) holds.   □

The work in Sect. 4.4 also holds for one-sided limits.

Example 4.5.4. Let f be the function given by

$$f(x) = e^{1/x}$$

for all $x \neq 0$. Then $\lim_{x \to 0+} f(x)$ does not exist, for if we set $s_n = 1/n$ for all $n \in \mathbb{N}$, then $s_n \to 0$ as $n \to \infty$ but

$$f(s_n) = e^n \to \infty$$

as $n \to \infty$.

On the other hand, for each $x > 0$ we have $x < e^x$. Thus if $t < 0$, then

$$0 < -\frac{1}{t} < e^{-1/t},$$

whence

$$0 < e^{1/t} < -t.$$

It therefore follows from the sandwich theorem that

$$\lim_{t \to 0^-} e^{1/t} = 0. \qquad\qquad \triangle$$

Exercises 4.3. 1. Find the following limits if they exist, giving reasons for any that do not:

(a) $\lim_{x \to 1-} \frac{x^2 + 2x - 3}{|x - 1|}$;   (e) $\lim_{x \to 0+} \frac{x - \sqrt{x}}{\sqrt{x}}$;

(b) $\lim_{x \to 1-} \frac{1 - x}{\sqrt{1 - x}}$;   (f) $\lim_{x \to 0+} \frac{\sin \sqrt{x}}{x}$;

(c) $\lim_{x \to 0+} \frac{1}{1 + 2^{-1/x}}$;   (g) $\lim_{x \to 0+} \frac{(x+1)^{1/3} - 1}{x}$;

(d) $\lim_{x \to 0-} \frac{1}{1 + 2^{-1/x}}$;   (h) $\lim_{x \to 1-} \frac{1}{\lfloor x \rfloor - 1}$.

4.6   Infinite Limits

Some functions increase or decrease without bound as their arguments approach a given number. We now define this concept formally.

Definition 4.6.1. Let f be a function and c an accumulation point of $\mathcal{D}_f$. We write

$$\lim_{x \to c} f(x) = \infty$$

if $f(s_n) \to \infty$ as $n \to \infty$ for every sequence $\{s_n\}$ in $\mathcal{D}_f - \{c\}$ converging to c.
The equation

$$\lim_{x \to c} f(x) = -\infty$$

may be defined similarly.

Thus

$$\lim_{x \to c} f(x) = \infty$$

if and only if for every N there exists $\delta > 0$ such that $f(x) > N$ for all $x \in \mathcal{D}_f$ satisfying $0 < |x - c| < \delta$. Similarly,

$$\lim_{x \to c} f(x) = -\infty$$

if and only if this condition holds with the inequality $f(x) > N$ replaced by $f(x) < N$.

We now discuss the behavior of a function f as its argument approaches $\pm\infty$.

Definition 4.6.2. Let f be a function and L a number. Suppose that $\mathcal{D}_f$ is not bounded above. We write

$$\lim_{x \to \infty} f(x) = L$$

if $f(s_n) \to L$ as $n \to \infty$ for every sequence $\{s_n\}$ in $\mathcal{D}_f$ such that $s_n \to \infty$ as $n \to \infty$.

The notation

$$\lim_{x \to -\infty} f(x) = L$$

may be defined in a similar manner.

Note that if $\mathcal{D}_f$ is not bounded above, then

$$\lim_{x \to \infty} f(x) = L$$

if and only if for every $\varepsilon > 0$ there exists M such that $|f(x) - L| < \varepsilon$ for all $x \in \mathcal{D}_f$ satisfying $x > M$. Moreover

$$\lim_{x \to -\infty} f(x) = L$$

if and only if this condition holds with the inequality $x > M$ replaced by $x < M$, provided that $\mathcal{D}_f$ is not bounded below (though it may be bounded above).

Of course, we also have such notations as

$$\lim_{x \to \infty} f(x) = \infty \tag{4.8}$$

and

$$\lim_{x \to \infty} f(x) = -\infty,$$

and so forth. We leave it to the reader to list the possibilities and to formulate appropriate definitions for them. For example, Eq. (4.8) holds if and only if $\mathcal{D}_f$ is not bounded above and for every sequence $\{s_n\}$ in $\mathcal{D}_f$ such that

$$\lim_{n\to\infty} s_n = \infty,$$

we have

$$\lim_{n\to\infty} f(s_n) = \infty.$$

Thus it follows from Theorem 2.7.10(1) that

$$\lim_{x\to\infty} e^x = \infty.$$

Similarly,

$$\lim_{x\to-\infty} e^x = 0,$$

by Example 2.8.2.

Example 4.6.1. Prove that

$$\lim_{x\to\infty} \frac{1}{x} = 0.$$

Proof. Let $\{s_n\}$ be a sequence of positive numbers such that $\lim_{n\to\infty} s_n = \infty$. Then

$$\lim_{n\to\infty} \frac{1}{s_n} = 0,$$

by Theorem 2.8.4(2). The result follows.   △

Example 4.6.2. We have seen in Example 2.7.8 that

$$\lim_{n\to\infty} \left(1 + \frac{1}{n}\right)^n = e.$$

We now show that $\lim_{x\to\pm\infty} f(x) = e$, where

$$f(x) = \left(1 + \frac{1}{x}\right)^x$$

for all $x \in \mathbb{Q} - [-1, 0]$. (Later we shall extend this result to the set $\mathbb{R} - [-1, 0]$.) First let $\{s_n\}$ be any sequence in $\mathbb{Q}$ such that $s_n \geq 1$ for all n and

$$\lim_{n\to\infty} s_n = \infty.$$

For each k let $n_k = \lfloor s_k \rfloor$. Then

$$0 < n_k \le s_k < n_k + 1.$$

Thus

$$\frac{1}{n_k + 1} < \frac{1}{s_k} \le \frac{1}{n_k},$$

and so

$$\left(1 + \frac{1}{n_k + 1}\right)^{n_k} < \left(1 + \frac{1}{s_k}\right)^{s_k} < \left(1 + \frac{1}{n_k}\right)^{n_k + 1}.$$

Letting $n_k \to \infty$ and using Lemma 2.2.1, we obtain

$$\left(1 + \frac{1}{n_k + 1}\right)^{n_k} = \left(1 + \frac{1}{n_k + 1}\right)^{n_k + 1}\left(1 + \frac{1}{n_k + 1}\right)^{-1} \to e \cdot 1 = e$$

and

$$\left(1 + \frac{1}{n_k}\right)^{n_k + 1} = \left(1 + \frac{1}{n_k}\right)^{n_k}\left(1 + \frac{1}{n_k}\right) \to e \cdot 1 = e.$$

Hence

$$\lim_{k \to \infty} \left(1 + \frac{1}{s_k}\right)^{s_k} = e.$$

We conclude that

$$\lim_{x \to \infty} \left(1 + \frac{1}{x}\right)^{x} = e.$$

In order to show that

$$\lim_{x \to -\infty} \left(1 + \frac{1}{x}\right)^{x} = e,$$

choose a sequence $\{x_n\}$ in $\mathbb{Q}$ such that

$$\lim_{n \to \infty} x_n = -\infty$$

and $x_n < -1$ for all n. Let $y_n = -x_n$ for all n. Then

$$\lim_{n \to \infty} y_n = \infty$$

and $y_n > 1$ for all n, and so

$$
\begin{aligned}
\left(1 + \frac{1}{x_n}\right)^{x_n} &= \left(1 - \frac{1}{y_n}\right)^{-y_n} \\
&= \left(\frac{y_n - 1}{y_n}\right)^{-y_n} \\
&= \left(\frac{y_n}{y_n - 1}\right)^{y_n} \\
&= \left(1 + \frac{1}{y_n - 1}\right)^{y_n} \\
&= \left(1 + \frac{1}{y_n - 1}\right)^{y_n - 1} \left(1 + \frac{1}{y_n - 1}\right) \\
&\to e
\end{aligned}
$$

as $n \to \infty$. $\quad\triangle$

Theorem 4.6.1. *Let f and g be functions, let L be a number, and let a be an accumulation point of both $\{x \in \mathcal{D}_f \mid x > a\}$ and $\{x \in \mathcal{D}_g \mid x > a\}$. Suppose that*

$$
\lim_{x \to a+} f(x) = L > 0
$$

and

$$
\lim_{x \to a+} g(x) = 0.
$$

Suppose also that there exists $\delta > 0$ such that $g(x) > 0$ for all $x \in \mathcal{D}_g$ satisfying $0 < x - a < \delta$. Then

$$
\lim_{x \to a+} \frac{f(x)}{g(x)} = \infty.
$$

Proof. Choose $N > 0$. The first limit given shows the existence of $\delta_1 > 0$ such that

$$
f(x) > \frac{L}{2}
$$

for all $x \in \mathcal{D}_f$ satisfying $0 < x - a < \delta_1$. It also follows from the second hypothesis that there exists $\delta_2 > 0$ such that

$$
|g(x)| < \frac{L}{2N}
$$

for all $x \in \mathcal{D}_g$ for which $0 < x - a < \delta_2$.

Now let $\delta_3 = \min\{\delta, \delta_1, \delta_2\}$, so that $\delta_3 > 0$, and choose $x \in \mathcal{D}_f \cap \mathcal{D}_g$ such that $0 < x - a < \delta_3$. Then

$$f(x) > \frac{L}{2} > 0.$$

Moreover

$$0 < g(x) = |g(x)| < \frac{L}{2N},$$

so that

$$\frac{1}{g(x)} > \frac{2N}{L} > 0.$$

Hence

$$\frac{f(x)}{g(x)} > \frac{L}{2} \cdot \frac{2N}{L} = N,$$

and the result follows.   □

Example 4.6.3. It follows immediately from Theorem 4.6.1 that

$$\lim_{x \to 0^+} \frac{1}{x} = \infty. \qquad \triangle$$

Theorem 4.6.2. *Let f and g be functions and let a be an accumulation point of $\mathcal{D}_{f \circ g}$. If*

$$\lim_{x \to \infty} f(x) = L,$$

for some number L, and

$$\lim_{x \to a} g(x) = \infty,$$

then

$$\lim_{x \to a} f(g(x)) = L.$$

Proof. Note that a is an accumulation point of $\mathcal{D}_g$, since $\mathcal{D}_{f \circ g} \subseteq \mathcal{D}_g$.

Choose $\varepsilon > 0$. By hypothesis, there exists M such that $|f(x) - L| < \varepsilon$ for all $x \in \mathcal{D}_f$ satisfying $x > M$. Similarly, there exists $\delta > 0$ such that $g(x) > M$ for all $x \in \mathcal{D}_g$ for which $0 < |x - a| < \delta$. Choose $x \in \mathcal{D}_{f \circ g}$ such that $0 < |x - a| < \delta$. Then $x \in \mathcal{D}_g$, so that $g(x) > M$, and $g(x) \in \mathcal{D}_f$. Therefore

$$|f(g(x)) - L| < \varepsilon,$$

and the result follows.   □

Clearly, a and L may each be replaced by ∞ or $-\infty$. Moreover the limits as x approaches a may be replaced with one-sided limits. Thus Example 4.6.3 yields the following corollary.

Corollary 4.6.3. *If* $\lim_{x\to\infty} f(x) = L$, *then*

$$\lim_{x\to 0^+} f\left(\frac{1}{x}\right) = L.$$

Corollary 4.6.4. *If* $\lim_{x\to a} g(x) = \infty$, *where a is an accumulation point of*

$$\{x \in \mathcal{D}_g \mid g(x) \neq 0\},$$

then

$$\lim_{x\to a} \frac{1}{g(x)} = 0.$$

Proof. Define $f(x) = 1/x$ for all $x \neq 0$. Then $\lim_{x\to\infty} f(x) = 0$, by Example 4.6.1. It therefore follows from Theorem 4.6.2 that

$$\lim_{x\to a} \frac{1}{g(x)} = \lim_{x\to a} f(g(x)) = 0. \qquad\qquad □$$

Once again a may be replaced by ∞ or $-\infty$ and the limits may be replaced by one-sided limits.

Example 4.6.4. From Example 4.6.3 and Theorem 4.6.2 we have

$$\lim_{x\to 0^+} e^{1/x} = \infty.$$

Consequently

$$\lim_{x\to 0^+} \frac{1}{e^{1/x} + 1} = 0,$$

using Corollary 4.6.4. On the other hand, Example 4.5.4 shows that

$$\lim_{x\to 0^-} \frac{1}{e^{1/x} + 1} = \frac{1}{0 + 1} = 1. \qquad\qquad △$$

Example 4.6.5. We show that $\lim_{x\to 0} f(x) = e$, where

$$f(x) = (1+x)^{1/x}$$

for all nonzero rational numbers $x > -1$. Since

$$\lim_{x\to\infty} \left(1 + \frac{1}{x}\right)^x = e$$

by Example 4.6.2, it follows from Corollary 4.6.3 that

$$\lim_{x\to 0^+} (1+x)^{1/x} = e.$$

A similar argument shows that

$$\lim_{x\to 0^-} (1+x)^{1/x} = e.$$

The result now follows from Theorem 4.5.2.   △

Exercises 4.4. 1. Find the following limits:

(a) $\lim_{x\to\infty} \frac{\cos 2x}{x}$;   (e) $\lim_{x\to\infty} \frac{\sqrt{x}-x}{\sqrt{x}+x}$;

(b) $\lim_{x\to\infty} \frac{x}{\sqrt{4x^2+5x+1}}$;   (f) $\lim_{x\to 0} \frac{1}{\sqrt{|x|}}$;

(c) $\lim_{x\to\infty}(\sqrt{x^2+x+1}-1)$;   (g) $\lim_{x\to-\infty}\left(1+\frac{1}{x}\right)^x$;

(d) $\lim_{x\to\infty} x\sin\frac{1}{x}$;   (h) $\lim_{x\to 1^-}\frac{\sqrt{1-x}}{x-1}$.

2. Give examples of functions f and g to show that if

$$\lim_{x\to 0} f(x) = \lim_{x\to 0} g(x) = 0$$

then

$$\lim_{x\to 0} \frac{f(x)}{g(x)}$$

may be any real number or $\pm\infty$ or may not exist.

Chapter 5
Continuity

5.1 Definition and Examples

We now come to a most important concept in analysis, one that has many applications, notably in optimization. Roughly speaking, we are talking about those functions that have no gaps in their graphs. Such functions are said to be continuous. For example, the functions f and g, such that

$$f(x) = g(x) = \frac{2x^2 - 2x}{x - 1}$$

for all $x \neq 1$ and $g(1) = 1$, are not continuous at 1 since the graphs of both functions have a gap where $x = 1$. On the other hand, the function h, such that $h(x) = f(x)$ for all $x \neq 1$ and $h(1) = 2$, is continuous (see Fig. 5.1).

The function given by

$$k(x) = e^{1/x},$$

for all $x \neq 0$, is not continuous at 0, for Example 4.5.4 shows that $\lim_{x \to 0+} e^{1/x} = \infty$ but $\lim_{x \to 0-} e^{1/x} = 0$, and so the graph of k will have a gap no matter how we define $k(0)$ (see Fig. 5.2).

Similarly, the function given by

$$\frac{1}{e^{1/x} + 1},$$

for all $x \neq 0$, is not continuous at 0 no matter how we define its value at 0 because Example 4.6.4 shows that

$$\lim_{x \to 0+} \frac{1}{e^{1/x} + 1} = 0$$

© Springer Science+Business Media New York 2015
C.H.C. Little et al., *Real Analysis via Sequences and Series*, Undergraduate
Texts in Mathematics, DOI 10.1007/978-1-4939-2651-0_5

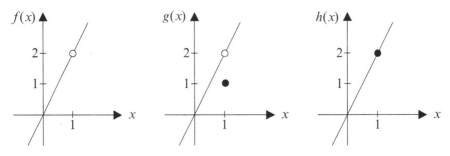

Fig. 5.1 Graphs of f, g and h

Fig. 5.2 Graph of $e^{1/x}$

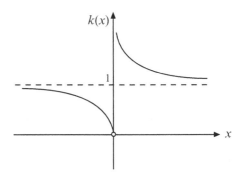

but

$$\lim_{x \to 0^-} \frac{1}{e^{1/x} + 1} = 1$$

(see Fig. 5.3).

A function f fails to be continuous at a certain point if its limit at that point is not equal to its value there. In particular, f is not continuous at a given point if either it or its limit is not defined at that point. We now make the concept of continuity precise.

Definition 5.1.1. A function f is **continuous** at an accumulation point c of $\mathcal{D}_f$ if

$$\lim_{z \to c} f(z) = f(c).$$

In view of the fact that $|f(z) - f(c)| = 0$ when $z = c$, the following theorem is immediate from Theorem 4.3.1.

Fig. 5.3 Graph of
$1/(e^{1/x} + 1)$

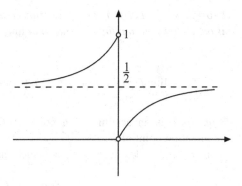

Theorem 5.1.1. *A function f is continuous at an accumulation point c of $\mathcal{D}_f$ if and only if for all $\varepsilon > 0$ there exists $\delta > 0$ such that*

$$|f(z) - f(c)| < \varepsilon$$

for each $z \in \mathcal{D}_f$ satisfying $|z - c| < \delta$.

Of course, the notion of continuity can also be formulated in terms of sequences.

Theorem 5.1.2. *Let f be a function and c an accumulation point of $\mathcal{D}_f$. Then f is continuous at c if and only if $\{f(s_n)\}$ converges to $f(c)$ for every sequence $\{s_n\}$ in $\mathcal{D}_f$ converging to c.*

The next theorem follows immediately from Theorem 4.4.7.

Theorem 5.1.3. *Let u and v be real-valued functions, and define*

$$f(z) = u(z) + iv(z)$$

for each $z \in \mathcal{D}_u \cap \mathcal{D}_v$. Then f is continuous at an accumulation point c of $\mathcal{D}_u \cap \mathcal{D}_v$ if and only if u and v are continuous at c.

It is an immediate consequence of Propositions 4.2.2 and 4.2.3 that constant functions and the identity function (the function that maps each number to itself) are continuous everywhere, as is the absolute value function, by Proposition 4.2.4. It follows from Theorem 4.4.1 that the sum or product of functions f and g that are continuous at a point c is also continuous at c if c is an accumulation point of $\mathcal{D}_f \cap \mathcal{D}_g$. Thus polynomial functions are continuous everywhere. The quotient f/g is also continuous at c provided that in addition $g(x) \neq 0$ for all x in some neighborhood of c. Theorem 2.3.13 shows that for each $m \in \mathbb{N}$ the function given by $x^{1/m}$ for all $x \geq 0$ is continuous at each such x.

Sometimes a function is not continuous but its product with another function is continuous. For example, the functions given by $e^{1/x}$ and $e^{-1/x}$, for all $x \neq 0$, are not continuous at 0. However, $e^{1/x} \cdot e^{-1/x}$ gives the constant function 1 for all $x \neq 0$, and if we define its value at 0 to be 1, then the resulting function is continuous everywhere.

Theorem 5.1.4. *Let f and g be functions and let c and L be numbers. Suppose that c is an accumulation point of $\mathcal{D}_{f \circ g}$, that $\lim_{z \to c} g(z) = L$, and that f is continuous at L. Then*

$$\lim_{z \to c} f(g(z)) = f(L) = f\left(\lim_{z \to c} g(z)\right) = \lim_{g(z) \to L} f(g(z)).$$

Proof. As c is an accumulation point of $\mathcal{D}_{f \circ g}$ and $\mathcal{D}_{f \circ g} \subseteq \mathcal{D}_g$, it is also an accumulation point of $\mathcal{D}_g$.

Choose $\varepsilon > 0$. There exists $\delta_1 > 0$ such that

$$|f(z) - f(L)| < \varepsilon$$

for all $z \in \mathcal{D}_f$ satisfying $|z - L| < \delta_1$. Similarly, there exists $\delta > 0$ such that

$$|g(z) - L| < \delta_1$$

for all $z \in \mathcal{D}_g$ satisfying $0 < |z - c| < \delta$. For each such z we therefore have

$$|f(g(z)) - f(L)| < \varepsilon$$

if $g(z) \in \mathcal{D}_f$, and the first equation follows. The proof is completed with the observation that $f(L)$ is equal to the two rightmost expressions, respectively, by substitution and the continuity hypothesized for f. $\square$

Remark. Clearly, the limits as z approaches c may be replaced by one-sided limits. Moreover the accumulation point c of $\mathcal{D}_{f \circ g}$ may be replaced by ∞ or $-\infty$.

Corollary 5.1.5. *If g is continuous at an accumulation point c of $\mathcal{D}_{f \circ g}$ and f is continuous at $g(c)$, then $f \circ g$ is continuous at c.*

Proof. Apply the theorem with $L = g(c)$. $\square$

Corollary 5.1.6. *Let f be a function and L and c numbers. Suppose that c is an accumulation point of $\mathcal{D}_f$. If $\lim_{z \to c} f(z) = L$, then*

$$\lim_{z \to c} |f(z)| = |L|.$$

Corollary 5.1.7. *Let f be a real function and L and a real numbers. Suppose that a is an accumulation point of $\mathcal{D}_f$ and*

$$\lim_{x \to a} f(x) = L.$$

Then, for every positive integer m,

$$\lim_{x \to a} (f(x))^{1/m} = L^{1/m}$$

if for each $\delta > 0$ there exists $x \in N_\delta^*(a) \cap \mathcal{D}_f$ such that $f(x) \geq 0$.

The next result is a corollary of Theorem 4.4.6.

Theorem 5.1.8. *Let f be a real-valued function that is continuous at a number c.*

1. *If $f(c) > 0$, then there exists $\delta > 0$ such that $f(z) > 0$ for all $z \in N_\delta(c) \cap \mathcal{D}_f$.*
2. *If $f(c) < 0$, then there exists $\delta > 0$ such that $f(z) < 0$ for all $z \in N_\delta(c) \cap \mathcal{D}_f$.*

Example 4.2.1 establishes the continuity of the exponential function.

Theorem 5.1.9. *The function e^x is continuous at all x.*

We now give a famous example of a discontinuous function that is due to Dirichlet.

Example 5.1.1. Let

$$f(x) = \begin{cases} 1 & \text{if } x \text{ is rational,} \\ 0 & \text{otherwise.} \end{cases}$$

We show that f is not continuous anywhere. First let c be a rational number. By the density theorem, for each $n \in \mathbb{N}$ we can find an irrational number

$$s_n \in \left(c - \frac{1}{n}, c + \frac{1}{n} \right).$$

Thus $f(s_n) = 0$, and so

$$\lim_{n \to \infty} f(s_n) = 0 \neq f(c),$$

yet the sequence $\{s_n\}$ converges to c. Consequently, f is not continuous at c.

The case where c is an irrational number can be handled in a similar manner. $\triangle$

The following function is a modification of the Dirichlet function. It is known as **Thomae's function** or the **popcorn function**.

Example 5.1.2. Let $f : (0, 1) \to \mathbb{R}$ be defined by

$$f(x) = \begin{cases} 0 & \text{if } x \text{ is irrational,} \\ \frac{1}{n} & \text{if } x = \frac{m}{n}, \text{ where } m \text{ and } n \text{ are relatively prime positive integers.} \end{cases}$$

We show that f is continuous at irrational points but not at rational points.

Let c be a rational number in $(0, 1)$. As in the previous example, we can construct a sequence $\{s_n\}$ of irrational numbers in $(0, 1)$ such that $s_n \to c$. Hence $f(s_n) = 0$ for all n. But $f(c) \neq 0$; hence f is not continuous at c.

Now let c be an irrational number in $(0, 1)$. Choose any $\varepsilon > 0$ and let

$$B = \left\{ \frac{m}{n} \in (0, 1) \,\middle|\, m \in \mathbb{N}, n \in \mathbb{N}, \frac{1}{n} \geq \varepsilon \right\}.$$

Then B is finite, as $m < n \leq 1/\varepsilon$, and $c \notin B$. Choose $\delta > 0$ such that $N_\delta(c) \subseteq (0, 1)$ and

$$N_\delta(c) \cap B = \emptyset.$$

Hence $f(x) < \varepsilon$ for all $x \in N_\delta(c)$. Consequently,

$$\lim_{x \to c} f(x) = 0 = f(c).$$

We conclude that f is continuous at c. $\triangle$

However, we can prove the following theorem. The proof is due to Volterra (1881).

Theorem 5.1.10. *There is no function $f: (0, 1) \to \mathbb{R}$ that is continuous at all rational numbers in $(0, 1)$ but not at any irrational number.*

Proof. Let $g: (0, 1) \to \mathbb{R}$ be defined by

$$g(x) = \begin{cases} 0 & \text{if } x \text{ is irrational,} \\ \frac{1}{n} & \text{if } x = \frac{m}{n}, \text{ where } m \text{ and } n \text{ are relatively prime positive integers.} \end{cases}$$

We have shown in Example 5.1.2 that g is discontinuous on $A = \mathbb{Q} \cap (0, 1)$ and continuous on $B = (0, 1) - A$.

Suppose there is a function $f: (0, 1) \to \mathbb{R}$ that is continuous on A but discontinuous on B. Let $c \in A$. Then f is continuous at c. Hence there exists $\delta > 0$ such that $(c - \delta, c + \delta) \subseteq (0, 1)$ and

$$|f(x) - f(c)| < \frac{1}{4}$$

for all $x \in (c - \delta, c + \delta)$. We may assume that $\delta \leq 1/4$. Choose $a_1, b_1 \in (c - \delta, c + \delta)$ such that $a_1 < b_1$; hence $b_1 - a_1 < 1/2$. Then for all $x, y \in (a_1, b_1)$ we have

$$|f(x) - f(y)| \leq |f(x) - f(c)| + |f(y) - f(c)| < \frac{1}{4} + \frac{1}{4} = \frac{1}{2}. \tag{5.1}$$

Now g is continuous at some point in (a_1, b_1). Therefore by the preceding argument with $(0, 1)$ replaced by (a_1, b_1), there exist $a_1', b_1' \in (a_1, b_1)$ such that

$$|g(x) - g(y)| < \frac{1}{2} \tag{5.2}$$

for all $x, y \in (a_1', b_1')$.

In summary, inequalities (5.1) and (5.2) both hold for all $x, y \in (a_1', b_1')$. Moreover (a_1', b_1') contains a point at which f is continuous and a point at which g is continuous. Therefore we may repeat the argument inductively to obtain a sequence of nested intervals

$$(a_1', b_1') \supset (a_2', b_2') \supset \ldots$$

in which each interval (a_n', b_n') has length less than $1/2^n$ and satisfies the condition that

$$|f(x) - f(y)| < \frac{1}{2^n}$$

and

$$|g(x) - g(y)| < \frac{1}{2^n}$$

whenever $x, y \in (a_n', b_n')$. Moreover the sequence $\{a_n'\}$ is increasing and bounded above. Therefore it converges. Similarly, $\{b_n'\}$ converges. Since

$$0 < b_n' - a_n' < \frac{1}{2^n}$$

for all n, it follows from the sandwich theorem that $\{a_n'\}$ and $\{b_n'\}$ both converge to some number d that belongs to every interval in the sequence. Therefore for every $n \in \mathbb{N}$ and every $x \in (a_n', b_n')$, it follows that

$$|f(x) - f(d)| < \frac{1}{2^n}$$

and

$$|g(x) - g(d)| < \frac{1}{2^n}.$$

It is now easy to show that f is continuous at d. Choose $\varepsilon > 0$. Since $1/2^n \to 0$ as $n \to \infty$, there exists N such that $1/2^n < \varepsilon$ for all $n \geq N$. Choose x such that

$$|x - d| < \min\{d - a_N', b_N' - d\}.$$

Then $x \in (a'_N, b'_N)$, and so

$$|f(x) - f(d)| < \frac{1}{2^N} < \varepsilon,$$

as required. A similar argument shows that g is continuous at d. Hence we reach the contradiction that $d \in A \cap B = \emptyset$.   □

Exercises 5.1.

1. Study the continuity of each of the following functions:

   (a) $f(x) = \begin{cases} \frac{|x|}{x} & \text{if } x \neq 0, \\ 0 & \text{if } x = 0. \end{cases}$

   (b) $f(x) = \begin{cases} x^2 & \text{if } x \leq 1, \\ 2 - x & \text{if } x > 1. \end{cases}$

   (c) $f(x) = \begin{cases} \frac{i(e^{ix} - 1)}{x} & \text{if } x \neq 0, \\ -1 & \text{if } x = 0. \end{cases}$

   (d) $f(x) = \begin{cases} x & \text{if } x \text{ is rational}, \\ 0 & \text{if } x \text{ is irrational}. \end{cases}$

2. Find c for which the following functions are continuous:

   (a) $f(x) = \begin{cases} x \sin \frac{1}{x} & \text{if } x \neq 0, \\ c & \text{if } x = 0. \end{cases}$

   (b) $f(x) = \begin{cases} x^2 - x + 1 & \text{if } x \leq 1, \\ cx^2 + 1 & \text{if } x > 1. \end{cases}$

3. Let f be continuous on an interval I. Suppose that $f(r) = r^2$ for every rational number r in I. Prove that $f(x) = x^2$ for all $x \in I$.

4. Let f and g be continuous functions. Show that the functions $f \vee g$ and $f \wedge g$ are also continuous, where

$$(f \vee g)(x) = \max\{f(x), g(x)\}$$

and

$$(f \wedge g)(x) = \min\{f(x), g(x)\}$$

for all x.
(Hint:

$$\max\{a, b\} = \frac{a + b + |a - b|}{2}$$

and

$$\min\{a, b\} = \frac{a + b - |a - b|}{2}.)$$

5. A subset S of $\mathbb{C}$ is **open** if for every $c \in S$ there exists $\delta > 0$ such that $N_\delta(c) \subseteq S$. Let $f: \mathbb{C} \to \mathbb{C}$ be continuous. Show that if S is open in $\mathbb{C}$, then

$$\{z \in \mathbb{C} \mid f(z) \in S\}$$

is open.

5.2 One-Sided Continuity

The concept of one-sided continuity is analogous to that of a one-sided limit.

Definition 5.2.1. A function f is **continuous on the right** at an accumulation point c of $\{x \in \mathcal{D}_f \mid x > c\}$ if

$$\lim_{x \to c^+} f(x) = f(c).$$

The definition of a function that is **continuous on the left** at c is analogous.

Example 5.2.1. If

$$f(x) = \frac{1}{e^{1/x} + 1}$$

for all $x \neq 0$, then we have already seen that

$$\lim_{x \to 0^+} f(x) = 0$$

and

$$\lim_{x \to 0^-} f(x) = 1.$$

Therefore, if we were to define $f(0) = 0$, then f would be continuous on the right at 0, and if we were to define $f(0) = 1$, then f would be continuous on the left at 0. △

Example 5.2.2. Let f be the function given by

$$f(x) = \sqrt{x}$$

for all $x \geq 0$. Then f is continuous on the right at 0. It is not continuous on the left at 0 since it is undefined at each $x < 0$. Nevertheless f is continuous at 0 according to our definition.   △

5.3 Continuity over an Interval

Definition 5.3.1. A function is **continuous** over a given set if it is continuous at all points in the set.

We shall show that if a function is continuous over a closed set, then it reaches a maximum and a minimum somewhere in that set.

In many applications the domain of a function is assumed to be closed and bounded. A closed bounded set is said to be **compact**.

Theorem 5.3.1. *Let $f : X \to \mathbb{C}$ be a continuous function, where X is a compact subset of $\mathbb{C}$. Then $f(X)$ is compact.*

Proof. Suppose that $f(X)$ is not bounded. Choose $w \in f(X)$. For every $n \in \mathbb{N}$, we can find $w_n \in f(X)$ such that

$$|w_n - w| > n.$$

There exists $z_n \in X$ such that $f(z_n) = w_n$. As X is bounded, Theorem 2.6.10 shows that $\{z_n\}$ contains an injective convergent subsequence $\{z_{k_n}\}$. Let

$$z = \lim_{n \to \infty} z_{k_n}.$$

Since X is closed, we have $z \in X$.

The continuity of f shows that

$$\lim_{n \to \infty} w_{k_n} = \lim_{n \to \infty} f(z_{k_n}) = f(z),$$

by the remark following Theorem 5.1.4. Thus there exists N such that

$$|w_{k_n} - f(z)| < 1$$

for all $n \geq N$. For each such n it follows that

$$\begin{aligned}
n &\leq k_n \\
&< |w_{k_n} - w| \\
&\leq |w_{k_n} - f(z)| + |f(z) - w| \\
&< 1 + |f(z) - w|.
\end{aligned}$$

Fig. 5.4 Graph of f in
Example 5.3.1

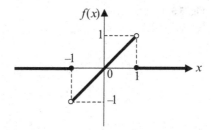

However this conclusion is impossible, as the right-hand side of the inequality above is a constant. Consequently $f(X)$ is bounded.

In order to show that $f(X)$ is closed, choose a limit point w of $f(X)$. There exists a sequence $\{w_n\}$ in $f(X)$ such that $\lim_{n \to \infty} w_n = w$. For each n there exists $z_n \in X$ such that $f(z_n) = w_n$. As before, $\{z_n\}$ contains a subsequence $\{z_{k_n}\}$ converging to a number $z \in X$. Theorem 2.4.1 and the continuity of f show that

$$w = \lim_{n \to \infty} w_{k_n} = f(z).$$

Thus $w \in f(X)$, and we conclude that $f(X)$ is closed.   □

Corollary 5.3.2 (Maximum- and Minimum-Value Theorem). *Let $f : X \to \mathbb{R}$ be a continuous function, where X is a compact set. Then $f(X)$ contains a maximum and a minimum value.*

Proof. The result follows from the theorem since the supremum and infimum of $f(X)$ are necessarily limit points.   □

In particular, a function that is continuous on a closed interval reaches a maximum and a minimum value in that interval. If a function is not continuous on a closed interval, then it may or may not have a maximum or a minimum on that interval.

Example 5.3.1. Let

$$f(x) = \begin{cases} x & \text{if } -1 < x < 1, \\ 0 & \text{otherwise.} \end{cases}$$

Here f is not continuous on the closed interval $[-1, 1]$ although it is continuous on the open interval $(-1, 1)$. In this case f has neither a maximum nor a minimum on $[-1, 1]$, since there is no largest number less than 1 and no smallest number greater than -1. Note that neither -1 nor 1 is a value of the function (see Fig. 5.4).   △

Example 5.3.2. Let

$$f(x) = \begin{cases} x & \text{if } -1 < x < 1, \\ 2 & \text{if } |x| \geq 1. \end{cases}$$

Fig. 5.5 Graph of f in
Example 5.3.2

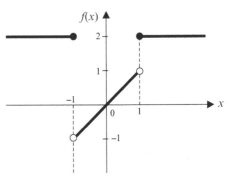

Fig. 5.6 Graph of f in
Example 5.3.3

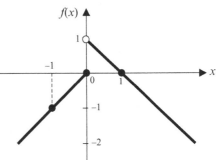

Here f is not continuous on the closed interval $[-1, 1]$, though it is continuous on
the open interval $(-1, 1)$. It has no minimum on $[-1, 1]$, but it has a maximum of 2,
attained at 1 (see Fig. 5.5).   △

It is not only discontinuities at endpoints of intervals that may invalidate the
conclusion of the theorem.

Example 5.3.3. Let

$$f(x) = \begin{cases} x & \text{if } x \le 0, \\ 1 - x & \text{if } x > 0. \end{cases}$$

This function is not continuous on $[-1, 1]$ as it is not continuous at 0. It has a
minimum of -1 on $[-1, 1]$, attained at -1, but no maximum (see Fig. 5.6).   △

Our next theorem asserts the existence of a zero of a continuous function where
the values of the function are not all of the same sign.

Theorem 5.3.3 (Bolzano). *Suppose that $f: [a, b] \to \mathbb{R}$ is a continuous function
such that $f(a) f(b) < 0$. Then there exists $\xi \in (a, b)$ such that $f(\xi) = 0$.*

Proof. We assume that $f(a) < 0$ and $f(b) > 0$, as the proof in the other case is
similar.

Suppose there is no $\xi \in (a, b)$ such that $f(\xi) = 0$. Let $a_0 = a$ and $b_0 = b$. Suppose now that a_j and b_j have been defined for some nonnegative integer j, that they are both in $[a, b]$, that $f(a_j) < 0$, that $f(b_j) > 0$, and that $b_j - a_j = (b - a)/2^j$. If

$$f\left(\frac{a_j + b_j}{2}\right) > 0,$$

then let $a_{j+1} = a_j$ and $b_{j+1} = (a_j + b_j)/2$; otherwise define $a_{j+1} = (a_j + b_j)/2$ and $b_{j+1} = b_j$. Then $\{a_n\}$ is a nondecreasing sequence in $[a, b]$, $\{b_n\}$ is a nonincreasing sequence in $[a, b]$, and for all n we have $f(a_n) < 0$, $f(b_n) > 0$ and

$$b_n - a_n = \frac{b - a}{2^n}.$$

As the sequences $\{a_n\}$ and $\{b_n\}$ are bounded and monotonic, they both converge. Moreover their limits are equal, since $b_n - a_n \to 0$. Let

$$\lim_{n \to \infty} a_n = \lim_{n \to \infty} b_n = \xi.$$

Since f is continuous, we have

$$0 \geq \lim_{n \to \infty} f(a_n) = f(\xi) = \lim_{n \to \infty} f(b_n) \geq 0.$$

Hence $f(\xi) = 0$. Moreover $\xi \in (a, b)$ since $f(a)f(b) \neq 0$.   □

Corollary 5.3.4 (Intermediate-Value Theorem). *Let $f : [a, b] \to \mathbb{R}$ be a continuous function such that $f(a) \neq f(b)$. For each number k between $f(a)$ and $f(b)$ there exists $\xi \in (a, b)$ such that $f(\xi) = k$.*

Proof. Apply Bolzano's theorem to the function

$$g(x) = f(x) - k$$

for all $x \in [a, b]$.   □

Example 5.3.4. We show that the equation

$$e^x = 4x$$

has a solution between 0 and 1. Let

$$f(x) = e^x - 4x$$

for all x. Then $f(0) = 1 > 0$ and $f(1) = e - 4 < 0$. By Bolzano's theorem, the equation $f(x) = 0$ has a solution between 0 and 1, and the result follows.   △

Example 5.3.5. If $a > 0$ and $n \in \mathbb{N}$, then we can use the intermediate-value theorem to prove the existence of a positive nth root of a. Choose $c > \max\{1, a\}$, and let $f(x) = x^n$ for all $x \in \mathbb{R}$. This function is continuous on $[0, c]$ and

$$f(0) = 0 < c^n = f(c).$$

As $0 < a < c < c^n$, the intermediate-value theorem confirms the existence of $\xi > 0$ such that $\xi^n = f(\xi) = a$. Thus ξ is an nth root of a. $\triangle$

A **fixed point** for a function f is a number $x \in \mathcal{D}_f$ such that $f(x) = x$.

Corollary 5.3.5 (Fixed-Point Theorem). *Let $f : [a, b] \rightarrow [a, b]$ be a continuous function. Then f has a fixed point.*

Proof. If either $f(a) = a$ or $f(b) = b$, then f has a fixed point. The remaining possibility is that $f(a) - a > 0$ and $f(b) - b < 0$. Let

$$g(x) = f(x) - x$$

for all $x \in [a, b]$. Then g is continuous on $[a, b]$, $g(a) > 0$ and $g(b) < 0$. Therefore, by Bolzano's theorem, there exists $\xi \in (a, b)$ such that $f(\xi) - \xi = 0$. The result follows.   □

By combining the maximum-value theorem, the minimum-value theorem, and the intermediate-value theorem, we obtain the following corollary.

Corollary 5.3.6. *If $f : [a, b] \rightarrow \mathbb{R}$ is continuous and m and M are its minimum and maximum values, respectively, then*

$$f([a, b]) = [m, M].$$

Theorem 5.3.7. *A continuous real function is injective if and only if it is strictly monotonic.*

Proof. Clearly, every strictly monotonic function is injective.

Let f be a continuous injective function that is not strictly monotonic. We shall complete the proof by finding a contradiction. Since f is injective but not monotonic, there exist $\alpha, \beta, \gamma \in \mathcal{D}_f$ such that $\alpha < \beta < \gamma$ and either

$$f(\beta) > \max\{f(\alpha), f(\gamma)\}$$

or

$$f(\beta) < \min\{f(\alpha), f(\gamma)\}.$$

We may suppose that the former inequality obtains, the argument in the other case being similar. Choose k such that

$$\max\{f(\alpha), f(\gamma)\} < k < f(\beta).$$

Then the intermediate-value theorem establishes the existence of c and d such that $\alpha < c < \beta < d < \gamma$ and $f(c) = k = f(d)$, contradicting the assumption that f is injective. $\qquad\square$

Theorem 5.3.8. *Let $f \colon [a, b] \to \mathbb{R}$ be an increasing (respectively, decreasing) continuous function. Then f^{-1} exists and is also increasing (respectively, decreasing) and continuous.*

Proof. We confine our attention to the case where f is increasing, as the argument in the other case is analogous. Our previous results (Corollary 5.3.6 and Theorem 5.3.7) show that f^{-1} exists with domain $[m, M]$, where $m = f(a)$ and $M = f(b)$. Now choose $y_1, y_2 \in [m, M]$, where $y_1 < y_2$. Then $y_1 = f(x_1)$ and $y_2 = f(x_2)$ for some $x_1, x_2 \in [a, b]$. Thus $x_1 = f^{-1}(y_1)$ and $x_2 = f^{-1}(y_2)$. If $x_1 \geq x_2$, then $y_1 = f(x_1) \geq f(x_2) = y_2$, a contradiction. Hence f^{-1} is increasing.

Let $c \in [m, M]$ and choose $\varepsilon > 0$. We want to find $\delta > 0$ such that

$$|f^{-1}(y) - f^{-1}(c)| < \varepsilon \tag{5.3}$$

for all $y \in N_\delta(c) \cap [m, M]$. If we set $x = f^{-1}(y)$ and $\beta = f^{-1}(c)$, then inequality (5.3) becomes

$$|x - \beta| < \varepsilon,$$

which is equivalent to

$$\beta - \varepsilon < x < \beta + \varepsilon. \tag{5.4}$$

As f is increasing, inequality (5.4) is equivalent to

$$f(\beta - \varepsilon) < f(x) < f(\beta + \varepsilon)$$

and therefore to

$$f(\beta - \varepsilon) - f(\beta) < f(x) - f(\beta) < f(\beta + \varepsilon) - f(\beta).$$

Since

$$f(x) - f(\beta) = y - c$$

whenever $y \in [m, M]$, the proof is completed by choosing

$$\delta = \min\{f(\beta) - f(\beta - \varepsilon), f(\beta + \varepsilon) - f(\beta)\} :$$

For each $y \in [m, M]$ such that $|y - c| < \delta$ we have

$$f(\beta - \varepsilon) - f(\beta) < 0 \le y - c < f(\beta + \varepsilon) - f(\beta)$$

if $y - c \ge 0$, whereas if $y - c < 0$, then

$$c - y < f(\beta) - f(\beta - \varepsilon),$$

so that

$$f(\beta - \varepsilon) - f(\beta) < y - c < 0 < f(\beta + \varepsilon) - f(\beta). \qquad \square$$

Exercises 5.2.

1. Let $f : [a, b] \to \mathbb{R}$ be continuous. Show that

$$M = \max\{|f(x)| \mid x \in [a, b]\}$$

   exists. Show also that for every $\varepsilon > 0$ there is an interval $[\alpha, \beta]$, included in $[a, b]$, such that

$$|f(x)| > M - \varepsilon$$

   for all $x \in [\alpha, \beta]$.
2. Let $f, g : [a, b] \to \mathbb{R}$ be continuous functions. Suppose that $f(a) < g(a)$ and $f(b) > g(b)$. Show that there exists $c \in (a, b)$ such that $f(c) = g(c)$.
3. Let $f : (a, b) \to \mathbb{R}$ be continuous. Show that for any $c_1, c_2, \ldots, c_n \in (a, b)$ there exists $c \in (a, b)$ such that

$$f(c) = \frac{1}{n} \sum_{j=1}^{n} f(c_j).$$

4. Let p be a nonzero polynomial. Show that $e^x = |p(x)|$ has a real solution.
5. Let $f : [a, \infty) \to \mathbb{R}$ be continuous and suppose that $\lim_{x \to \infty} f(x)$ exists and is finite. Show that f is bounded on $[a, \infty)$.
6. Let f be continuous on $\mathbb{R}$ and suppose that

$$\lim_{x \to \infty} f(x) = \lim_{x \to -\infty} f(x) = 0.$$

   Show that f attains a maximum value or a minimum value.
7. Let f be a function and c a number. We say that $f(c)$ is a **local maximum** (respectively, **local minimum**) of f if there exists $\delta > 0$ such that $f(c) \ge f(x)$ [respectively, $f(c) \le f(x)$] for all $x \in N_\delta(c)$.

Let $f : [a, b] \to \mathbb{R}$ be continuous. Suppose that $f(\alpha)$ and $f(\beta)$ are local maxima of f, where $\alpha < \beta$. Show that there exists $c \in (\alpha, \beta)$ such that $f(c)$ is a local minimum.
8. Let $f : [a, b] \to \mathbb{R}$ be continuous. Show that if f does not have a local maximum or a local minimum, then it must be monotonic on $[a, b]$.

5.4 The Logarithm Function

Recall that the exponential function is increasing (Theorem 2.7.12) and continuous everywhere (Theorem 5.1.9). Therefore it has a continuous increasing inverse. This inverse is called the **logarithm** function and is denoted by log. (The reader is cautioned that some authors use a different notation.) The logarithm of a number x is often written as $\log x$ rather than $\log(x)$. Since $\lim_{x \to \infty} e^x = \infty$ and $\lim_{x \to -\infty} e^x = 0$, the continuity of the exponential function, together with the intermediate-value theorem, shows that $\log x$ is defined for all $x > 0$. Theorem 5.3.8 shows that it is increasing and continuous on $(0, \infty)$.

As the logarithm and exponential functions are inverses, $\log x = y$ if and only if $e^y = x$. Therefore

$$e^{\log x} = x$$

and

$$\log e^y = y.$$

In particular, $\log e = 1$, so that $e^x = e^{x \log e}$. We use this observation to motivate the following definition for every $a > 0$ and every real x:

$$a^x = e^{x \log a}. \tag{5.5}$$

In particular,

$$a^0 = e^0 = 1$$

and

$$a^1 = e^{\log a} = a.$$

We also define $0^x = 0$ for all $x > 0$.

If a, x, y are real numbers with $a > 0$, then

$$a^x a^y = e^{x \log a} e^{y \log a} = e^{x \log a + y \log a} = e^{(x+y) \log a} = a^{x+y}.$$

Thus, for every number w we have

$$1 = a^0 = a^{w-w} = a^w a^{-w}.$$

Therefore

$$a^{-w} = \frac{1}{a^w}.$$

Applying this result with a and w replaced by e and $y \log a$, respectively, we find that

$$\frac{a^x}{a^y} = \frac{e^{x \log a}}{e^{y \log a}} = e^{x \log a} e^{-y \log a} = e^{(x-y) \log a} = a^{x-y}.$$

Moreover Eq. (5.5) shows that

$$\log a^x = x \log a$$

for all $a > 0$ and all real x. If y is also a real number, it follows that

$$\log(a^x)^y = y \log a^x = xy \log a = \log a^{xy},$$

and we deduce that

$$(a^x)^y = a^{xy}.$$

In addition,

$$\log 1 = \log a^0 = 0 \log a = 0.$$

Hence

$$1^x = e^{x \log 1} = e^0 = 1$$

for all real x.

Note that

$$a^{x+1} = a^x \cdot a^1 = a^x a.$$

Since we also have $a^0 = 1$, it follows by induction that our definition of a^x agrees with our previous understanding in the case where x is a nonnegative integer. As $a^{-x} = 1/a^x$, the same can be said if x is a negative integer. Note also that if m and n are integers and $n > 0$, then $(a^{1/n})^m = a^{m/n}$. In particular,

$$(a^{1/n})^n = a^1 = a.$$

Thus our definition also agrees with our previous understanding in the case where x is rational.

Now suppose that $a > 0$ and $b > 0$. Since

$$e^{\log a + \log b} = e^{\log a} e^{\log b} = ab,$$

it follows that

$$\log ab = \log a + \log b.$$

One consequence is that

$$a^x b^x = e^{x \log a} e^{x \log b} = e^{x(\log a + \log b)} = e^{x \log ab} = (ab)^x$$

for every real x. Similarly,

$$\log \frac{a}{b} = \log a - \log b,$$

since

$$e^{\log a - \log b} = e^{\log a} e^{-\log b} = \frac{e^{\log a}}{e^{\log b}} = \frac{a}{b}.$$

Hence

$$\frac{a^x}{b^x} = \frac{e^{x \log a}}{e^{x \log b}} = e^{x(\log a - \log b)} = e^{x \log \frac{a}{b}} = \left(\frac{a}{b}\right)^x.$$

We also observe that the result of Example 4.6.2 holds for all real $x \notin [-1, 0]$. Indeed, the argument used in that example extends to the case of such an x. The result of Example 3.6.1 may similarly be extended to all real p.

Since $\log 1 = 0$ and $\log$ is increasing and continuous on $(0, \infty)$, we see that $\log x$ is negative for all $x \in (0, 1)$ and positive for all $x > 1$. Moreover if we choose $M > 0$ and $x > e^M$, then

$$\log x > \log e^M = M.$$

Hence

$$\lim_{x \to \infty} \log x = \infty.$$

Also, $\log x < -M$ for all x such that $0 < x < e^{-M}$, and so

$$\lim_{x \to 0^+} \log x = -\infty.$$

Example 5.4.1. By continuity we see that

$$\lim_{n \to \infty} \frac{\log n}{n} = \lim_{n \to \infty} \log n^{1/n} = \log \left(\lim_{n \to \infty} n^{1/n} \right) = \log 1 = 0.$$

In other words, $\log n \ll n$. $\triangle$

Example 5.4.2. We show that

$$\sum_{j=2}^{\infty} \frac{1}{j \log j}$$

is divergent. This series is known as **Abel's series**.

We observe that $\{1/(n \log n)\}$ is a decreasing sequence of positive terms. Hence Cauchy's condensation test is applicable. The condensed series is

$$\sum_{k=1}^{\infty} 2^k \frac{1}{2^k \log 2^k} = \sum_{k=1}^{\infty} \frac{1}{k \log 2}$$

$$= \frac{1}{\log 2} \sum_{k=1}^{\infty} \frac{1}{k},$$

which is divergent. Hence Abel's series is also divergent. $\triangle$

Logarithms are used in the proof of a test, due to Gauss, for the convergence of a series.

Theorem 5.4.1 (Gauss's Test). *Let $\sum_{j=1}^{\infty} a_j$ be a series of positive terms. Suppose there exist a bounded sequence $\{s_n\}$ and a constant c such that*

$$\frac{a_{n+1}}{a_n} = 1 - \frac{c}{n} + \frac{s_n}{n^2} \tag{5.6}$$

for all $n > 0$. Then the series is convergent if $c > 1$ and divergent if $c \le 1$.

Proof.   Case 1: Suppose $c \ne 1$. From Eq. (5.6) we see that

$$n \left(1 - \frac{a_{n+1}}{a_n} \right) = c - \frac{s_n}{n},$$

and the expression on the right-hand side approaches c as n approaches infinity since $\{s_n\}$ is bounded. The required result therefore follows from Raabe's test.

Case 2: Suppose $c = 1$. We apply the Kummer-Jensen test to the sequence

$$\{(n - 1) \log(n - 1)\}.$$

We already know that Abel's series diverges, and so it remains only to investigate the limiting behavior of

$$c_n = (n-1)\log(n-1) - \left(1 - \frac{1}{n} + \frac{s_n}{n^2}\right) n \log n$$

$$= (n-1)\log(n-1) - \left(n - 1 + \frac{s_n}{n}\right)\log n$$

$$= (n-1)(\log(n-1) - \log n) - \frac{s_n \log n}{n}.$$

Since $\{s_n\}$ is bounded, we have

$$\lim_{n\to\infty} \frac{s_n \log n}{n} = 0.$$

(See Theorem 2.5.3 and recall that $\log n << n$.) Moreover, using the analog of Theorem 5.1.4 with c replaced by ∞, we obtain

$$\lim_{n\to\infty} c_n = \lim_{n\to\infty} (n-1)(\log(n-1) - \log n)$$

$$= \lim_{n\to\infty} \log \left(\frac{n-1}{n}\right)^{n-1}$$

$$= \log \lim_{n\to\infty} \left(\frac{n}{n-1} \cdot \left(1 - \frac{1}{n}\right)^n\right)$$

$$= \log \left(1 \cdot e^{-1}\right)$$

$$= -1.$$

As the result of this calculation is negative, there exists N such that $c_n < 0$ for all $n \geq N$. Consequently, the theorem follows from the Kummer-Jensen test. □

Example 5.4.3. Test the series

$$\sum_{j=1}^{\infty} \left(\frac{(2j-1)!}{2^{2j-1}(j-1)!j!}\right)^2$$

for convergence.

Solution. As in Example 3.9.1 we write

$$a_n = \left(\frac{(2n-1)!}{2^{2n-1}(n-1)!n!}\right)^2$$

for all $n > 0$. Then

$$\frac{a_{n+1}}{a_n} = \left(\frac{2n+1}{2n+2}\right)^2$$

$$= 1 - \frac{1}{n} + \frac{1}{n^2}\left(\frac{5n^2+4n}{4n^2+8n+4}\right),$$

the last step being obtained upon division of $4n^2 + 4n + 1$ by $4n^2 + 8n + 4$ to give a quotient of $1 - 1/n$ and a remainder of $5 + 4/n$. Since the sequence

$$\left\{\frac{5n^2+4n}{4n^2+8n+4}\right\}$$

converges (to $5/4$), it is bounded. The divergence of the given series now follows from Gauss's test with $c = 1$.

$\triangle$

Exercises 5.3.

1. Let $a_n = \log^{\log n} n$ for all $n > 1$. Show that $a_n = n^{\log \log n}$ and hence that $\sum_{j=2}^{\infty} 1/a_j$ converges.
2. Determine the convergence of the following series:

   (a) $\sum_{j=1}^{\infty}(\log(j+1) - \log j)^p$ where $p > 0$;

   (b) $\sum_{j=2}^{\infty} \frac{\log j}{j}$;

   (c) $\sum_{j=1}^{\infty} \frac{1}{j}\log\left(1 + \frac{1}{j}\right)$;

   (d) $\sum_{j=2}^{\infty} \frac{1}{j \log^p j}$ where $p \in \mathbb{R}$;

   (e) $\sum_{j=2}^{\infty} \frac{1}{j \log j \log \log j}$.

   (Hint: Try Cauchy's condensation test.)
3. Find the interval of convergence of the power series

$$\sum_{j=2}^{\infty} \frac{\log j}{(j+1)^2}(x-1)^j.$$

4. Let $\{a_n\}$ be a sequence of positive terms and $\{s_n\}$ a bounded sequence. Let c be a constant such that

$$\frac{a_n}{a_{n+1}} = 1 + \frac{c}{n} + \frac{s_n}{n^2}$$

for all $n > 0$. Prove that the series $\sum_{j=1}^{\infty} a_j$ converges if $c > 1$ and diverges if $c \leq 1$.

5. (Bertrand's test) Let $\sum_{j=1}^{\infty} a_j$ be a series of positive terms, and for each $n \in \mathbb{N}$ let

$$\beta_n = \left(n \left(\frac{a_{n+1}}{a_n} - 1 \right) - 1 \right) \log n.$$

(a) Show that if $\lim_{n \to \infty} \beta_n > 1$, then the series converges.
(b) Show that if $\lim_{n \to \infty} \beta_n < 1$, then the series diverges.

(Hint: Take

$$b_n = (n-1) \log(n-1)$$

in the Kummer-Jensen test.)

5.5   Uniformly Continuous Functions

Let f be a continuous function and let c be an accumulation point of $\mathcal{D}_f$. Then for each $\varepsilon > 0$ there exists $\delta > 0$ such that

$$|f(z) - f(c)| < \varepsilon$$

for all $z \in N_\delta(c) \cap \mathcal{D}_f$. It may be that δ depends on c. If not, then we say that f is uniformly continuous.

Definition 5.5.1. A function f is **uniformly continuous** if for each $\varepsilon > 0$ there exists $\delta > 0$ such that

$$|f(z_1) - f(z_2)| < \varepsilon$$

for all $z_1, z_2 \in \mathcal{D}_f$ satisfying $|z_1 - z_2| < \delta$.

Remark. Thus a uniformly continuous function must be continuous at every point in its domain.

We can prove the following sequential characterisation of uniform continuity.

Proposition 5.5.1. *A function f is uniformly continuous if and only if for each pair of sequences $\{s_n\}$ and $\{t_n\}$ in $\mathcal{D}_f$ such that*

$$\lim_{n \to \infty} (s_n - t_n) = 0,$$

we have

$$\lim_{n \to \infty} (f(s_n) - f(t_n)) = 0.$$

Proof. Suppose there exist sequences $\{s_n\}$ and $\{t_n\}$ in $\mathcal{D}_f$ such that $\{s_n - t_n\}$ is null but $\{f(s_n) - f(t_n)\}$ is not. Some $\varepsilon > 0$ therefore has the property that for every M there exists $n \geq M$ satisfying

$$|f(s_n) - f(t_n)| \geq \varepsilon, \tag{5.7}$$

but on the other hand, for every $\delta > 0$ there exists N such that

$$|s_n - t_n| < \delta$$

for all $n \geq N$. Some such n must satisfy inequality (5.7), and so f cannot be uniformly continuous.

Conversely, suppose f is not uniformly continuous. Then for some $\varepsilon > 0$ and all $\delta > 0$ there exist $s, t \in \mathcal{D}_f$ such that $|s - t| < \delta$ and $|f(s) - f(t)| \geq \varepsilon$. Thus for each $n \in \mathbb{N}$ there exist $s_n, t_n \in \mathcal{D}_f$ such that

$$|s_n - t_n| < \frac{1}{n}$$

and

$$|f(s_n) - f(t_n)| \geq \varepsilon.$$

The sandwich theorem shows that the sequence $\{s_n - t_n\}$ is null, but $\{f(s_n) - f(t_n)\}$ is evidently not.   □

Example 5.5.1. Consider the function $f : (0, 2] \to \mathbb{R}$ defined by

$$f(x) = \frac{1}{x}.$$

Let $\varepsilon = 1/2$ and, for each $n \in \mathbb{N}$, define $s_n = 1/n$ and $t_n = 2/n$. Then

$$|s_n - t_n| = \frac{1}{n} \to 0$$

as $n \to \infty$, but

$$|f(s_n) - f(t_n)| = \frac{n}{2} \geq \varepsilon$$

for all $n \in \mathbb{N}$. We conclude that f is not uniformly continuous.   △

We also deduce the following theorem.

Theorem 5.5.2. *Every function that is continuous over a compact set is uniformly continuous.*

Proof. Suppose that a function f is continuous over a compact set X but not uniformly. Then we can find $\varepsilon > 0$ with the property that for each $n \in \mathbb{N}$ there exist s_n and t_n in X satisfying

$$\lim_{n \to \infty} |s_n - t_n| = 0$$

and

$$|f(s_n) - f(t_n)| \geq \varepsilon. \tag{5.8}$$

Since $\{s_n\}$ is bounded, by Corollary 2.6.8 it has a subsequence $\{s_{k_n}\}$ that converges to some number s. If $s_{k_m} = s$ for some number m, then $s \in X$. In the remaining case s is a limit point of X by Theorem 2.6.11, and once again $s \in X$ since X is closed. Now

$$|t_{k_n} - s| \leq |t_{k_n} - s_{k_n}| + |s_{k_n} - s| \to 0$$

as $n \to \infty$. Hence

$$\lim_{n \to \infty} t_{k_n} = s.$$

As f is continuous at $s \in X$, we obtain

$$\lim_{n \to \infty} f(s_{k_n}) = \lim_{n \to \infty} f(t_{k_n}) = f(s).$$

Therefore

$$|f(s_{k_n}) - f(t_{k_n})| \leq |f(s_{k_n}) - f(s)| + |f(t_{k_n}) - f(s)| \to 0$$

as $n \to \infty$. We now have a contradiction to inequality (5.8). $\qquad\square$

Theorem 5.5.3. *A continuous function f is uniformly continuous on an open interval (a, b) if and only if $\lim_{x \to a+} f(x)$ and $\lim_{x \to b-} f(x)$ exist and are finite.*

Proof. Suppose that f is uniformly continuous on (a, b). Choose $\varepsilon > 0$. There exists $\delta > 0$ such that

$$|f(x) - f(y)| < \varepsilon \tag{5.9}$$

for all x and y in (a, b) satisfying $|x - y| < \delta$. Choose x and y in $N^*_{\delta/2}(a) \cap (a, b)$. Then

$$|x - y| \leq |x - a| + |a - y| < \frac{\delta}{2} + \frac{\delta}{2} = \delta,$$

so that inequality (5.9) holds. Thus f satisfies Cauchy's condition at a, and we conclude from Theorem 4.3.2 that $\lim_{x \to a+} f(x)$ exists and is finite. The limit as $x \to b^-$ is dealt with in a similar manner.

Conversely, suppose that both limits exist and are finite. Then f can be extended to a continuous function on $[a, b]$ that is uniformly continuous by Theorem 5.5.2. Hence f is also uniformly continuous.   □

Example 5.5.2. Let $f(x) = e^{-1/x}$ for all $x \neq 0$. Then $\lim_{x \to 0+} f(x) = 0$. Hence f is uniformly continuous on $(0, b]$ for each $b > 0$.   △

A function is uniformly continuous over a closed interval if and only if it is continuous over that interval. Thus if functions f and g are uniformly continuous over an interval $[a, b]$, then so are $f + g$, fg, and αf for all $\alpha \in \mathbb{R}$. In view of Theorem 5.5.3, this result is also true if we replace the interval by (a, b). However, it is in general not true if we replace the interval by $[a, \infty)$.

Example 5.5.3. Let $f(x) = g(x) = x$ for all x. Then, clearly, f and g are uniformly continuous over $[0, \infty)$. However, we show that the function given by x^2 is not uniformly continuous over that interval. For all $n \geq 0$ let $s_n = \sqrt{n + 1}$ and $t_n = \sqrt{n}$. Then

$$\begin{aligned}
s_n - t_n &= \sqrt{n + 1} - \sqrt{n} \\
&= \frac{(\sqrt{n + 1} - \sqrt{n})(\sqrt{n + 1} + \sqrt{n})}{\sqrt{n + 1} + \sqrt{n}} \\
&= \frac{1}{\sqrt{n + 1} + \sqrt{n}} \\
&\to 0.
\end{aligned}$$

However,

$$s_n^2 - t_n^2 = 1$$

for all n. Hence according to Proposition 5.5.1, the function x^2 is not uniformly continuous over $[0, \infty)$.   △

The family of uniformly continuous functions includes an important subfamily, which we define below.

Definition 5.5.2. A function f is said to be **Lipschitz continuous** if there exists a positive constant M such that

$$|f(z_1) - f(z_2)| \leq M |z_1 - z_2|$$

for all z_1 and z_2 in $\mathcal{D}_f$.

Example 5.5.4. The modulus function is Lipschitz continuous since

$$||z_1| - |z_2|| \leq |z_1 - z_2|$$

for all $z_1, z_2 \in \mathbb{C}$.   △

It follows from Proposition 5.5.1 and the sandwich theorem that every function that is Lipschitz continuous is also uniformly continuous. However, the converse is not always true:

Example 5.5.5. The function f given by

$$f(x) = \sqrt{x}$$

is continuous and therefore uniformly continuous over $[0, 1]$. We show that it is not Lipschitz continuous. Suppose there exists $M > 0$ such that

$$|f(x_1) - f(x_2)| \leq M|x_1 - x_2|$$

for all $x_1, x_2 \in [0, 1]$. In particular,

$$|f(x) - f(0)| \leq M|x|$$

for every $x \in (0, 1]$. Thus

$$\sqrt{x} \leq Mx,$$

so that $1 \leq M\sqrt{x}$. However, this result is contradicted for x such that $0 < x < 1/M^2$.   △

Exercises 5.4.

1. Show that the following functions are uniformly continuous:

   (a) $\sqrt{x}$, where $x \in [0, \infty)$;
   (b) $x \sin \frac{1}{x}$, where $x \in (0, 1)$.

2. Show that the following functions are not uniformly continuous:

   (a) $\log x$, where $x \in (0, 1)$;
   (b) e^x, where $x \in [0, \infty)$.

3. Suppose that f is uniformly continuous on $(a, b]$ and on $[b, c)$. Show that f is uniformly continuous on (a, c).
4. Let f be continuous on $[a, \infty)$ and suppose that $\lim_{x \to \infty} f(x)$ exists and is finite. Show that f is uniformly continuous on $[a, \infty)$.

Chapter 6
Differentiability

The notion of a derivative is motivated by studying two kinds of problems: finding instantaneous velocities and determining slopes of tangents to curves. Here we shall not dwell on these problems. Rather, we undertake a study of the general mathematical properties of derivatives. We assume all functions to be of a complex variable and complex-valued unless an indication to the contrary is given.

6.1 Derivatives

Definition 6.1.1. Let f be a function and c an accumulation point of $\mathcal{D}_f$. Define

$$f'(c) = \lim_{z \to c} \frac{f(z) - f(c)}{z - c}.$$

If the limit exists and is a number, then f is said to be **differentiable** at c and $f'(c)$ is the **derivative** of f at c.

Remark 1. We sometimes write

$$\frac{d}{dz} f(z) = f'(z)$$

if f is differentiable at z.

Remark 2. In the case where f is a real-valued function of a real variable x, the quotient

$$\frac{f(x) - f(c)}{x - c}$$

© Springer Science+Business Media New York 2015

C.H.C. Little et al., *Real Analysis via Sequences and Series*, Undergraduate Texts in Mathematics, DOI 10.1007/978-1-4939-2651-0_6

gives the slope of the line joining the points $(c, f(c))$ and $(x, f(x))$ on the graph of f. We may interpret $f'(c)$ geometrically as the slope of the graph of f at the point $(c, f(c))$.

Example 6.1.1. Let $f(z) = az + b$ for all $z \in \mathbb{C}$, where $a, b \in \mathbb{C}$. Then, for each $c \in \mathbb{C}$ and $z \neq c$,

$$\frac{f(z) - f(c)}{z - c} = \frac{(az + b) - (ac + b)}{z - c}$$

$$= \frac{a(z - c)}{z - c}$$

$$= a.$$

Hence $f'(c) = a$. △

In particular, if $f(z) = b$ for all $z \in \mathbb{C}$, then $f'(c) = 0$ for all c. If $f(z) = z$ for all $z \in \mathbb{C}$, then $f'(c) = 1$ for all c.

Example 6.1.2. Let $f(z) = z^2$ for all $z \in \mathbb{C}$. Then, for each c and $z \neq c$,

$$\frac{f(z) - f(c)}{z - c} = \frac{z^2 - c^2}{z - c}$$

$$= z + c$$

$$\to 2c$$

as $z \to c$. Hence $f'(c) = 2c$. △

Definition 6.1.1 may be rewritten using the limit in the next theorem.

Theorem 6.1.1. *Let f be a function that is differentiable at a number c. Then*

$$f'(c) = \lim_{h \to 0} \frac{f(c + h) - f(c)}{h}.$$

Proof. Let $k(z) = c + z$ for all z. Then $\lim_{z \to 0} k(z) = c$. Define

$$g(z) = \frac{f(z) - f(c)}{z - c}$$

for all $z \in \mathcal{D}_f - \{c\}$, and let $g(c) = f'(c)$. Then 0 is an accumulation point of $\mathcal{D}_{g \circ k}$ and g is continuous at c. Notice also that

$$g(k(z)) = \frac{f(c + z) - f(c)}{z}.$$

It therefore follows from Theorem 5.1.4 that

$$f'(c) = \lim_{z \to c} g(z) = \lim_{k(z) \to c} g(k(z)) = \lim_{z \to 0} \frac{f(c+z) - f(c)}{z},$$

as required. □

Example 6.1.3. Recall that $\exp(z) = e^z$ for all complex z. We have seen that Theorem 2.7.10 holds also for complex numbers. Therefore so does Corollary 2.7.11, and we can use the latter, together with Theorem 6.1.1, to show that

$$\exp'(c) = \lim_{h \to 0} \frac{e^{c+h} - e^c}{h} = e^c$$

for all c. In other words, the exponential function is its own derivative. △

The following theorem is clear from the sequential formulation of limits.

Theorem 6.1.2. *Let f be a function. If there exist sequences $\{s_n\}$ and $\{t_n\}$ in $\mathcal{D}_f$ such that*

$$\lim_{n \to \infty} s_n = \lim_{n \to \infty} t_n = c$$

and

$$\lim_{n \to \infty} \frac{f(s_n) - f(c)}{s_n - c} \neq \lim_{n \to \infty} \frac{f(t_n) - f(c)}{t_n - c},$$

then $f'(c)$ does not exist.

Example 6.1.4. Consider the function $f(z) = \bar{z}$, defined on $\mathbb{C}$. We show that $f'(0)$ does not exist. For each $n \in \mathbb{N}$, take $s_n = 1/n$ and $t_n = i/n$. Clearly,

$$\lim_{n \to \infty} s_n = \lim_{n \to \infty} t_n = 0.$$

Now

$$\frac{f(s_n) - f(0)}{s_n - 0} = 1$$

and

$$\frac{f(t_n) - f(0)}{t_n - 0} = -1.$$

Hence

$$\lim_{n \to \infty} \frac{f(s_n) - f(0)}{s_n - 0} \neq \lim_{n \to \infty} \frac{f(t_n) - f(0)}{t_n - 0}.$$

Therefore $f'(0)$ does not exist. △

Example 6.1.5.   Consider the real function

$$f(x) = \begin{cases} x^2 & \text{if } x \leq 0, \\ x & \text{if } x > 0. \end{cases}$$

We show that $f'(0)$ does not exist. For each $x \neq 0$ let

$$\begin{aligned} Q(x) &= \frac{f(x) - f(0)}{x - 0} \\ &= \frac{f(x)}{x} \\ &= \begin{cases} x & \text{if } x \leq 0, \\ 1 & \text{if } x > 0. \end{cases} \end{aligned}$$

Thus

$$\lim_{x \to 0^-} Q(x) = 0$$

and

$$\lim_{x \to 0^+} Q(x) = 1.$$

Hence $\lim_{x \to 0} Q(x)$ does not exist. In other words, $f'(0)$ does not exist. Note that the function is continuous but its graph has a "corner" at 0 (see Fig. 6.1). $\triangle$

We now show that if f is differentiable at c, then f is continuous at c. Examples 6.1.4 and 6.1.5 show that the converse is not always true.

Theorem 6.1.3. *If a function f is differentiable at a number c, then f is continuous at c.*

Fig. 6.1 Graph of f
in Example 6.1.5

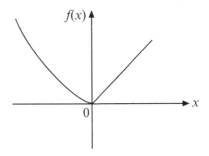

Proof. Let f be differentiable at c. Then, for each $z \in \mathcal{D}_f - \{c\}$,

$$f(z) - f(c) = \frac{f(z) - f(c)}{z - c}(z - c).$$

As z approaches c, the right-hand side approaches $f'(c) \cdot 0 = 0$. Hence

$$\lim_{z \to c} f(z) = f(c).$$

$\square$

Our next example shows that, even for a real function f, the existence of

$$\lim_{x \to c} f'(x)$$

does not guarantee the existence of $f'(c)$.

Example 6.1.6. Let

$$f(x) = \begin{cases} x + 1 & \text{if } x \le 0, \\ x & \text{if } x > 0. \end{cases}$$

Then $f'(x) = 1$ for all $x \ne 0$. Thus

$$\lim_{x \to 0} f'(x) = 1.$$

But f is not continuous at 0, since $\lim_{x \to 0^-} f(x) = 1$ and $\lim_{x \to 0^+} f(x) = 0$. Therefore, by Theorem 6.1.3, $f'(0)$ does not exist. $\triangle$

Our next theorem gives an idea of the behavior of a function in the vicinity of a point where its derivative exists. The result is stronger than Theorem 6.1.3.

Theorem 6.1.4. *Let f be differentiable at c. Then there exist $\delta > 0$ and $M > 0$ such that*

$$|f(z) - f(c)| < M|z - c|$$

for all $z \in N_\delta^*(c) \cap \mathcal{D}_f$.

Proof. There exists $\delta > 0$ such that

$$\left| \frac{f(z) - f(c)}{z - c} - f'(c) \right| < 1$$

for all $z \in N_\delta^*(c) \cap \mathcal{D}_f$. Hence

$$\left| \frac{f(z) - f(c)}{z - c} \right| - |f'(c)| < 1.$$

Setting $M = |f'(c)| + 1 > 0$, we obtain

$$|f(z) - f(c)| < M|z - c|$$

for all $z \in N_\delta^*(c) \cap \mathcal{D}_f$.   □

The condition given in the conclusion of Theorem 6.1.4 is called the **Lipschitz condition** at c. Functions that satisfy it are continuous at c but not necessarily differentiable there.

Example 6.1.7.   The absolute value function $|x|$ satisfies the Lipschitz condition at 0 but can be shown to be not differentiable there. Hence the converse of Theorem 6.1.4 is in general not true.   △

We end this section by defining $f^{(0)} = f$, where f is a function, and if $f^{(n)}$ has been defined for some nonnegative integer n, then $f^{(n+1)} = (f^{(n)})'$. We call $f^{(n)}$ the nth **derivative** of f. If f is a function of a variable z, then its nth derivative is sometimes written as $\frac{d^n}{dz^n} f(z)$. These higher-order derivatives will be used later to study approximations of functions by polynomials.

Exercises 6.1.

1. Find the derivatives of the following functions:
   (a) $\frac{1}{x}$, where $x \neq 0$;
   (b) x^n, where $n \in \mathbb{N}$;
   (c) $x|x|$.

2. Show that the absolute value function is not differentiable at 0.
3. Show that the function $\sqrt{x}$, where $x \geq 0$, is differentiable at all $x > 0$.
4. Use the properties of the sine and cosine functions to prove that $\sin' x = \cos x$ and $\cos' x = -\sin x$ for all x.
5. Let

$$f(x) = \begin{cases} x^2 & \text{if } x \in \mathbb{Q}, \\ 0 & \text{if } x \notin \mathbb{Q}. \end{cases}$$

   Show that f is differentiable only at 0.
6.

$$f(x) = \begin{cases} x^2 & \text{if } x < 0, \\ x^3 & \text{if } x \geq 0. \end{cases}$$

   Find $f'(x)$ and $f''(x)$ for all x.
7. Let $f(x) = |x|^3$ for all x. Find $f'(x)$ and $f''(x)$ for all x and show that $f'''(0)$ does not exist.

6.2   Differentiation Formulas

We now present some results that are helpful in finding derivatives of functions.

Theorem 6.2.1. *If f and g are functions and c is an accumulation point of $\mathcal{D}_f \cap \mathcal{D}_g$, where f and g are differentiable, then*

1.

$$(f + g)'(c) = f'(c) + g'(c),$$

2.

$$(fg)'(c) = f(c)g'(c) + f'(c)g(c),$$

3.

$$\left(\frac{f}{g}\right)'(c) = \frac{f'(c)g(c) - f(c)g'(c)}{g^2(c)}$$

if there is a neighborhood $N_\delta(c)$ such that $g(z) \neq 0$ for all $z \in N_\delta(c)$.

Proof. 1. We compute

$$
\begin{aligned}
(f + g)'(c) &= \lim_{z \to c} \frac{(f + g)(z) - (f + g)(c)}{z - c} \\
&= \lim_{z \to c} \left(\frac{f(z) - f(c)}{z - c} + \frac{g(z) - g(c)}{z - c} \right) \\
&= f'(c) + g'(c).
\end{aligned}
$$

2. Since

$$
\begin{aligned}
(fg)(z) - (fg)(c) &= f(z)g(z) - f(c)g(c) \\
&= f(z)g(z) - f(z)g(c) + f(z)g(c) - f(c)g(c) \\
&= f(z)(g(z) - g(c)) + (f(z) - f(c))g(c),
\end{aligned}
$$

we have

$$
\begin{aligned}
(fg)'(c) &= \lim_{z \to c} f(z) \lim_{z \to c} \frac{g(z) - g(c)}{z - c} + \lim_{z \to c} \frac{f(z) - f(c)}{z - c} g(c) \\
&= f(c)g'(c) + f'(c)g(c)
\end{aligned}
$$

because f is continuous at c.

3. We deal first with the function $1/g$, recalling that g is continuous at c:

$$
\left(\frac{1}{g}\right)'(c) = \lim_{z \to c} \frac{\frac{1}{g(z)} - \frac{1}{g(c)}}{z - c}
$$

$$
= \lim_{z \to c} \left(\frac{1}{z - c} \cdot \frac{g(c) - g(z)}{g(z)g(c)}\right)
$$

$$
= \lim_{z \to c} \frac{1}{g(z)g(c)} \cdot \lim_{z \to c} \frac{g(c) - g(z)}{z - c}
$$

$$
= \frac{-g'(c)}{g^2(c)}.
$$

An application of part (2) therefore gives

$$
\left(\frac{f}{g}\right)'(c) = \left(f \cdot \frac{1}{g}\right)'(c)
$$

$$
= f'(c)\frac{1}{g(c)} + f(c)\left(\frac{1}{g}\right)'(c)
$$

$$
= \frac{f'(c)}{g(c)} - \frac{f(c)g'(c)}{g^2(c)}
$$

$$
= \frac{f'(c)g(c) - f(c)g'(c)}{g^2(c)}.
$$

□

Remark. Again, the hypothesis that c be an accumulation point of $\mathcal{D}_f \cap \mathcal{D}_g$ is essential (see Remark 2 after Theorem 4.4.1).

Corollary 6.2.2. *Let f be a function and a and c numbers. If f is differentiable at c, then*

$$
(af)'(c) = af'(c).
$$

Proof. Let $g(z) = a$ for all z. Thus $g'(c) = 0$ for all c. Hence

$$
(af)'(c) = af'(c) + 0 \cdot f(c) = af'(c),
$$

by Theorem 6.2.1(2).   □

Example 6.2.1. Let $f(z) = z^n$ for all z, where n is a positive integer. We show by induction that

$$
f'(z) = nz^{n-1}
$$

for all z, where $0^0 = 1$. Certainly, $f'(z) = 1$ for all z if $n = 1$. Assume that $n > 1$ and that the result holds with n replaced by $n - 1$. Then $z^n = z \cdot z^{n-1}$, and so

$$f'(z) = z(n - 1)z^{n-2} + 1 \cdot z^{n-1}$$
$$= nz^{n-1}$$

for all z, by Theorem 6.2.1(2), as required.

Suppose now that $z \neq 0$. Then the result just proved holds also when $n = 0$. In fact, we can extend it to the case where n is a negative integer, using Theorem 6.2.1(3) and the fact that $z^n = 1/z^{-n}$. Thus if $f(z) = z^n$ for all $z \neq 0$ and n is a negative integer, then $-n > 0$, and so

$$f'(z) = \frac{-(-nz^{-n-1})}{z^{-2n}} = nz^{n-1}$$

for all $z \neq 0$. △

Our next theorem, known as the chain rule, deals with compositions of functions.

Theorem 6.2.3 (Chain Rule). *Suppose that f and g are functions such that both $g'(c)$ and $f'(g(c))$ exist, where $c \in \mathcal{D}_{f \circ g}$. Suppose also that c is an accumulation point of $\mathcal{D}_{f \circ g}$. Then*

$$(f \circ g)'(c) = f'(g(c))g'(c).$$

Proof. Define

$$u(z) = \frac{g(z) - g(c)}{z - c} - g'(c)$$

for all $z \in \mathcal{D}_g - \{c\}$. Thus

$$\lim_{z \to c} u(z) = g'(c) - g'(c) = 0.$$

Define $u(c) = 0$. Then u is continuous at c.

Similarly, let $b = g(c)$ and define

$$v(z) = \begin{cases} \frac{f(z) - f(b)}{z - b} - f'(b) & \text{if } z \in \mathcal{D}_f - \{b\}, \\ 0 & \text{if } z = b. \end{cases}$$

Then v is continuous at b.

From the definitions of $u(z)$ and $v(z)$, we have

$$g(z) - b = (z - c)(g'(c) + u(z))$$

for all $z \in \mathcal{D}_g$ (even for $z = c$) and

$$f(z) - f(b) = (z - b)(f'(b) + v(z))$$

for all $z \in \mathcal{D}_f$. Hence

$$\begin{aligned}
f(g(z)) - f(g(c)) &= f(g(z)) - f(b) \\
&= (g(z) - b)(f'(b) + v(g(z))) \\
&= (z - c)(g'(c) + u(z))(f'(b) + v(g(z)))
\end{aligned}$$

for all $z \in \mathcal{D}_{f \circ g}$, so that

$$\begin{aligned}
(f \circ g)'(c) &= \lim_{z \to c}(g'(c) + u(z))(f'(b) + v(g(z))) \\
&= g'(c)f'(b),
\end{aligned}$$

for $v \circ g$ is continuous at c by Corollary 5.1.5 since g is continuous at c and v is continuous at $b = g(c)$.   □

Example 6.2.2. Since

$$\sin z = \frac{e^{iz} - e^{-iz}}{2i}$$

for all $z \in \mathbb{C}$, we have

$$\sin' z = \frac{ie^{iz} + ie^{-iz}}{2i} = \frac{e^{iz} + e^{-iz}}{2} = \cos z,$$

by Theorems 6.2.1 and 6.2.3. Similarly,

$$\cos' z = \frac{ie^{iz} - ie^{-iz}}{2} = \frac{i^2(e^{iz} - e^{-iz})}{2i} = -\sin z.$$

Thus the sine and cosine functions are continuous everywhere.

Recall also that $\sec x = 1/\cos x$ and $\tan x = (\sin x)/(\cos x)$ whenever $\cos x \neq 0$. We therefore have

$$\tan' x = \frac{\cos^2 x + \sin^2 x}{\cos^2 x} = \frac{1}{\cos^2 x} = \sec^2 x.$$

It follows that the tangent function is continuous wherever it is defined. Note also that

$$\sec' x = -\frac{1}{\cos^2 x}(-\sin x) = \sec x \tan x$$

for all x such that $\cos x \neq 0$.   △

Theorem 6.2.4 (Inverse Function Theorem). *Let* $f : [a, b] \to \mathbb{R}$ *be an increasing differentiable function. Let* c *be a number in* $[a, b]$ *such that* $f'(c) \neq 0$. *Then* f^{-1} *is differentiable at* $f(c)$ *and*

$$\left(f^{-1}\right)'(f(c)) = \frac{1}{f'(c)}.$$

Proof. Note first that f is continuous, by Theorem 6.1.3. Therefore f^{-1} exists and is continuous and increasing on its domain, by Theorem 5.3.8. Let $g = f^{-1}$. Thus $\mathcal{D}_g = [f(a), f(b)]$.

According to Theorem 6.1.1, we must show that

$$\lim_{k \to 0} \frac{g(f(c) + k) - c}{k} = \frac{1}{f'(c)},$$

since $g(f(c)) = c$. It is therefore enough to prove that

$$f'(c) = \lim_{k \to 0} \frac{k}{g(f(c) + k) - c}.$$

For each k such that $f(c) + k \in \mathcal{D}_g$, define

$$h(k) = g(f(c) + k) - c.$$

Thus $k \in \mathcal{D}_h$ if and only if $f(a) \leq f(c) + k \leq f(b)$. Consequently,

$$\mathcal{D}_h = [f(a) - f(c), f(b) - f(c)].$$

Therefore $0 \in \mathcal{D}_h$ since f is increasing, and

$$h(0) = g(f(c)) - c = c - c = 0.$$

Furthermore,

$$f^{-1}(f(c) + k) = c + h(k);$$

hence

$$f(c) + k = f(c + h(k)),$$

so that

$$k = f(c + h(k)) - f(c).$$

Thus it suffices to show that

$$f'(c) = \lim_{k \to 0} \frac{f(c + h(k)) - f(c)}{h(k)}.$$

But

$$f'(c) = \lim_{h(k) \to 0} \frac{f(c + h(k)) - f(c)}{h(k)}.$$

We therefore introduce the function j defined by $j(0) = f'(c)$ and

$$j(x) = \frac{f(c + x) - f(c)}{x}$$

for all $x \in [a - c, b - c] - \{0\}$, so that

$$f'(c) = \lim_{h(k) \to 0} j(h(k)).$$

We now check that the hypotheses of Theorem 5.1.4 are satisfied by the functions j and h. As h is evidently continuous on its domain, we see that

$$\lim_{k \to 0} h(k) = h(0) = 0.$$

Moreover j is continuous at 0 by definition. Finally, we have $k \in \mathcal{D}_{j \circ h}$ if and only if $k \in \mathcal{D}_h$ and $h(k) \in [a - c, b - c]$. But if $k \in \mathcal{D}_h$, then $h(k) \in [a - c, b - c]$ since $\mathcal{R}_g = [a, b]$. We conclude that $\mathcal{D}_{j \circ h} = \mathcal{D}_h$ and therefore that 0 is an accumulation point of $\mathcal{D}_{j \circ h}$. We can thus apply Theorem 5.1.4 to show that

$$f'(c) = \lim_{k \to 0} j(h(k)),$$

as required.   □

The inverse function theorem also holds for functions of a complex variable with continuous derivatives (see [11]).

Since the logarithm and exponential functions are inverses, we have

$$x = \exp(\log x) \tag{6.1}$$

for all $x > 0$. Moreover the logarithm function is differentiable at all $x > 0$ by Theorem 6.2.4, since the exponential function is nonzero, differentiable, and increasing everywhere. Differentiation of Eq. (6.1) therefore yields

$$1 = \exp(\log x) \log' x = x \log' x,$$

and so

$$\log' x = \frac{1}{x}$$

for all $x > 0$. This calculation provides another confirmation that the logarithm function is continuous at all $x > 0$.

Let us now define $f(x) = x^a = e^{a \log x} > 0$ for all $x > 0$, where a is any real number. Then

$$f'(x) = e^{a \log x} \cdot \frac{a}{x} = x^a \cdot \frac{a}{x} = ax^{a-1}.$$

Similarly, let $g(x) = a^x = e^{x \log a}$ for all x, where $a > 0$. Then

$$g'(x) = e^{x \log a} \log a = a^x \log a.$$

Exercises 6.2.

1. Show that $\log' x = 1/x$ for all $x > 0$ by evaluating

$$\lim_{h \to 0} \frac{\log(x + h) - \log x}{h}.$$

2. Let $f : [a, b] \to \mathbb{R}$ be differentiable at $c \in (a, b)$. Let $\{a_n\}$ and $\{b_n\}$ be sequences such that

$$a < a_n < c < b_n < b$$

for all positive integers n. If

$$\lim_{n \to \infty} a_n = \lim_{n \to \infty} b_n = c,$$

prove that

$$f'(c) = \lim_{n \to \infty} \frac{f(b_n) - f(a_n)}{b_n - a_n}.$$

[Hint: Use the fact that

$$\frac{f(b_n) - f(a_n)}{b_n - a_n} - f'(c)$$

$$= \frac{b_n - c}{b_n - a_n} \left(\frac{f(b_n) - f(c)}{b_n - c} - f'(c) \right) - \frac{a_n - c}{b_n - a_n} \left(\frac{f(a_n) - f(c)}{a_n - c} - f'(c) \right)$$

for all $n \in \mathbb{N}$.]

3. Let f and g be functions having nth derivatives. Prove that

$$(fg)^{(n)} = \sum_{j=0}^{n} \binom{n}{j} f^{(n-j)} g^{(j)}.$$

(This result is known as Leibniz's rule.)

4. Let $F = f_1 f_2 \cdots f_n$ and $F_j = F/f_k$ whenever $f_j(x) \neq 0$. Show by induction that we then have

$$F'(x) = \sum_{j=1}^{n} F_j(x) f_j'(x).$$

5. Let

$$f(x) = \begin{cases} x & \text{if } x \in \mathbb{Q}, \\ -x & \text{if } x \notin \mathbb{Q}. \end{cases}$$

Show that $(f \circ f)(x) = x$ for all x. What can you say about the chain rule?

6. Let f be as in Exercise 6.5 and let $g = -f$. Show that $(fg)(x) = -x^2$ for all x. What can you say about the product rule?

7. Let m and n be positive integers. An $m \times n$ matrix is defined as an array of numbers arranged in m rows and n columns. The array is usually enclosed in parentheses. The determinant $\begin{vmatrix} p & q \\ r & s \end{vmatrix}$ of the 2×2 matrix $\begin{pmatrix} p & q \\ r & s \end{pmatrix}$ is defined by the equation

$$\begin{vmatrix} p & q \\ r & s \end{vmatrix} = ps - rq.$$

Let (a, b) be an open interval, and for all $x \in (a, b)$ define

$$F(x) = \begin{vmatrix} f_1(x) & f_2(x) \\ g_1(x) & g_2(x) \end{vmatrix},$$

where f_1, f_2, g_1, g_2 are functions that are differentiable on (a, b). Show that

$$F'(x) = \begin{vmatrix} f_1'(x) & f_2'(x) \\ g_1(x) & g_2(x) \end{vmatrix} + \begin{vmatrix} f_1(x) & f_2(x) \\ g_1'(x) & g_2'(x) \end{vmatrix}$$

for all $x \in (a, b)$.

6.3   The Mean-Value Theorem for Derivatives

The first theorem to be discussed in this section is important in its own right but is even more important as the basis for a number of other theorems that are among the most useful in the theory of functions.

Theorem 6.3.1 (Rolle). *Let* $f : [a, b] \to \mathbb{R}$ *be continuous on* $[a, b]$ *and differentiable on* (a, b), *and suppose that* $f(a) = f(b)$. *Then* $f'(\xi) = 0$ *for some* $\xi \in (a, b)$.

Proof. If f is a constant function, then $f'(x) = 0$ for all $x \in [a, b]$. Hence we assume that $f(x_1) \neq f(a)$ for some $x_1 \in (a, b)$. We may also assume that

$$f(x_1) > f(a),$$

as the argument for the other case is similar.

By the maximum-value theorem, there exists $\xi \in [a, b]$ such that $f(x) \leq f(\xi)$ for all $x \in [a, b]$. Since

$$f(\xi) \geq f(x_1) > f(a) = f(b),$$

we have $\xi \notin \{a, b\}$. Thus $\xi \in (a, b)$.

We claim that $f'(\xi) = 0$. Define

$$Q(x) = \frac{f(x) - f(\xi)}{x - \xi}$$

for all $x \in [a, b] - \{\xi\}$. Then

$$\lim_{x \to \xi} Q(x) = f'(\xi).$$

As $f(x) \leq f(\xi)$ for all $x \in [a, b]$, it follows that $f(x) - f(\xi) \leq 0$ for all such x. Hence $Q(x) \geq 0$ for all $x \in [a, \xi)$ and $Q(x) \leq 0$ for all $x \in (\xi, b]$. Thus

$$\lim_{x \to \xi^-} Q(x) \geq 0$$

and

$$\lim_{x \to \xi^+} Q(x) \leq 0;$$

consequently, $\lim_{x \to \xi} Q(x) = 0$, as required.   □

Remark 1. The number ξ in Rolle's theorem need not be unique. For example, let $f(x) = x^3 - x$ for all $x \in [-1, 1]$. Then f satisfies the hypotheses of the theorem.

However, $f'(x) = 3x^2 - 1$ for all $x \in (-1, 1)$, so that

$$f'\left(\frac{1}{\sqrt{3}}\right) = f'\left(-\frac{1}{\sqrt{3}}\right) = 0.$$

Remark 2. The conditions in Rolle's theorem cannot be relaxed. For instance, for all $x \in [0, 1]$ let

$$f(x) = x,$$
$$g(x) = x - \lfloor x \rfloor,$$

and

$$h(x) = |2x - 1|.$$

The function f is continuous and differentiable everywhere, yet $f'(x) = 1 \neq 0$ for all x. Note that $f(0) \neq f(1)$. The function g satisfies $g(x) = x$ for all $x \in [0, 1)$ and $g(1) = 1 - 1 = 0 = g(0)$. It is differentiable on $(0, 1)$, but $g'(x) = 1 \neq 0$ for all $x \in (0, 1)$. Note that g is not continuous at 1. Finally, h is continuous everywhere and $h(0) = h(1) = 1$. Its derivative is $-2 \neq 0$ at all $x \in (0, 1/2)$ and $2 \neq 0$ at all $x \in (1/2, 1)$. It is not differentiable at $1/2$.

Remark 3. The converse of Rolle's theorem is not, in general, true. In fact, the conclusion of the theorem does not imply any of its hypotheses. Take, for example, the function f such that

$$f(x) = x^2 - \lfloor x^2 \rfloor$$

for each $x \in [-1, 3/2]$. Here we have $f(3/2) = 9/4 - 2 = 1/4 \neq 0 = f(-1)$. Moreover f is not continuous (and therefore not differentiable) at 1. Nevertheless, $f(x) = x^2$ for all $x \in (-1, 1)$, and so $f'(0) = 0$. Note that $0 \in (-1, 3/2)$.

By maneuvering the x-axis, we can generalize Rolle's theorem to one of the most fundamental theorems of real analysis—the mean-value theorem.

Theorem 6.3.2 (Mean-Value Theorem). *Let f be a function of a real variable and suppose that f is continuous on a closed interval $[a, b]$ and differentiable on (a, b). Then there is a number $\xi \in (a, b)$ such that*

$$f'(\xi) = \frac{f(b) - f(a)}{b - a}. \tag{6.2}$$

Discussion: Notice that the right-hand side of Eq. (6.2) is the slope of the chord of the graph of f joining the points $(a, f(a))$ and $(b, f(b))$. The theorem asserts that some tangent to the graph between these points is parallel to the chord. We may rotate and translate the chord, if necessary, until it is superimposed on the x-axis. The mean-value theorem then becomes identical to Rolle's theorem.

Proof. The equation of the chord joining $(a, f(a))$ and $(b, f(b))$ is

$$y - f(a) = \frac{f(b) - f(a)}{b - a}(x - a).$$

Thus

$$y = f(a) + \frac{f(b) - f(a)}{b - a}(x - a).$$

Define

$$g(x) = f(x) - y = f(x) - f(a) - \frac{f(b) - f(a)}{b - a}(x - a)$$

for each $x \in [a, b]$. Then g is continuous on $[a, b]$ and differentiable on (a, b). Moreover $g(a) = 0$ and

$$g(b) = f(b) - f(a) - (f(b) - f(a)) = 0.$$

Thus Rolle's theorem may be applied to find a number $\xi \in (a, b)$ such that $g'(\xi) = 0$. But

$$g'(\xi) = f'(\xi) - \frac{f(b) - f(a)}{b - a}.$$

The result follows. □

Example 6.3.1. If $f(x) = e^x$ for all x, then $f'(x) = e^x$ for all x. We can use this result and the mean-value theorem to show that

$$e^x \geq 1 + x$$

for all real x. Indeed, equality holds if $x = 0$. Suppose that $x > 0$. By the mean-value theorem, there exists $\xi \in (0, x)$ such that

$$e^x - e^0 = e^\xi(x - 0).$$

Since $\xi > 0$, we conclude that

$$e^x - 1 = e^\xi x > x,$$

and the result follows.

The case where $x < 0$ is handled in a similar way. △

We now establish some theorems that are plausible intuitively but difficult to prove without invoking the mean-value theorem.

Theorem 6.3.3. *Let $f : [a, b] \to \mathbb{R}$ be a function that is continuous on $[a, b]$ and differentiable on (a, b). If $f'(x) = 0$ for all $x \in (a, b)$, then f is a constant function.*

Proof. Suppose f is not a constant function. Then there exist c and d in $[a, b]$ such that $f(c) \neq f(d)$. Suppose without loss of generality that $c < d$. By the mean-value theorem, there exists $\xi \in (c, d)$ such that

$$\frac{f(d) - f(c)}{d - c} = f'(\xi) = 0.$$

Thus we reach the contradiction that $f(d) = f(c)$.   □

Corollary 6.3.4. *Let $f, g : [a, b] \to \mathbb{R}$ be functions that are continuous on $[a, b]$ and differentiable on (a, b). Suppose that $f'(x) = g'(x)$ for all $x \in (a, b)$. Then there is a constant c such that*

$$f(x) = g(x) + c$$

for all $x \in [a, b]$.

Proof. Apply Theorem 6.3.3 to the function $f - g$.   □

The next two theorems relate the sign of the derivative of a function to the monotonicity of the function.

Theorem 6.3.5. *Let $f : [a, b] \to \mathbb{R}$ be a function that is continuous on $[a, b]$ and differentiable on (a, b). If $f'(x) > 0$ for all $x \in (a, b)$, then f is increasing on $[a, b]$.*

Proof. Choose c and d in $[a, b]$ such that $c < d$. The function f satisfies the hypotheses of the mean-value theorem on $[c, d]$. Therefore there exists $\xi \in (c, d)$ such that

$$\frac{f(d) - f(c)}{d - c} = f'(\xi) > 0.$$

Consequently, $f(d) > f(c)$, since $d - c > 0$. We conclude that f is increasing on $[a, b]$.   □

Theorem 6.3.6. *Let $f : [a, b] \to \mathbb{R}$ be a function that is continuous on $[a, b]$ and differentiable on (a, b). If $f'(x) < 0$ for all $x \in (a, b)$, then f is decreasing on $[a, b]$.*

Proof. Apply Theorem 6.3.5 to the function $-f$.   □

Corollary 6.3.7. *If f' is continuous at some $c \in (a,b)$ and $f'(c) \neq 0$, then f is strictly monotonic in some neighborhood of c.*

Proof. Suppose first that $f'(c) > 0$. Since f' is continuous at c, Theorem 5.1.8 confirms the existence of $\delta > 0$ such that $f'(x) > 0$ for all $x \in N_\delta(c)$. By Theorem 6.3.5, f is increasing on $N_\delta(c)$.

The argument is similar if $f'(c) < 0$.   □

Example 6.3.2. Let $f(x) = x^a$ for all $x > 0$, where $a \in \mathbb{R}$. Then $f'(x) = ax^{a-1}$ for all $x > 0$. Thus f is increasing if $a > 0$ and decreasing if $a < 0$.   △

Example 6.3.3. Let $f(x) = a^x$ for all $x \in \mathbb{R}$, where $a > 0$. Then $f'(x) = a^x \log a$ for all x, so that f is increasing if $a > 1$ and decreasing if $0 < a < 1$.   △

Example 6.3.4. Let

$$f(x) = ax^2 + bx + c$$

for all real x, where a, b, c are real and $a > 0$. Then

$$f'(x) = 2ax + b$$

for all x, and so $f'(x) = 0$ if and only if $x = -b/(2a)$. Moreover $f'(x) < 0$ for all $x < -b/(2a)$ and $f'(x) > 0$ for all $x > -b/(2a)$, since $a > 0$. It follows from Theorems 6.3.6 and 6.3.5 that f is decreasing on $(-\infty, -b/(2a)]$ and increasing on $[-b/(2a), \infty)$. Therefore, by Theorem 5.3.7, the equation $f(x) = 0$ can have at most two real solutions, one in the former interval and one in the latter. In fact, since

$$f(x) = a\left(x^2 + \frac{bx}{a} + \frac{c}{a}\right)$$

$$= a\left(\left(x + \frac{b}{2a}\right)^2 - \frac{b^2}{4a^2} + \frac{c}{a}\right),$$

we have

$$\lim_{x \to -\infty} f(x) = \lim_{x \to \infty} f(x) = \infty,$$

and so there are two real solutions if $f(-b/(2a)) < 0$, none if $f(-b/(2a)) > 0$, and just the solution $x = -b/(2a)$ if $f(-b/(2a)) = 0$. Note that

$$f\left(-\frac{b}{2a}\right) = a \cdot \frac{b^2}{4a^2} - \frac{b^2}{2a} + c$$

$$= \frac{4ac - b^2}{4a}.$$

As we saw in Sect. 1.4, the number $b^2 - 4ac$ is called the discriminant of $f(x)$. Since $a > 0$, the equation has two distinct real solutions if the discriminant Δ is positive, just one if $\Delta = 0$, and none if $\Delta < 0$. We also observe that the equation has two distinct real solutions if and only if there exists a number ξ such that $f(\xi) < 0$.

If $a < 0$, similar arguments show that once again the equation has two distinct real solutions if $\Delta > 0$, just one if $\Delta = 0$, and none if $\Delta < 0$. In this case, however, two distinct real solutions exist if and only if there is a number ξ such that $f(\xi) > 0$.   △

Definition 6.3.1. Let f be a real-valued function. The value of f at a number $c \in \mathcal{D}_f$ is called a **local maximum** if there exists a neighborhood $N_\delta(c)$ such that $f(z) \leq f(c)$ for all $z \in N_\delta(c)$. A **local minimum** of f is defined analogously. A number c is an **extremal point** or **extremum** of f if $f(c)$ is a local maximum or local minimum, and $f(c)$ is then an **extremal value** of f.

Our next result gives a sufficient condition for the existence of an extremum.

Theorem 6.3.8 (First Derivative Test). *Let* $f : [a, b] \to \mathbb{R}$ *be a continuous function and let* $c \in (a, b)$. *Suppose there exists* $\delta > 0$ *such that* f *is differentiable at all* $x \in N_\delta^*(c)$.

1. *If* $f'(x) \geq 0$ *whenever* $c - \delta < x < c$ *and* $f'(x) \leq 0$ *whenever* $c < x < c + \delta$, *then* f *has a local maximum at* c.
2. *If* $f'(x) \leq 0$ *whenever* $c - \delta < x < c$ *and* $f'(x) \geq 0$ *whenever* $c < x < c + \delta$, *then* f *has a local minimum at* c.

Proof.

1. The mean-value theorem shows that for each $x \in (c - \delta, c)$ there exists $\xi \in (x, c)$ such that

$$f(c) - f(x) = f'(\xi)(c - x) \geq 0.$$

   Hence $f(c) \geq f(x)$. Likewise, $f(c) \geq f(x)$ for each $x \in (c, c + \delta)$. Therefore $f(c)$ is a local maximum.
2. The proof of part (2) is similar.   □

Example 6.3.5. This theorem shows that the function $|x|$, for all $x \in \mathbb{R}$, has a local minimum at 0. Note that this function is not differentiable at 0.   △

The following theorem is often of assistance in locating local maxima and minima.

Theorem 6.3.9. *Let* $f : [a, b] \to \mathbb{R}$ *and* $c \in (a, b)$. *If* $f(c)$ *is an extremal value of* f *and* $f'(c)$ *exists, then* $f'(c) = 0$.

Theorem 6.3.9 is an immediate consequence of the following lemma, which asserts that an extremal point of a function f is a "turning" point of the graph of f.

Lemma 6.3.10. *Let* $f : [a, b] \to \mathbb{R}$ *and* $c \in (a, b)$.

1. *If* $f'(c) > 0$, *then there exists* $\delta > 0$ *such that* $f(x) < f(c)$ *for all* $x \in \mathcal{D}_f$ *for which* $c - \delta < x < c$ *and* $f(x) > f(c)$ *for all* $x \in \mathcal{D}_f$ *for which* $c < x < c + \delta$.
2. *If* $f'(c) < 0$, *then there exists* $\delta > 0$ *such that* $f(x) > f(c)$ *for all* $x \in \mathcal{D}_f$ *for which* $c - \delta < x < c$ *and* $f(x) < f(c)$ *for all* $x \in \mathcal{D}_f$ *for which* $c < x < c + \delta$.

Proof. If $f'(c) > 0$, then by Theorem 4.4.6 there exists $\delta > 0$ such that

$$\frac{f(x) - f(c)}{x - c} > 0$$

for all $x \in N_\delta^*(c)$. Thus $f(x) > f(c)$ if $x > c$, but $f(x) < f(c)$ if $x < c$.

The case where $f'(c) < 0$ follows by considering the function $-f$. $\square$

A number c such that $f'(c) = 0$ is sometimes called a **critical point** of the function f.

Our next theorem shows that derivatives satisfy the conclusion of the intermediate-value theorem even though they may not be continuous. First, we establish a special case.

Lemma 6.3.11. *If* $f : [a, b] \to \mathbb{R}$ *is differentiable and* $f'(a) f'(b) < 0$, *then there exists* $\xi \in (a, b)$ *such that* $f'(\xi) = 0$.

Proof. We may assume that $f'(a) > 0$ [and therefore that $f'(b) < 0$] as the argument is similar if $f'(a) < 0$. The differentiable function f is continuous and therefore has a maximum value at some $\xi \in [a, b]$. By Lemma 6.3.10, there exists $\delta > 0$ such that $f(x) > f(a)$ for each x such that $a < x < a + \delta$. Therefore $\xi \neq a$. Similarly, $\xi \neq b$, so that $\xi \in (a, b)$. Finally, $f'(\xi) = 0$ by Theorem 6.3.9. $\square$

Theorem 6.3.12 (Darboux). *If* $f : [a, b] \to \mathbb{R}$ *is differentiable and* $f'(a) \neq f'(b)$, *then for each* v *between* $f'(a)$ *and* $f'(b)$ *there exists* $\xi \in (a, b)$ *such that* $f'(\xi) = v$.

Proof. Apply Lemma 6.3.11 to the function g such that

$$g(x) = f(x) - vx$$

for all $x \in [a, b]$. $\square$

Remark. If the derivative f' is continuous, then Darboux's theorem follows immediately from the intermediate-value theorem.

Corollary 6.3.13. *Let* f *be continuous on* $[a, b]$ *and differentiable on* (a, b), *and suppose that* $f'(x) \neq 0$ *for all* $x \in (a, b)$. *Then* f *is strictly monotonic on* $[a, b]$.

Proof. If there exist x and y such that $a < x < y < b$ and $f'(x) f'(y) < 0$, then we may apply Darboux's theorem to the interval $[x, y]$ to produce a number $\xi \in (x, y)$ for which $f'(\xi) = 0$. This finding contradicts the hypothesis. Therefore either $f'(x) > 0$ for all $x \in (a, b)$ or $f'(x) < 0$ for all $x \in (a, b)$. The conclusion is immediate in both cases, by Theorem 6.3.5 or Theorem 6.3.6, respectively. □

Remark. The condition hypothesized for the continuity of f in this corollary may be relaxed provided that a corresponding change is made to the conclusion. For example, suppose that the interval $[a, b]$ is replaced by $[a, b)$, and choose y such that $a < y < b$. Then f is continuous on $[a, y]$ and therefore strictly monotonic on that interval. As y is any number in (a, b), we conclude that f is strictly monotonic on $[a, b)$.

Exercises 6.3.

1.(a) Show that

$$x - \frac{x^2}{2} < \log(1 + x) < x - \frac{x^2}{2} + \frac{x^3}{3}$$

for all $x > 0$.

(b) Let

$$s_n = \frac{\left(1 + \frac{1}{n}\right)^{n^2}}{e^n}$$

for each $n > 0$. Show that

$$-\frac{1}{2} < \log s_n < -\frac{1}{2} + \frac{1}{3n}.$$

(c) Compute $\lim_{n \to \infty} s_n$.

2.(a) Show that

$$\frac{2x}{x + 2} < \log(1 + x)$$

for any $x > 0$. (Hint: Try using Theorem 6.3.5 or 6.3.6.)

(b) Show that

$$\lim_{n \to \infty} n(a^{1/n} - 1) = \log a.$$

3. Prove that

$$\log(1 + x) < x - \frac{x^2}{2(1 + x)}$$

if $x > 0$ and that the inequality is reversed if $-1 < x < 0$.

4. For all $x > 0$ let

$$f(x) = \left(1 + \frac{1}{x}\right)^x$$

and

$$g(x) = \left(1 - \frac{1}{x}\right)^{-x}.$$

Show that f is increasing on $(0, \infty)$ and g is decreasing on $(1, \infty)$. [Hint: Study the signs of the derivatives of $\log f(x)$ and $\log g(x)$.]

5. Use the mean-value theorem to show that the following functions are uniformly continuous on $[0, \infty)$:

   (a) $\cos kx$ for every real k.
   (b) $\log(1 + x)$.

6. Use the mean-value theorem to show that

$$\frac{1}{2\sqrt{n+1}} < \sqrt{n+1} - \sqrt{n} < \frac{1}{\sqrt{n}}$$

   for every $n \in \mathbb{N}$.

7. Show that the equation

$$x^4 + 4x + c = 0$$

   has at most two real roots for every real c and exactly two if $c < 0$.

8. Show that $x^5 + 7x - 2 = 0$ has exactly one real root.

9. Suppose that f is differentiable on $\mathbb{R}$ and has two real roots. Show that f' has at least one root.

10. Let $p(x)$ be a polynomial of degree $n \geq 2$. Suppose that the equation $p(x) = 0$ has n real roots (which may be repeated). Show that $p'(x) = 0$ has $n - 1$ real roots.

11. Let f be a function such that f' is continuous on $[a, b]$ and differentiable on (a, b), and suppose that

$$f(a) = f(b) = f'(a) = 0.$$

   Show that there exists $\alpha \in (a, b)$ such that $f''(\alpha) = 0$.

12. Suppose that f'' exists and is bounded on (a, b). Show by the mean-value theorem that f' is also bounded on (a, b) and hence that f is uniformly continuous on (a, b). [Note that the function $\sqrt{x}$ is uniformly continuous and differentiable on $(0, 1)$, but its derivative is not bounded on $(0, 1)$.]

13. Let f be a function that is continuous on $[0, 1]$ and differentiable on $(0, 1)$, and
    suppose that $f(0) = f(1) = 0$. Show that there exists $c \in (0, 1)$ for which
    $f'(c) = f(c)$. [Hint: Apply the mean-value theorem to $f(x)/e^x$.]
14. Let f be a differentiable function. Suppose that $|f'(x)| \leq M$ for all $x \in \mathcal{D}_f$.
    Show that f is Lipschitz continuous.
15. Show that $\sqrt{x^2 + 1}$ and $\sin x$ are Lipschitz continuous on $\mathbb{R}$.

6.4   Periodicity of Sine and Cosine

The sine and cosine functions may be used to define another important mathematical
constant. First, we derive the following proposition.

Proposition 6.4.1. *There exists $\phi \in (1, 2)$ such that $\cos \phi = 0$.*

Proof. Proposition 3.16.1 shows that

$$\cos x \geq 1 - \frac{x^2}{2}$$

for every real x. Hence

$$\cos 1 \geq 1 - \frac{1}{2} = \frac{1}{2} > 0.$$

Similarly,

$$\cos x \leq 1 - \frac{x^2}{2} + \frac{x^4}{24},$$

so that

$$\cos 2 \leq 1 - \frac{4}{2} + \frac{16}{24} = -\frac{1}{3} < 0.$$

An appeal to continuity and the intermediate-value theorem completes the proof. $\square$

For the sine function we have the following result.

Proposition 6.4.2. *If $x \in (0, 2)$, then $\sin x > 0$.*

Proof. The proposition follows from the facts that

$$\sin x = \sum_{j=0}^{\infty} (-1)^j \frac{x^{2j+1}}{(2j + 1)!}$$

$$= \sum_{j=0}^{\infty} \left(\frac{x^{4j+1}}{(4j+1)!} - \frac{x^{4j+3}}{(4j+3)!} \right)$$

$$= \sum_{j=0}^{\infty} \frac{x^{4j+1}}{(4j+1)!} \left(1 - \frac{x^2}{(4j+3)(4j+2)} \right)$$

for all real x and

$$(4j+3)(4j+2) \geq 6 > x^2$$

for all $x \in (0, 2)$. □

Thus

$$\cos' x = -\sin x < 0$$

for all $x \in (0, 2)$, so that cos is decreasing on the interval $[0, 2]$. We infer that the ϕ of Proposition 6.4.1 must be unique. Since $\sin \phi > 0$, it follows that

$$\sin \phi = \sqrt{1 - \cos^2 \phi} = 1.$$

Furthermore,

$$\sin 2\phi = 2 \sin \phi \cos \phi = 0.$$

We also have the following proposition.

Proposition 6.4.3. *If $x \in (0, 2\phi)$, then $\sin x > 0$.*

Proof. Since $\phi \in (1, 2)$, we already know that $\sin x > 0$ for all $x \in (0, \phi]$ by Proposition 6.4.2. It remains to consider the case where $\phi < x < 2\phi$. Let $y = x - \phi$; hence $0 < y < \phi$. Moreover

$$\sin x = \sin(y + \phi) = \sin y \cos \phi + \cos y \sin \phi = \cos y.$$

We therefore need to show that $\cos y > 0$. But this result is immediate from the facts that $\cos \phi = 0$, $0 < y < \phi$, and cos is decreasing on $[0, \phi]$. □

We now define $\pi = 2\phi$. This is the new mathematical constant whose introduction was foreshadowed earlier. Our results therefore show that $2 < \pi < 4$, $\cos(\pi/2) = 0$, $\sin(\pi/2) = 1$, and $\sin \pi = 0$. In fact, we see from Proposition 6.4.3 that π is the smallest positive number whose sine is 0. Moreover it follows from Proposition 6.4.3 and the formula $\cos' x = -\sin x$ that cos is decreasing on $[0, \pi]$. In addition,

$$\sin\left(x + \frac{\pi}{2}\right) = \sin x \cos\frac{\pi}{2} + \cos x \sin\frac{\pi}{2} = \cos x \qquad (6.3)$$

and

$$\cos\left(x + \frac{\pi}{2}\right) = \cos x \cos\frac{\pi}{2} - \sin x \sin\frac{\pi}{2} = -\sin x \qquad (6.4)$$

for all x. Therefore

$$\sin(x + \pi) = \sin\left(x + \frac{\pi}{2} + \frac{\pi}{2}\right) = \cos\left(x + \frac{\pi}{2}\right) = -\sin x.$$

It follows that $\sin x < 0$ for all $x \in (\pi, 2\pi)$ and hence that $\cos$ is increasing on $[\pi, 2\pi]$. Furthermore,

$$\cos(x + \pi) = -\sin\left(x + \frac{\pi}{2}\right) = -\cos x$$

and

$$\sin(x + 2\pi) = -\sin(x + \pi) = \sin x$$

for all x. Similarly,

$$\cos(x + 2\pi) = \cos x.$$

Since $\cos 0 = 1$ and $\cos\pi = -\cos 0 = -1$, an appeal to the continuity of the cosine and the intermediate-value theorem shows that the cosine, restricted to the interval $[0, \pi]$, is a bijection between that interval and the interval $[-1, 1]$. Note also that $\cos(3\pi/2) = -\cos(\pi/2) = 0$. Since the cosine is decreasing on $[0, \pi]$ but increasing on $[\pi, 2\pi]$, we see that $\cos x < 0$ if $\pi/2 < x < 3\pi/2$. Therefore $\cos x > 0$ if $-\pi/2 < x < \pi/2$. We deduce that the sine function is increasing on $[-\pi/2, \pi/2]$ and decreasing on $[\pi/2, 3\pi/2]$. Since $\sin(\pi/2) = 1$ and $\sin(-\pi/2) = -\sin(\pi/2) = -1$, it follows that the sine, restricted to the interval $[-\pi/2, \pi/2]$, is a bijection between that interval and the interval $[-1, 1]$.

A function f defined for all x and not constant is said to be **periodic** if there exists $\theta > 0$ such that

$$f(x + \theta) = f(x)$$

for all x. The smallest such θ is called the **period** of f. We have now shown that the sine and cosine functions are both periodic with period 2π.

We have observed that $\sin$ is continuous everywhere, and it is also increasing on $[-\pi/2, \pi/2]$. Consequently, if its domain were restricted to that interval, then the resulting function would have an inverse. This inverse is called the **inverse sine** function and is denoted by arcsin. Its domain is $[-1, 1]$, since $\sin(-\pi/2) = -1$ and

$\sin(\pi/2) = 1$. As $\cos x \neq 0$ for all $x \in (-\pi/2, \pi/2)$, we see from the inverse function theorem (Theorem 6.2.4) that arcsin is differentiable on $(-1, 1)$. In fact, if $y \in (-1, 1)$, then $y = \sin x$ for some $x \in (-\pi/2, \pi/2)$, and

$$\arcsin'(y) = \frac{1}{\sin' x} = \frac{1}{\cos x} = \frac{1}{\sqrt{1 - y^2}}$$

since $\cos x > 0$.

A similar argument shows that if the domain of cos were restricted to the interval $[0, \pi]$, then the resulting function would have an inverse. This inverse is called the **inverse cosine** function and denoted by arccos. It is also differentiable on $(-1, 1)$. If $y \in (-1, 1)$, then $y = \cos x$ for some $x \in (0, \pi)$, and

$$\arccos'(y) = -\frac{1}{\sin x} = -\frac{1}{\sqrt{1 - y^2}}.$$

Since $\tan x = \sin x / \cos x$ for all x for which $\cos x \neq 0$, we have

$$\tan' x = \frac{\cos^2 x + \sin^2 x}{\cos^2 x} = 1 + \tan^2 x > 0$$

for all such x. We infer that tan is continuous and increasing on each interval over which it is defined. An example of such an interval is $(-\pi/2, \pi/2)$. Since

$$\lim_{x \to \frac{\pi}{2}^-} \tan x = \lim_{x \to \frac{\pi}{2}^-} \frac{\sin x}{\cos x} = \infty$$

and, similarly,

$$\lim_{x \to -\frac{\pi}{2}^+} \tan x = -\infty,$$

it follows that for all y there exists a unique $x \in (-\pi/2, \pi/2)$ for which $\tan x = y$. We write $x = \arctan(y)$ and refer to x as the **inverse tangent** of y. For instance, $\arctan(0) = 0$. This function is differentiable, by the inverse function theorem. Since

$$\tan \arctan(y) = y,$$

differentiation yields

$$1 = (1 + \tan^2 \arctan(y)) \arctan'(y) = (1 + y^2) \arctan'(y),$$

so that

$$\arctan'(y) = \frac{1}{1 + y^2}.$$

The parentheses around the arguments of the inverse sine, inverse cosine, and inverse tangent functions are usually omitted.

Let $\theta = \arcsin(-x)$, where $x \in [-1, 1]$. Then $\sin \theta = -x$ and $-\pi/2 \le \theta \le \pi/2$. Hence $-\pi/2 \le -\theta \le \pi/2$ and $\sin(-\theta) = -\sin \theta = x$, so that $-\theta = \arcsin x$. We deduce that

$$\arcsin(-x) = -\arcsin x.$$

On the other hand, let $\theta = \arccos(-x)$, where $x \in [-1, 1]$. Then $\cos \theta = -x$ and $0 \le \theta \le \pi$. Hence $0 \le \pi - \theta \le \pi$ and

$$\cos(\pi - \theta) = -\cos(-\theta) = -\cos \theta = x,$$

so that $\pi - \theta = \arccos x$. Therefore

$$\arccos(-x) = \pi - \arccos x.$$

Example 6.4.1. We now show that

$$\lim_{x \to 0} \cos \frac{1}{x}$$

does not exist. For each positive integer n take

$$s_n = \frac{1}{2n\pi}$$

and

$$t_n = \frac{2}{(2n + 1)\pi}.$$

Then

$$\lim_{n \to \infty} s_n = \lim_{n \to \infty} t_n = 0.$$

However,

$$\cos \frac{1}{s_n} = \cos 2n\pi = 1$$

and

$$\cos \frac{1}{t_n} = \cos \left(n\pi + \frac{\pi}{2} \right) = 0 \neq 1.$$

Therefore the limit in question does not exist. In fact, this argument shows that

$$\lim_{x \to 0^+} \cos \frac{1}{x}$$

does not exist. Neither does $\lim_{x \to 0^-} \cos(1/x)$, by a similar argument. $\triangle$

In the next example, we use the sine and cosine functions to show that the derivative of a differentiable function might not be continuous.

Example 6.4.2. Let f be the real function defined by

$$f(x) = \begin{cases} x^2 \sin \frac{1}{x} & \text{if } x \neq 0, \\ 0 & \text{if } x = 0. \end{cases}$$

By the product and chain rules, f is differentiable at all $x \neq 0$. Moreover for all $x \neq 0$ we have

$$0 \leq \left| \frac{f(x) - f(0)}{x - 0} \right| = \left| \frac{f(x)}{x} \right| = \left| x \sin \frac{1}{x} \right| \leq |x|.$$

It therefore follows from the sandwich theorem that

$$f'(0) = \lim_{x \to 0} \frac{f(x) - f(0)}{x - 0} = 0.$$

Hence f is differentiable at all x.

On the other hand, for all $x \neq 0$ we have

$$f'(x) = 2x \sin \frac{1}{x} + x^2 \left(-\frac{1}{x^2} \right) \cos \frac{1}{x}$$

$$= 2x \sin \frac{1}{x} - \cos \frac{1}{x}.$$

As

$$\lim_{x \to 0} 2x \sin \frac{1}{x} = 0$$

by Example 4.4.1 but $\lim_{x \to 0} \cos(1/x)$ does not exist, we conclude that $\lim_{x \to 0} f'(x)$ does not exist. Therefore f' is not continuous at 0. $\triangle$

Our next example shows that a function may have a local extremum at some point c, yet the values of its derivative might not have equal sign throughout the left or right half of any neighborhood of c.

Example 6.4.3. Let

$$f(x) = \begin{cases} 2x^4 + x^4 \sin \frac{1}{x} & \text{if } x \neq 0, \\ 0 & \text{if } x = 0. \end{cases}$$

Since

$$2x^4 + x^4 \sin \frac{1}{x} = x^4 \left(2 + \sin \frac{1}{x} \right) \geq x^4 > 0$$

for all $x \neq 0$, we find that $f(0) = 0$ is a local minimum.

Now f is differentiable at all $x \neq 0$. In fact,

$$f'(x) = 8x^3 + 4x^3 \sin \frac{1}{x} - x^2 \cos \frac{1}{x}$$

for all $x \neq 0$. Thus for all integers $n \geq 2$ we have

$$f'\left(\frac{1}{2n\pi} \right) = \frac{1}{n^3\pi^3} \cdot \frac{1}{4n^2\pi^2} = \frac{4 - n\pi}{4n^3\pi^3} < 0,$$

whereas

$$f'\left(\frac{1}{2n\pi + \frac{\pi}{2}} \right) = \frac{12}{\left(2n\pi + \frac{\pi}{2} \right)^3} > 0.$$

Since

$$\frac{1}{2n\pi} \to 0$$

and

$$\frac{1}{2n\pi + \frac{\pi}{2}} \to 0$$

as $n \to 0$, we find that each neighborhood $N_\delta(0)$ of 0 contains positive numbers a and b such that $f'(a)f'(b) < 0$. We can also find negative numbers in $N_\delta(0)$ with the same property, because

$$f'\left(-\frac{1}{2n\pi} \right) < 0$$

and

$$f'\left(-\frac{1}{(2n+1)\pi}\right) = -\frac{8}{(2n+1)^3\pi^3} + \frac{1}{(2n+1)^2\pi^2}$$

$$= \frac{-8+(2n+1)\pi}{(2n+1)^3\pi^3}$$

$$> 0$$

for all integers $n > 1$.   △

We now give an example showing that $f'(c)$ may be nonzero, but f is neither increasing nor decreasing on any interval containing c.

Example 6.4.4. If

$$f(x) = \begin{cases} x^2 \sin\frac{1}{x} + \frac{x}{2} & \text{for } x \neq 0, \\ 0 & \text{for } x = 0, \end{cases}$$

then

$$f'(x) = 2x\sin\frac{1}{x} - \cos\frac{1}{x} + \frac{1}{2}$$

for all $x \neq 0$. In order to compute $f'(0)$, for each $x \neq 0$ define

$$Q(x) = \frac{f(x) - f(0)}{x - 0}$$

$$= \frac{f(x)}{x}$$

$$= x\sin\frac{1}{x} + \frac{1}{2}.$$

Thus

$$f'(0) = \lim_{x \to 0} Q(x) = \frac{1}{2}.$$

But for each $n \in \mathbb{Z} - \{0\}$ we have

$$f'\left(\frac{1}{2n\pi}\right) = -\frac{1}{2}$$

and

$$f'\left(\frac{1}{(2n+1)\pi}\right) = \frac{3}{2}.$$

Thus f is neither increasing nor decreasing on any interval containing 0.   △

The mean-value theorem may be used to sharpen inequality (3.27).

Example 6.4.5. We shall show that $\sin x < x$ for all $x > 0$. This inequality is certainly true if $x > 1$, and so we may suppose that $0 < x \leq 1$. As the sine function is differentiable everywhere, we may apply the mean-value theorem to it on the interval $[0, x]$ to find $\xi \in (0, x)$ such that

$$\sin x - \sin 0 = (x - 0)\cos \xi.$$

Therefore

$$\sin x = x \cos \xi < x,$$

for $\cos \xi < 1$ since $0 < \xi < x \leq 1 < 2\pi$.   △

The next example presents a result known as Jordan's inequality.

Example 6.4.6. For all $x \in [-\pi/2, \pi/2] - \{0\}$ we shall show that

$$\frac{2}{\pi} \leq \frac{\sin x}{x}.$$

Define

$$f(x) = \pi \sin x - 2x$$

for all such x. Then

$$f'(x) = \pi \cos x - 2$$

and

$$f''(x) = -\pi \sin x.$$

The last equation shows that f' is decreasing on $[0, \pi/2]$. Since $f'(0) = \pi - 2 > 0$ and $f'(\pi/2) = -2 < 0$, it follows from the intermediate-value theorem that there is a unique $\xi \in (0, \pi/2)$ such that $f'(\xi) = 0$. Thus $f'(x) > 0$ for all $x \in [0, \xi)$ and $f'(x) < 0$ for all $x \in (\xi, \pi/2]$, so that f is increasing on $[0, \xi]$ and decreasing on $[\xi, \pi/2]$. As $f(0) = f(\pi/2) = 0$, we conclude that $f(x) > 0$ for all $x \in (0, \pi/2)$. Thus

$$\pi \sin x > 2x$$

for all such x and the desired inequality follows in this case upon division by $\pi x > 0$.

If $x \in (-\pi/2, 0)$, then $-x \in (0, \pi/2)$. The previous result therefore shows that

$$\pi \sin(-x) > -2x,$$

whence $\pi \sin x < 2x$ and the desired result again holds.

Clearly, equality holds if $x = \pm\pi/2$.

Jordan's inequality spawns several other results. For instance, let $0 < x < \pi/2$. Then

$$0 < \frac{\pi}{2} - x < \frac{\pi}{2}.$$

Moreover

$$\sin\left(\frac{\pi}{2} - x\right) = -\sin\left(x - \frac{\pi}{2}\right) = \cos x$$

by Eq. (6.4). Jordan's inequality therefore shows that

$$\frac{2}{\pi} \leq \frac{\sin\left(\frac{\pi}{2} - x\right)}{\frac{\pi}{2} - x} = \frac{\cos x}{\frac{\pi}{2} - x}.$$

Thus

$$\cos x \geq \frac{2}{\pi}\left(\frac{\pi}{2} - x\right) = 1 - \frac{2x}{\pi}$$

for all $x \in (0, \pi/2)$. This inequality is due to Kober. Since

$$1 - \cos x \leq \frac{2x}{\pi},$$

Kober's inequality may be rewritten as

$$\frac{1 - \cos x}{x} \leq \frac{2}{\pi}.$$

Note that equality holds for $x = \pi/2$.

Further information about these inequalities may be found in [18].   △

As an illustration of the computation of a sine or a cosine of a real number, we offer the following example.

Example 6.4.7.

$$\sin \frac{\pi}{4} = \frac{1}{\sqrt{2}}.$$

Proof. First, observe that

$$1 = \sin \frac{\pi}{2} = 2 \sin \frac{\pi}{4} \cos \frac{\pi}{4}.$$

Since $0 < \pi/4 < \pi/2$, we also know that $\cos(\pi/4) > 0$. Hence

$$\frac{1}{2} = \sin \frac{\pi}{4} \cos \frac{\pi}{4} = \left(\sin \frac{\pi}{4} \right) \sqrt{1 - \sin^2 \frac{\pi}{4}},$$

so that

$$\frac{1}{4} = \left(\sin^2 \frac{\pi}{4} \right) \left(1 - \sin^2 \frac{\pi}{4} \right) = \sin^2 \frac{\pi}{4} - \sin^4 \frac{\pi}{4}.$$

Thus

$$0 = 4 \sin^4 \frac{\pi}{4} - 4 \sin^2 \frac{\pi}{4} + 1 = \left(2 \sin^2 \frac{\pi}{4} - 1 \right)^2,$$

whence

$$\sin^2 \frac{\pi}{4} = \frac{1}{2}$$

and the result follows.

$\triangle$

Hence

$$\cos \frac{\pi}{4} = \sqrt{1 - \frac{1}{2}} = \frac{1}{\sqrt{2}}.$$

Note that $\cos x = 0$ if and only if $x = (2k + 1)\pi/2$ for some integer k. For all other values of x we have

$$\tan x = \frac{\sin x}{\cos x}.$$

For instance,

$$\tan 0 = 0$$

and

$$\tan \frac{\pi}{4} = \frac{1}{\sqrt{2}} \cdot \sqrt{2} = 1.$$

Theorem 6.4.4. *If $\cos x = \cos y$ and $\sin x = \sin y$, then $x = y + 2k\pi$ for some integer k.*

Proof. First we reduce the problem to values of x and y in $[0, 2\pi)$. There exists an integer l such that

$$2l\pi \leq x < 2(l+1)\pi = 2l\pi + 2\pi.$$

Let $x_1 = x - 2l\pi$; hence $0 \leq x_1 < 2\pi$. Similarly, let m be the integer such that

$$2m\pi \leq y < 2(m+1)\pi$$

and let $y_1 = y - 2m\pi$, so that $0 \leq y_1 < 2\pi$. Note also that $\cos x_1 = \cos y_1$ and $\sin x_1 = \sin y_1$.

   Suppose that $\sin x_1 = \sin y_1 = 0$. Then $\{x_1, y_1\} \subseteq \{0, \pi\}$. Since $\cos 0 \neq \cos \pi$, it follows that $x_1 = y_1$.
   Suppose $\sin x_1 > 0$. Then x_1 and y_1 are both in $(0, \pi)$. Since $\cos$ is decreasing on that interval and $\cos x_1 = \cos y_1$, we deduce that $x_1 = y_1$.
   Suppose $\sin x_1 < 0$. Now $\{x_1, y_1\} \subset (\pi, 2\pi)$. Since $\cos$ is increasing on $(\pi, 2\pi)$ and $\cos x_1 = \cos y_1$, we again have $x_1 = y_1$.
   Thus $x_1 = y_1$ in every case. Hence

$$x - 2l\pi = y - 2m\pi,$$

and so

$$x = y + 2l\pi - 2m\pi = y + 2(l - m)\pi,$$

as required. $\qquad\qquad\square$

   We now give a geometric interpretation of the sine and cosine functions for real numbers. The idea is to establish a bijection between the interval $[0, 2\pi)$ and the unit circle

$$x^2 + y^2 = 1,$$

the circle that is centered at the origin and has radius 1. Given a point (x, y) on the circle, we observe that $|x| \leq 1$. Therefore there is a unique number $\alpha \in [0, \pi]$ for which $\cos \alpha = x$. Moreover

$$y^2 = 1 - x^2 = 1 - \cos^2 \alpha = \sin^2 \alpha,$$

Fig. 6.2 $\sin \varphi = y$ and
$\cos \varphi = x$

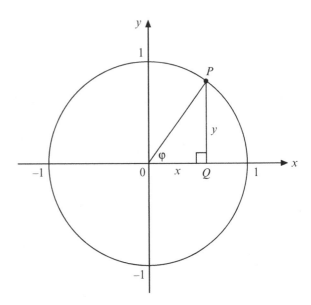

so that $y = \pm \sin \alpha$. We define $f(x, y) = \alpha$ if $y = \sin \alpha$. Otherwise $y = -\sin \alpha \neq 0$. In this case $\alpha \neq 0$, $x = \cos(-\alpha) = \cos(2\pi - \alpha)$, and $y = \sin(-\alpha) = \sin(2\pi - \alpha)$, and we define $f(x, y) = 2\pi - \alpha$. In each case we find that $f(x, y)$ is the unique (see Theorem 6.4.4) number in $[0, 2\pi)$ with cosine x and sine y. The function f is a surjection onto $[0, 2\pi)$: Given φ such that $0 \leq \varphi < 2\pi$, if we put $x = \cos \varphi$ and $y = \sin \varphi$, then $f(x, y) = \varphi$ and $x^2 + y^2 = 1$. We see that it is also injective, for if $f(u, v) = f(x, y) = \varphi$, then $u = \cos \varphi = x$ and similarly $v = y$.

The bijection f manifests itself geometrically as the measurement of an angle by means of a number in the interval $[0, 2\pi)$. Thus for every point $P = (x, y)$ on the circle $x^2 + y^2 = 1$ there is a unique angle $\varphi \in [0, 2\pi)$, measured from the positive x-axis in the counterclockwise sense to the line joining the origin O to P, such that $x = \cos \varphi$ and $y = \sin \varphi$. If $0 < \varphi < \pi/2$, then $x > 0$ and $y > 0$. In this case let Q be the foot of the perpendicular from P to the positive x-axis. Then $\sin \varphi$ and $\cos \varphi$ are the lengths of PQ and OQ, respectively (see Fig. 6.2). If $\varphi = \pi/2$, then P is on the positive y-axis and φ is a right angle.

If we now multiply all coordinates by some factor $r > 0$, then the unit circle is replaced by a circle C of radius r but still centered at the origin. Its equation is

$$x^2 + y^2 = r^2.$$

If $0 < \varphi < \pi/2$, then the triangle OPQ is replaced by a right triangle with hypotenuse of length r extending from the origin to a point (x, y) on C. The remaining vertex of the triangle is the point $(x, 0)$ on the positive x-axis. The angle from the positive x-axis to the hypotenuse, measured in the counterclockwise sense, is still $\varphi > 0$. The length of the side coincident with the x-axis is $x = r \cos \varphi$, and

Fig. 6.3 $\sin \varphi = s/r$ and $\cos \varphi = t/r$

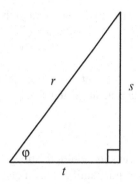

that of the side parallel to the y-axis is $y = r \sin \varphi$. We therefore perceive that, for an angle $\varphi < \pi/2$ in a given right triangle, $\sin \varphi$ is the ratio of the length s of the side opposite φ to the length r of the hypotenuse. Similarly, $\cos \varphi = t/r$, where t is the length of the side adjacent to the right angle and φ (see Fig. 6.3). In fact, the triangle may be positioned anywhere in the plane and oriented in any manner, so that the angle φ may be measured, geometrically, from any line. However, since $\varphi > 0$, we use the convention that the angle is measured in the counterclockwise sense. Negative numbers may similarly be perceived as angles measured in the clockwise sense. Thus the numbers φ and $-\varphi$ measure the same angle but in opposite senses. The numbers φ and $\varphi + 2k\pi$, for each integer k, also measure the same angle.

For each $\varphi \in [0, 2\pi)$ we define $\theta = \varphi - 2\pi$ if $\pi < \varphi < 2\pi$ and $\theta = \varphi$ otherwise. Then $\theta \in (-\pi, \pi]$, $\cos \theta = x/r$, and $\sin \theta = y/r$. Moreover $\theta \in (-\pi/2, \pi/2)$ if and only if $x > 0$, whereas $\theta \in (0, \pi)$ if and only if $y > 0$. If $x = 0$, then $\theta = \pi/2$ if $y > 0$ and $\theta = -\pi/2$ if $y < 0$. On the other hand, for $y = 0$ we have $\theta = 0$ if $x > 0$ and $\theta = \pi$ if $x < 0$.

We now give a geometric interpretation of the multiplication of complex numbers. Suppose that $z = x + iy$, where x and y are real and $|z| = 1$. Geometrically, z is therefore a point on the unit circle centered at the origin. There exists a unique number $\theta \in (-\pi, \pi]$, which we will call the argument of z, such that $x = \cos \theta$ and $y = \sin \theta$. Thus

$$z = \cos \theta + i \sin \theta = e^{i\theta}.$$

For example,

$$e^{i\pi} = \cos \pi + i \sin \pi = -1.$$

Note also that

$$z = e^{i(\theta + 2k\pi)}$$

for every integer k, and that

$$\bar{z} = \cos\theta - i\sin\theta = \cos(-\theta) + i\sin(-\theta) = e^{-i\theta}.$$

Now let us multiply two complex numbers w and z with modulus 1 and respective arguments α and β:

$$wz = e^{i\alpha}e^{i\beta} = e^{i(\alpha+\beta)}.$$

Thus the multiplication of two complex numbers on the unit circle centered at the origin manifests itself geometrically as a rotation about the origin. We easily obtain a corresponding result for division:

$$\frac{w}{z} = \frac{e^{i\alpha}}{e^{i\beta}} = e^{i(\alpha-\beta)}.$$

Furthermore, since

$$(e^{i\alpha})^n = e^{in\alpha}$$

for every integer n, we obtain the following theorem, which is due to de Moivre.

Theorem 6.4.5 (de Moivre). *For each integer n and each real α,*

$$(\cos\alpha + i\sin\alpha)^n = \cos n\alpha + i\sin n\alpha.$$

More generally, for every $z \neq 0$ we have

$$\left|\frac{z}{|z|}\right| = \frac{|z|}{|z|} = 1,$$

and so there is a unique number $\theta \in (-\pi, \pi]$ for which

$$\frac{z}{|z|} = e^{i\theta}.$$

Hence

$$z = re^{i\theta},$$

where $r = |z|$. This expression is called the **polar** form of z. The number θ is the **argument** of z and is written as $\arg z$.

There are various formulas for $\arg z$. Let $z = (x, y) \neq 0$, where x and y are real. If $\theta \in [0, \pi/2]$, then $\sin\theta = y/r$, so that

$$\arg z = \theta = \arcsin\frac{y}{r}.$$

in this case. If $\theta \in (\pi/2, \pi]$, then $0 \leq \pi - \theta < \pi/2$ and

$$\sin(\pi - \theta) = -\sin(-\theta) = \sin \theta = \frac{y}{r}.$$

Hence

$$\pi - \theta = \arcsin \frac{y}{r},$$

so that

$$\arg z = \theta = \pi - \arcsin \frac{y}{r}.$$

If $\theta \in [-\pi/2, 0)$, then $0 < -\theta \leq \pi/2$ and $\sin(-\theta) = -y/r$, whence

$$-\theta = \arcsin \left(-\frac{y}{r}\right) = -\arcsin \frac{y}{r},$$

so that

$$\arg z = \arcsin \frac{y}{r}.$$

If $\theta \in (-\pi, -\pi/2)$, then $0 < \theta + \pi < \pi/2$ and $\sin(\theta + \pi) = -y/r$, so that

$$\arg z = \arcsin \left(-\frac{y}{r}\right) - \pi = -\pi - \arcsin \frac{y}{r}.$$

Note that if z is real, then $\arg z = 0$ if $z > 0$ and $\arg z = \pi$ if $z < 0$. We have already observed that

$$z = |z|e^{i\theta} = |z|(\cos \theta + i \sin \theta)$$

for all $z \neq 0$, where $\theta = \arg z$. If we also have

$$z = |z|(\cos \gamma + i \sin \gamma)$$

for some γ, then

$$|z|(\cos \theta + i \sin \theta) = |z|(\cos \gamma + i \sin \gamma).$$

Hence $\cos \theta = \cos \gamma$ and $\sin \theta = \sin \gamma$, and so

$$\gamma = \theta + 2k\pi = \arg z + 2k\pi$$

for some integer k.

If w is another nonzero complex number, then

$$wz = |wz|e^{i \arg wz} = |wz|(\cos \alpha + i \sin \alpha),$$

where $\alpha = \arg wz$. If $\arg w = \phi$, then we can also write

$$wz = |w|e^{i\phi}|z|e^{i\theta}$$
$$= |wz|e^{i(\phi+\theta)}$$
$$= |wz|(\cos(\phi + \theta) + i \sin(\phi + \theta)).$$

Hence

$$\cos \alpha = \cos(\phi + \theta)$$

and

$$\sin \alpha = \sin(\phi + \theta).$$

We conclude that $\alpha = \phi + \theta + 2k\pi$ for some integer k. In other words,

$$\arg wz = \arg w + \arg z + 2k\pi.$$

A similar argument shows that

$$\arg \frac{w}{z} = \arg w - \arg z + 2k\pi$$

for some integer k.

There are many expressions for π. Several can be obtained by using the argument of a complex number. For instance, starting with the equation

$$(2 + i)(3 + i) = (2, 1)(3, 1) = (5, 5) = 5(1, 1), \tag{6.5}$$

observe that the components of the complex numbers $(1, 1), (2, 1), (3, 1)$ are all positive. Therefore the arguments of these numbers are all in the interval $(0, \pi/2)$. In fact, the argument of the number on the right-hand side is equal to $\arg(1, 1)$, which is $\arcsin(1/\sqrt{2}) = \pi/4$. Letting

$$\theta = \arg(2, 1) = \arcsin \frac{1}{\sqrt{5}},$$

we find that $\sin \theta = 1/\sqrt{5}$, so that

$$\cos \theta = \sqrt{1 - \frac{1}{5}} = \frac{2}{\sqrt{5}}$$

and therefore $\tan \theta = 1/2$. Thus $\arg(2, 1) = \arctan 1/2$. Similarly, since

$$\arg(3, 1) = \arcsin \frac{1}{\sqrt{10}}$$

we have $\cos \arg(3, 1) = 3/\sqrt{10}$, and we deduce that $\arg(3, 1) = \arctan 1/3$. As

$$\arg(2, 1) + \arg(3, 1) < \pi,$$

it follows that

$$\arg((2, 1)(3, 1)) = \arctan \frac{1}{2} + \arctan \frac{1}{3}.$$

Hence

$$\arctan \frac{1}{2} + \arctan \frac{1}{3} = \frac{\pi}{4},$$

so that

$$\pi = 4 \left(\arctan \frac{1}{2} + \arctan \frac{1}{3} \right).$$

The power series expansion for $\arctan x$, which we will derive later, therefore provides a means of estimating π. More efficient formulas with which to begin can be found in the exercises. The approximate value of π is 3.14159.

In contrast to the injective nature of the exponential function for real variables, for complex arguments we have the following theorem.

Theorem 6.4.6. *For complex w and z we have $e^w = e^z$ if and only if $z = w + 2k\pi i$ for some integer k.*

Proof. If $z = w + 2k\pi i$, then, since

$$e^{2k\pi i} = \cos 2k\pi + i \sin 2k\pi = 1,$$

we have

$$e^z = e^{w+2k\pi i} = e^w e^{2k\pi i} = e^w.$$

To prove the converse, suppose first that $e^z = e^0 = 1$, where $z = x + iy$ for some real numbers x and y. Then

$$1 = e^z = e^{x+iy} = e^x e^{iy} = e^x (\cos y + i \sin y),$$

so that $e^x \cos y = 1$ and $e^x \sin y = 0$. Since $e^x \neq 0$, it follows that $\sin y = 0$, and so $y = n\pi$ for some integer n. Hence

$$1 = e^x \cos n\pi = (-1)^n e^x,$$

and since $e^x > 0$, it follows that n is even and $e^x = 1$. The former condition shows that $n = 2k$ for some integer k, and from the latter we have $x = 0$, so that

$$z = iy = n\pi i = 2k\pi i,$$

as required.

The general case now follows easily, for if $e^w = e^z$, then $e^{z-w} = 1$, so that

$$z - w = 2k\pi i$$

for some integer k. Hence $z = w + 2k\pi i$. $\square$

For all integers n we have $\cos 2n\pi = 1$ and $\sin n\pi = 0$. Thus $\{\cos nx\}$ is convergent if $x = 2k\pi$ for some $k \in \mathbb{Z}$, and $\{\sin nx\}$ is convergent if $x = k\pi$. We now show that these are the only cases where the sequences are convergent and x is real.

Theorem 6.4.7. *1. The sequence $\{\cos nx\}$ is convergent if and only if $x = 2k\pi$ for some $k \in \mathbb{Z}$.*
2. The sequence $\{\sin nx\}$ is convergent if and only if $x = k\pi$ for some $k \in \mathbb{Z}$.

Proof. Suppose that $x \neq 2k\pi$ for every integer k. Clearly, $\{\cos nx\}$ is divergent if $x = (2k+1)\pi$, for in that case we have $\cos nx = 1$ if n is even and $\cos nx = -1$ if n is odd. We therefore assume further that $x \neq k\pi$ for every $k \in \mathbb{Z}$.

Now

$$\cos(n+1)x - \cos nx = \cos nx \cos x - \sin nx \sin x - \cos nx$$

for all n. If $\{\cos nx\}$ were convergent, then the left-hand side would converge to 0. Therefore $\{\sin nx\}$ would converge as well ($\sin x \neq 0$ since $x \neq k\pi$ for each integer k), and therefore so would $\{e^{nxi}\}$ since

$$e^{nxi} = \cos nx + i \sin nx.$$

We now find a contradiction by showing that the difference between consecutive terms of this sequence does not approach 0 as $n \to \infty$. First,

$$|e^{(n+1)xi} - e^{nxi}| = |e^{nxi}(e^{xi} - 1)| = |e^{xi} - 1|.$$

But

$$e^{xi} = \cos x + i \sin x \neq 1,$$

for $\cos x \neq 1$ since $x \neq 2k\pi$ for each integer k. Therefore $|e^{xi} - 1|$ is a positive constant.

By Cauchy's principle, we have now reached the contradiction that the sequence $\{e^{nxi}\}$ does not converge. Part (1) of the theorem follows. The proof of part (2) is similar, using the identity

$$\sin(n+1)x - \sin nx = \sin nx \cos x + \cos nx \sin x - \sin nx$$

for all n. $\square$

We conclude this section with the observation that the mean-value theorem does not hold in general for functions of a complex variable. For example, let $f(z) = e^{iz}$ for all $z \in \mathbb{C}$. If $a = 0$ and $b = 2\pi$, then

$$\frac{f(b) - f(a)}{b - a} = \frac{e^{2\pi i} - e^0}{2\pi} = \frac{1 - 1}{2\pi} = 0,$$

but $f'(z) = ie^{iz} \neq 0$ for every z.

Exercises 6.4.

1. Use the equation

$$\cos x = \frac{e^{ix} + e^{-ix}}{2},$$

    for all x, to express $\cos 3x$ in terms of $\cos x$.

2. Use the result of Example 6.4.5 to show that

$$\cos x > 1 - \frac{x^2}{2}$$

    for all $x \neq 0$.

3. Show that

$$\frac{1}{\pi - x} \leq \frac{\tan \frac{x}{2}}{x}$$

    for all x such that $0 < x \leq \pi/2$.

4. Prove that the following strengthening of Kober's inequality holds for all x such that $0 < x \leq \pi/2$:

$$\frac{1 - \cos x}{x} \leq \frac{\tan \frac{x}{2}}{x} \leq \frac{2}{\pi}.$$

[Hint: $1 - \cos x = 2\sin^2(x/2)$.]

5. For each real α and β, show that

$$\sin x \cos(\alpha + \beta - x) + \cos x \sin(\alpha + \beta - x)$$

is a constant function. Deduce the addition rules for the sine and cosine functions from this fact.

6. Differentiate the identity

$$\sin\left(\frac{\pi}{2} - x\right) = \cos x$$

to deduce that

$$\cos\left(\frac{\pi}{2} - x\right) = \sin x.$$

7. Show that $\cos x + x \sin x$ is increasing on $[0, \pi/2]$ and hence deduce that

$$\cos x + x \sin x > 1$$

for all $x \in (0, \pi/2]$.

8. Let θ and r be real numbers and suppose that $|r| < 1$.

(a) Show that

i. $\sum_{j=0}^{\infty} r^j \cos j\theta = \frac{1 - r\cos\theta}{1 - 2r\cos\theta + r^2}$;

ii. $\sum_{j=1}^{\infty} r^j \cos j\theta = \frac{r\sin\theta}{1 - 2r\cos\theta + r^2}$.

(Hint: Write $z = re^{i\theta}$ and investigate $\sum_{j=0}^{\infty} z^j$.)

(b) Find the sums $\sum_{j=0}^{n} \cos j\theta$ and $\sum_{j=1}^{n} \sin j\theta$. Are they convergent? (Hint: See Theorem 6.4.7.)

9. Prove that

$$\sum_{j=1}^{\infty} \frac{\cos j\theta}{j}$$

converges if and only if θ is not an integer multiple of 2π.

10. (a) Establish the following identities:

i. $\cos\left(j \pm \frac{1}{2}\right)\theta = \cos j\theta \cos \frac{\theta}{2} \mp \sin j\theta \sin \frac{\theta}{2}$;

ii. $\cos\left(j - \frac{1}{2}\right)\theta - \cos\left(j + \frac{1}{2}\right)\theta = 2\sin j\theta \sin \frac{\theta}{2}$.

(b) Use part (a) to find the sum

$$S_n = \sum_{j=1}^{n} \sin j\theta.$$

Show also that the sequence $\{S_n\}$ is bounded.

11. Show that

$$\lim_{x \to 0} \frac{\sin |z|}{z}$$

does not exist.

12. Show that

$$\cos \frac{\pi}{6} = \frac{\sqrt{3}}{2}.$$

13. Compute $(5 - i)^4(1 + i)$ and hence show that

$$\frac{\pi}{4} = 4 \arctan \frac{1}{5} - \arctan \frac{1}{239}.$$

This formula is due to Machin.

14. Compute

$$\frac{(4 + i)^3(20 + i)}{1 + i}$$

and hence show that

$$\frac{\pi}{4} = 3 \arctan \frac{1}{4} + \arctan \frac{1}{20} + \arctan \frac{1}{1985}.$$

15. Show that $\sin z = 0$ if and only if z is an integer multiple of π.

16. Solve the following equations:

(a) $\cos z = 0$.
(b) $\cos z = 1$.

17. Let n be a positive integer.

(a) Express $\sin^{2n} x$ in terms of $\cos kx$, where $k \in \mathbb{Z}$.
(b) Express $\sin^{2n+1} x$ in terms of $\sin kx$, where $k \in \mathbb{Z}$.

6.5 L'Hôpital's Rule

The concept of a derivative may be applied to the calculation of limits that do not succumb to the methods of Chap. 4. First we need a generalized version of the mean-value theorem. This generalization is due to Cauchy.

Consider two functions f and g that satisfy the hypotheses of the mean-value theorem on some interval $[a, b]$. Then there exist numbers ξ_f and ξ_g in (a, b) such that

$$\frac{f(b) - f(a)}{b - a} = f'(\xi_f)$$

and

$$\frac{g(b) - g(a)}{b - a} = g'(\xi_g).$$

Thus

$$\frac{f(b) - f(a)}{g(b) - g(a)} = \frac{f'(\xi_f)}{g'(\xi_g)} \tag{6.6}$$

if $g(b) \neq g(a)$ and $g'(\xi_g) \neq 0$. It is natural to wonder whether there is a number $\xi \in (a, b)$ such that Eq. (6.6) holds with $\xi_f = \xi_g = \xi$. Cauchy's mean-value formula satisfies our curiosity in this regard.

Theorem 6.5.1 (Cauchy's Mean-Value Formula). *Let f and g be functions that are continuous on a closed interval $[a, b]$ and differentiable on (a, b). Suppose also that $g'(x) \neq 0$ for all $x \in (a, b)$. Then there exists $\xi \in (a, b)$ such that*

$$\frac{f(b) - f(a)}{g(b) - g(a)} = \frac{f'(\xi)}{g'(\xi)}.$$

Proof. Notice first that $g(b) - g(a) \neq 0$, for if $g(a) = g(b)$, then Rolle's theorem would reveal the existence of a number $c \in (a, b)$ for which $g'(c) = 0$. Such a c cannot exist, however, by hypothesis.

Now define a function F such that

$$F(x) = (f(b) - f(a))g(x) - (g(b) - g(a))f(x)$$
$$= f(b)g(x) - f(a)g(x) - f(x)g(b) + f(x)g(a)$$

for all $x \in [a, b]$. This function is continuous on $[a, b]$ and differentiable on (a, b). Since

$$F(a) = f(b)g(a) - f(a)g(b) = F(b),$$

it satisfies all the hypotheses of Rolle's theorem. Therefore there is a number $\xi \in (a,b)$ such that

$$0 = F'(\xi) = (f(b) - f(a))g'(\xi) - (g(b) - g(a))f'(\xi).$$

Hence

$$(f(b) - f(a))g'(\xi) = (g(b) - g(a))f'(\xi),$$

and the result follows upon division by the nonzero number $(g(b) - g(a))g'(\xi)$. $\square$

Remark. The mean-value theorem is the special case where $g(x) = x$ for all $x \in [a,b]$.

Corollary 6.5.2. *Let f and g be functions that are differentiable on an open interval (a,b) and suppose that $g'(x) \neq 0$ for all $x \in (a,b)$. Suppose also that*

$$\lim_{x \to a^+} \frac{f'(x)}{g'(x)} = L$$

for some real number L. Then for every $\varepsilon > 0$ there exists $\delta \in (0, b-a)$ such that

$$\left| \frac{f(y) - f(x)}{g(y) - g(x)} - L \right| < \varepsilon$$

for all x and y satisfying $a < x < y < a + \delta$.

Proof. Choose $\varepsilon > 0$. By hypothesis we have $(a,b) \subseteq \mathcal{D}_{f'/g'}$ and

$$\lim_{x \to a^+} \frac{f'(x)}{g'(x)} = L.$$

Therefore there exists $\delta > 0$ such that

$$\left| \frac{f'(y)}{g'(y)} - L \right| < \varepsilon \tag{6.7}$$

for each $y \in (a,b)$ satisfying $a < y < a + \delta$. For each such y we therefore have

$$\frac{f'(c)}{g'(c)} < L + \varepsilon \tag{6.8}$$

for all $c \in (a, y)$. We may assume that $\delta < b - a$.

Now fix x and y such that $a < x < y < a + \delta$. Functions f and g satisfy the hypotheses of Cauchy's mean-value formula on $[x, y]$. Therefore there exists $\xi \in (x, y)$ such that

$$\frac{f(y) - f(x)}{g(y) - g(x)} = \frac{f'(\xi)}{g'(\xi)}.$$

As $\xi \in (a, y)$, it follows from inequality (6.8) that

$$\frac{f(y) - f(x)}{g(y) - g(x)} < L + \varepsilon.$$

A similar argument shows that

$$\frac{f(y) - f(x)}{g(y) - g(x)} > L - \varepsilon,$$

and the conclusion follows. $\Box$

Remark. Corresponding results may be obtained by replacing L with $\pm\infty$. For instance, if

$$\lim_{x \to a^+} \frac{f'(x)}{g'(x)} = \infty,$$

then for each number M there exists $\delta \in (0, b - a)$ such that

$$\frac{f(y) - f(x)}{g(y) - g(x)} > M$$

for all x and y satisfying $a < x < y < a + \delta$.

We now discuss several closely related theorems known collectively as l'Hôpital's rule. Suppose first that f and g are functions of a real variable x and that they both approach 0 as x approaches some value (finite or infinite). The limit, which may be one- or two-sided, of $f(x)/g(x)$ is then described as being of the indeterminate form $0/0$. L'Hôpital's rule can sometimes be applied to find this limit. We deal first with the case of one-sided limits.

Theorem 6.5.3. *Let f and g be functions that are differentiable on an open interval (a, b) and suppose that $g'(x) \neq 0$ for all $x \in (a, b)$. Suppose also that*

$$\lim_{x \to a^+} f(x) = \lim_{x \to a^+} g(x) = 0$$

and

$$\lim_{x \to a^+} \frac{f'(x)}{g'(x)} = L$$

for some real number L. Then

$$\lim_{x \to a^+} \frac{f(x)}{g(x)} = L.$$

Proof. Choose $\varepsilon > 0$. By Corollary 6.5.2 there exists $\delta \in (0, b - a)$ such that

$$L - \varepsilon < \frac{f(x) - f(y)}{g(x) - g(y)} < L + \varepsilon \qquad (6.9)$$

for all x and y satisfying $a < y < x < a + \delta$. Thus $\delta > 0$. Choose x such that $0 < x - a < \delta$ and $g(x) \neq 0$. Then

$$a < x < a + \delta < a + b - a = b,$$

so that $x \in \mathcal{D}_{f/g}$. Moreover inequality (6.9) holds for each $y \in (a, x)$. Taking limits as $y \to a^+$, we find that

$$L - \varepsilon \le \frac{f(x)}{g(x)} \le L + \varepsilon,$$

so that

$$\left| \frac{f(x)}{g(x)} - L \right| \le \varepsilon < 2\varepsilon$$

and the result follows. □

Remark. A similar argument shows that L may be replaced by ∞ or $-\infty$. Similar results also hold for limits as $x \to b^-$, and therefore for two-sided limits as well.

Example 6.5.1. Evaluate

$$\lim_{x \to 0} \frac{\sin x}{x}.$$

Solution. Since $\sin x \to 0$ as $x \to 0$, we have

$$\lim_{x \to 0} \frac{\sin x}{x} = \lim_{x \to 0} \cos x = 1$$

by l'Hôpital's rule.

△

Example 6.5.2. Evaluate

$$\lim_{x \to 0} \frac{e^x - 1}{\sin x}.$$

Solution. Note first that

$$\lim_{x \to 0} (e^x - 1) = 0 = \lim_{x \to 0} \sin x.$$

Therefore l'Hôpital's rule may be applied. We conclude that

$$\lim_{x \to 0} \frac{e^x - 1}{\sin x} = \lim_{x \to 0} \frac{e^x}{\cos x} = 1.$$

$\triangle$

Example 6.5.3. L'Hôpital's rule is not applicable to the evaluation of

$$\lim_{x \to 0^+} \frac{\log x}{x}$$

because

$$\lim_{x \to 0^+} \log x = -\infty.$$

An attempt to apply l'Hôpital's rule would yield a wrong answer, since

$$\lim_{x \to 0^+} \frac{\frac{1}{x}}{1} = \lim_{x \to 0^+} \frac{1}{x} = \infty,$$

yet for each $x \in (0, 1)$ we find that $\log x$ is negative and x positive. $\triangle$

Example 6.5.4. Evaluate

$$\lim_{x \to 0} \frac{\sin x - \tan x}{x^2}.$$

Solution. Since

$$\lim_{x \to 0} (\sin x - \tan x) = 0 = \lim_{x \to 0} x^2,$$

we may apply l'Hôpital's rule. Therefore

$$\lim_{x \to 0} \frac{\sin x - \tan x}{x^2} = \lim_{x \to 0} \frac{\cos x - \sec^2 x}{2x}.$$

As

$$\lim_{x \to 0} (\cos x - \sec^2 x) = 0 = \lim_{x \to 0} 2x,$$

we must apply l'Hôpital's rule once again. We conclude that

$$\lim_{x \to 0} \frac{\sin x - \tan x}{x^2} = \lim_{x \to 0} \frac{\cos x - \sec^2 x}{2x}$$

$$= \lim_{x \to 0} \frac{-\sin x - 2\sec^2 x \tan x}{2}$$

$$= 0.$$

$\triangle$

The next theorem also concerns the indeterminate form $0/0$, but the limit is taken as $x \to \infty$.

Theorem 6.5.4. *Let f and g be functions that are differentiable at all $x > a$ for some positive number a, and suppose that $g'(x) \neq 0$ for all $x > a$. Suppose also that*

$$\lim_{x \to \infty} f(x) = \lim_{x \to \infty} g(x) = 0$$

and

$$\lim_{x \to \infty} \frac{f'(x)}{g'(x)} = L,$$

where L may be any real number, ∞, or $-\infty$. Then

$$\lim_{x \to \infty} \frac{f(x)}{g(x)} = L.$$

Proof. For each $x > a > 0$ we may define $t = 1/x$. Thus $0 < t < 1/a$. For each t satisfying these inequalities, define $F(t) = f(1/t)$ and $G(t) = g(1/t)$. By Corollary 4.6.3 we then have

$$\lim_{t \to 0+} F(t) = \lim_{t \to 0+} f\left(\frac{1}{t}\right) = \lim_{x \to \infty} f(x) = 0$$

and

$$\lim_{t \to 0+} G(t) = \lim_{t \to 0+} g\left(\frac{1}{t}\right) = \lim_{x \to \infty} g(x) = 0.$$

Moreover F and G are differentiable on $(0, 1/a)$, and for each $t \in (0, 1/a)$ we have $1/t > a$, so that

$$G'(t) = -\frac{1}{t^2} g'\left(\frac{1}{t}\right) \neq 0$$

by hypothesis. Therefore Theorem 6.5.3 is applicable. We conclude that

$$\lim_{x \to \infty} \frac{f(x)}{g(x)} = \lim_{t \to 0^+} \frac{f\left(\frac{1}{t}\right)}{g\left(\frac{1}{t}\right)}$$

$$= \lim_{t \to 0^+} \frac{F(t)}{G(t)}$$

$$= \lim_{t \to 0^+} \frac{F'(t)}{G'(t)}$$

$$= \lim_{t \to 0^+} \frac{-\frac{1}{t^2} f'\left(\frac{1}{t}\right)}{-\frac{1}{t^2} g'\left(\frac{1}{t}\right)}$$

$$= \lim_{x \to \infty} \frac{f'(x)}{g'(x)}$$

$$= L.$$

$\square$

Remark. A similar proof establishes a corresponding theorem in which $x \to -\infty$.

Sometimes an application of these theorems makes matters worse.

Example 6.5.5. Suppose we wish to evaluate

$$\lim_{x \to 0^+} x \log x.$$

We may convert this limit to the indeterminate form $0/0$ by writing

$$x \log x = \frac{x}{\frac{1}{\log x}}$$

for all $x > 0$ such that $x \neq 1$, because

$$\lim_{x \to 0^+} \frac{1}{\log x} = 0$$

since $\log x \to -\infty$ as $x \to 0^+$. An application of l'Hôpital's rule then shows that

$$\lim_{x \to 0^+} \frac{x}{\frac{1}{\log x}} = \lim_{x \to 0^+} \frac{1}{\frac{1}{x}\left(-\frac{1}{\log^2 x}\right)} = -\lim_{x \to 0^+} \frac{x}{\frac{1}{\log^2 x}}.$$

Although the limit on the right-hand side is still of the indeterminate form $0/0$, it is more complicated than the limit on the left-hand side. The appeal to Theorem 6.5.3 has therefore failed to solve the problem. However, we can also write

$$x \log x = \frac{\log x}{\frac{1}{x}},$$

for all $x > 0$, and note that

$$\lim_{x \to 0^+} \log x = -\infty$$

and

$$\lim_{x \to 0^+} \frac{1}{x} = \infty.$$

$\triangle$

The observation at the end of the example above motivates a study of other indeterminate forms. We prepare for the resulting theorem with a helpful lemma concerning limits.

Lemma 6.5.5. *Let f be a function defined on an interval (a, b) and let L be a number. Suppose that for each $q_1 > L$ there exists $c_1 \in (a, b)$ such that $f(x) < q_1$ whenever $x \in (a, c_1)$. Suppose similarly that for each $q_2 < L$ there exists $c_2 \in (a, b)$ such that $q_2 < f(x)$ whenever $x \in (a, c_2)$. Then*

$$\lim_{x \to a^+} f(x) = L.$$

Proof. Choose $\varepsilon > 0$. Then $L - \varepsilon < L < L + \varepsilon$. Consequently, there exist numbers c_1 and c_2 in (a, b) such that $f(x) < L + \varepsilon$ whenever $x \in (a, c_1)$ and $L - \varepsilon < f(x)$ whenever $x \in (a, c_2)$. If we choose x such that

$$0 < x - a < \min\{c_1 - a, c_2 - a\},$$

then $a < x < c_1$ and $a < x < c_2$, so that

$$L - \varepsilon < f(x) < L + \varepsilon.$$

Hence $|f(x) - L| < \varepsilon$, and the result follows. $\square$

Theorem 6.5.6. *Let f and g be functions that are differentiable on an open interval (a, b) and suppose that $g'(x) \neq 0$ for all $x \in (a, b)$. Suppose also that*

$$\lim_{x \to a^+} g(x) = \infty$$

and

$$\lim_{x \to a+} \frac{f'(x)}{g'(x)} = L$$

for some real number L. Then

$$\lim_{x \to a+} \frac{f(x)}{g(x)} = L.$$

Discussion: One obvious way to attempt a proof of this theorem is to try to show that $\lim_{x \to a+} f(x) = \infty$ and then to convert the limit to the indeterminate form $0/0$ by writing $f(x)/g(x)$ as

$$\frac{\frac{1}{g(x)}}{\frac{1}{f(x)}}.$$

Using Theorem 6.5.3 together with the chain rule, we should then obtain

$$\lim_{x \to a+} \frac{f(x)}{g(x)} = \lim_{x \to a+} \frac{\frac{1}{g(x)}}{\frac{1}{f(x)}}$$

$$= \lim_{x \to a+} \frac{-\frac{g'(x)}{g^2(x)}}{-\frac{f'(x)}{f^2(x)}}$$

$$= \lim_{x \to a+} \frac{f^2(x)}{g^2(x)} \cdot \lim_{x \to a+} \frac{g'(x)}{f'(x)}$$

$$= \left(\lim_{x \to a+} \frac{f(x)}{g(x)} \right)^2 \frac{1}{L}.$$

We now encounter several problems, the most obvious of which is that L might be 0. Furthermore, $\lim_{x \to a+} f(x)/g(x)$ might not exist. Even if it were to exist, it would have to be nonzero in order for us to be able to deduce the required result as a consequence of the calculation above. The following proof circumvents these difficulties.

Proof. Choose $q_1 > L$. The density property shows that there is a q satisfying $L < q < q_1$. By Corollary 6.5.2 there exists $\delta \in (0, b - a)$ such that

$$\left| \frac{f(x) - f(y)}{g(x) - g(y)} - L \right| < q - L$$

for all x and y satisfying $a < x < y < a + \delta$. Fix a y for which

$$a < y < a + \delta < b.$$

It follows that

$$\frac{f(x) - f(y)}{g(x) - g(y)} < L + q - L = q \tag{6.10}$$

for all $x \in (a, y)$.

Since

$$\lim_{x \to a^+} g(x) = \infty,$$

there exists $\delta_1 > 0$ such that

$$g(x) > \max\{0, g(y)\} \tag{6.11}$$

for all $x \in (a, b)$ satisfying $0 < x - a < \delta_1$. Let

$$c = \min\{y, a + \delta_1\} > a.$$

Thus $c \in (a, b)$, since $y < b$.

Now choose $x \in (a, c)$. Then $a < x < c \leq y$, so that inequality (6.10) holds. But we also have

$$0 < x - a < c - a \leq \delta_1,$$

so that $g(x) - g(y) > 0$ by inequality (6.11). Therefore

$$f(x) - f(y) < q(g(x) - g(y)),$$

and so

$$f(x) < f(y) + qg(x) - qg(y).$$

Moreover $g(x) > 0$. We conclude that

$$\frac{f(x)}{g(x)} < h(x)$$

for all $x \in (a, c)$, where

$$h(x) = \frac{f(y)}{g(x)} + q - q\frac{g(y)}{g(x)}$$

for all such x.

Now

$$\lim_{x \to a^+} h(x) = q$$

since

$$\lim_{x \to a^+} \frac{1}{g(x)} = 0$$

by Corollary 4.6.4. Therefore there exists $\delta_2 > 0$ such that

$$|h(x) - q| < q_1 - q \qquad (6.12)$$

for all $x \in (a, c)$ satisfying $0 < x - a < \delta_2$. Define

$$c_1 = \min\{c, a + \delta_2\} > a.$$

Thus $c_1 \in (a, b)$. Choose $x \in (a, c_1)$. Then $x \in (a, c)$ and

$$0 < x - a < c_1 - a \leq \delta_2.$$

Therefore inequality (6.12) holds, so that

$$h(x) < q + q_1 - q = q_1.$$

We conclude that

$$\frac{f(x)}{g(x)} < h(x) < q_1.$$

In summary, we have now shown that for each $q_1 > L$ there exists $c_1 \in (a, b)$ such that

$$\frac{f(x)}{g(x)} < q_1$$

whenever $x \in (a, c_1)$. A similar argument shows that for each $q_2 < L$ there exists $c_2 \in (a, b)$ such that

$$q_2 < \frac{f(x)}{g(x)}$$

whenever $x \in (a, c_2)$. An appeal to Lemma 6.5.5 therefore completes the proof. $\square$

Example 6.5.6. Evaluate

$$\lim_{x\to 0^+} x \log x.$$

Solution. We have

$$\lim_{x\to 0^+} x \log x = \lim_{x\to 0^+} \frac{\log x}{\frac{1}{x}}.$$

Since $\lim_{x\to 0^+} 1/x = \infty$, it follows from Theorem 6.5.6 that

$$\lim_{x\to 0^+} x \log x = \lim_{x\to 0^+} \frac{\frac{1}{x}}{-\frac{1}{x^2}} = - \lim_{x\to 0^+} x = 0.$$

$\triangle$

Theorem 6.5.6 may be extended to limits as x approaches a^-, a, ∞, or $-\infty$.

Remark. If $\lim f'(x)/g'(x)$ does not exist, then we cannot draw any conclusion about $\lim f(x)/g(x)$.

Example 6.5.7. If $f(x) = \sin x$ and $g(x) = x$ for all x, then $f'(x) = \cos x$ and $g'(x) = 1$, so that $\lim_{x\to\infty} f'(x)/g'(x)$ does not exist. $\triangle$

When applying l'Hôpital's rule to find $\lim_{x\to a} f(x)/g(x)$, we must make sure that all the hypotheses are satisfied, as we may get spurious results otherwise. For instance, if g' has a zero in each neighborhood of a, then we must not apply l'Hôpital's rule. Corresponding remarks hold if x approaches a^+, a^-, ∞, or $-\infty$. In a case where $f'(x) = \alpha(x)h(x)$, $g'(x) = \alpha(x)k(x)$, and $\alpha(x)$ does not approach a limit but $h(x)/k(x)$ does, then we must resist the temptation to cancel $\alpha(x)$ in $f'(x)/g'(x)$. In 1879, Stolz gave an example to illustrate this point. In 1956, Boas constructed infinitely many examples, including the one constructed by Stolz. We present an example here.

Example 6.5.8. For all x let

$$f(x) = \frac{\sin 2x + 2x}{4},$$

and let φ be any function such that the functions $\varphi(\sin x)$ and $\varphi'(\sin x)$ are positive and bounded for all x. For instance, the exponential function satisfies these conditions. Let

$$g(x) = f(x)\varphi(\sin x)$$

for all x. Thus

$$\lim_{x\to\infty} g(x) = \lim_{x\to\infty} f(x) = \infty.$$

Moreover $g(x) > 0$ for all $x > 1/2$.

Now let us attempt to apply l'Hôpital's rule to compute $\lim_{x\to\infty} f(x)/g(x)$. First we check that the hypotheses are satisfied. For all x we have

$$f'(x) = \frac{\cos 2x + 1}{2} = \cos^2 x$$

and

$$g'(x) = \varphi(\sin x)\cos^2 x + f(x)\varphi'(\sin x)\cos x$$
$$= (\varphi(\sin x)\cos x + f(x)\varphi'(\sin x))\cos x.$$

Thus $g'(x) = 0$ whenever $\cos x = 0$. Consequently, for every a there exist values of $x > a$ such that $g'(x) = 0$. We conclude that l'Hôpital's rule cannot in fact be applied.

Note that $\cos x$ is a common factor of $f'(x)$ and $g'(x)$. Canceling this factor, we obtain

$$\frac{f'(x)}{g'(x)} = \frac{\cos x}{\varphi(\sin x)\cos x + f(x)\varphi'(\sin x)}$$

for 1all x such that $g'(x) \neq 0$. The properties hypothesized for φ and the fact that $\lim_{x\to\infty} f(x) = \infty$ therefore show that

$$\lim_{x\to\infty} \frac{f'(x)}{g'(x)} = 0.$$

However,

$$\frac{f(x)}{g(x)} = \frac{1}{\varphi(\sin x)}1$$

for all $x > 1/2$, and this quotient does not approach 0 as $x \to \infty$. $\triangle$

It is easy to use l'Hôpital's rule to prove that $\log x \ll x$:

$$\lim_{x\to\infty} \frac{\log x}{x} = \lim_{x\to\infty} \frac{1}{x} = 0.$$

In fact, it can be proved by induction that $\log^k x \ll x$ for every positive integer k, for if we assume that

$$\lim_{x\to\infty} \frac{\log^k x}{x} = 0$$

for some fixed k, then, using l'Hôpital's rule, we obtain

$$\lim_{x\to\infty} \frac{\log^{k+1} x}{x} = \lim_{x\to\infty} \frac{(k+1)\log^k x}{x} = 0.$$

Exercises 6.5.

1. Show that the conclusion of Theorem 6.5.1 does not hold for the functions

$$f(x) = 4x^3 + 6x^2 - 12x$$

and

$$g(x) = 3x^4 + 4x^3 - 6x^2$$

on the interval $[0, 1]$. Which of the conditions of the theorem is not satisfied in this case?

2. Show that Theorem 6.5.6 may not be applied to evaluate

$$\lim_{x\to\infty} \frac{x + \cos x}{x + \sin x}$$

directly.

3. Let

$$f(x) = x + \cos x \sin x$$

and

$$g(x) = e^{\sin x} f(x)$$

for all x. Show that l'Hôpital's rule does not apply to the evaluation of

$$\lim_{x\to\infty} \frac{f(x)}{g(x)}.$$

Does the limit exist?

4. Show that l'Hôpital's rule does not apply to the evaluation of

$$\lim_{x\to\infty} \frac{2x + \sin 2x}{x \sin x + \cos x}.$$

Does the limit exist?

5. Evaluate the following limits:
   (a) $\lim_{x\to 0} \frac{e^x - 1}{\log(x+1)}$;
   (b) $\lim_{x\to\infty} \frac{x \log x}{x^2 + 1}$;

(c) $\lim_{x \to 1+} \frac{e^x - 1}{x - 1}$;

(d) $\lim_{x \to 0+} \frac{\log \sin x}{\log x}$;

(e) $\lim_{x \to \infty} x(\log(x + 2) - \log x)$;

(f) $\lim_{x \to \infty} x(a^{1/x} - 1)$, where $a > 0$.

(g) $\lim_{x \to (-1)+} (x + 1)^{1/3} \log(x + 1)$;

(h) $\lim_{x \to 0+} x^x$;

(i) $\lim_{x \to \infty} (\sqrt{x^2 + x} - x)$;

(j) $\lim_{x \to (\pi/2)-} \left(x - \frac{\pi}{2}\right) \tan x$;

(k) $\lim_{x \to \infty} \left(\frac{x+1}{x-2}\right)^x$.

6. The determinant

$$\begin{vmatrix} a_{11} & a_{12} & a_{13} \\ a_{21} & a_{22} & a_{23} \\ a_{31} & a_{32} & a_{33} \end{vmatrix}$$

of the 3×3 matrix

$$\begin{pmatrix} a_{11} & a_{12} & a_{13} \\ a_{21} & a_{22} & a_{23} \\ a_{31} & a_{32} & a_{33} \end{pmatrix}$$

is defined by the equation

$$\begin{vmatrix} a_{11} & a_{12} & a_{13} \\ a_{21} & a_{22} & a_{23} \\ a_{31} & a_{32} & a_{33} \end{vmatrix} = a_{11} \begin{vmatrix} a_{22} & a_{23} \\ a_{32} & a_{33} \end{vmatrix} - a_{12} \begin{vmatrix} a_{21} & a_{23} \\ a_{31} & a_{33} \end{vmatrix} + a_{13} \begin{vmatrix} a_{21} & a_{22} \\ a_{31} & a_{32} \end{vmatrix}.$$

Let $f, g: [a, b] \to \mathbb{R}$. Apply Rolle's theorem with

$$F(x) = \begin{vmatrix} 1 & 1 & 1 \\ f(a) & f(x) & f(b) \\ g(a) & g(x) & g(b) \end{vmatrix}$$

to prove Cauchy's mean-value formula. (See question 7 of the exercises for Sect. 6.2.)

7. Let f and g be functions that are continuous on $[a, b]$ and differentiable on (a, b). Then, as in the proof of Cauchy's mean-value formula, there exists $\xi \in (a, b)$ satisfying the equation

$$(f(b) - f(a))g'(\xi) = (g(b) - g(a))f'(\xi). \tag{6.13}$$

Show that Cauchy's mean-value formula does not apply in the case where

$$f(x) = 3x^4 - 2x^3 - x^2 + 1$$

and

$$g(x) = 4x^3 - 3x^2 - 2x$$

for all $x \in [0, 1]$ but that there exists $\xi \in (0, 1)$ satisfying Eq. (6.13).

8. [17] Let $\alpha_1, \alpha_2, \alpha_3$ be real numbers with sum equal to 1. Let f_1, f_2, f_3 be functions that are continuous on $[a, b]$ and differentiable on (a, b), and suppose that $f_k(a) \neq f_k(b)$ for each $k \in \{1, 2, 3\}$. Show that there exists $\xi \in (a, b)$ for which

$$\frac{\alpha_1}{f_1(b) - f_1(a)} f_1'(\xi) + \frac{\alpha_2}{f_2(b) - f_2(a)} f_2'(\xi) + \frac{\alpha_3}{f_3(b) - f_2(a)} f_3'(\xi) = 0.$$

[Hint: Apply Rolle's theorem with

$$F(x) = \alpha_1(f_2(b) - f_2(a))(f_3(b) - f_3(a))(f_1(x) - f_1(a))$$
$$+ \alpha_2(f_1(b) - f_1(a))(f_3(b) - f_3(a))(f_2(x) - f_2(a))$$
$$+ \alpha_3(f_1(b) - f_1(a))(f_2(b) - f_2(a))(f_3(x) - f_3(a)).]$$

Obtain an expression for $f_1'(\xi)$ by taking $\alpha_1 = -1$, and derive Cauchy's mean-value formula by taking $\alpha_1 = -1$, $\alpha_2 = 1$, and $\alpha_3 = 0$.

The result can be generalized to an arbitrary number of functions. Use it to show that the equation

$$f(x) = -3x + \frac{\pi}{2} \cos \frac{\pi x}{2} + \frac{e^x}{e - 1} + \frac{1}{(x + 1) \log 2} = 0$$

has at least one solution in $(0, 1)$. [Note that $f(0)$ and $f(1)$ are both positive and so the conclusion is not an obvious consequence of the intermediate-value theorem.]

9. [2] Let a and $b > a$ be real numbers and suppose that f and g are functions that are continuous on $[a, b]$ and differentiable on (a, b). Suppose also that $g'(x) \neq 0$ for all $x \in (a, b)$. Show that if $f'(x)/g'(x)$ is increasing (respectively, decreasing), then so are

$$\frac{f(x) - f(a)}{g(x) - g(a)}$$

and

$$\frac{f(x) - f(b)}{g(x) - g(b)}.$$

[Hint: Note that Corollary 6.3.13 implies that either $g'(x) < 0$ for all $x \in (a, b)$ or $g'(x) > 0$ for all $x \in (a, b)$. Assuming that $f'(x)/g'(x)$ is increasing, show that the derivative of

$$\frac{f(x) - f(a)}{g(x) - g(a)}$$

is nonnegative.]

10. This problem concerns the following question. Suppose that f and g are functions such that

$$\lim_{x \to a+} \frac{f(x)}{g(x)} \tag{6.14}$$

is of the indeterminate form $0/0$, where a is a real number. It follows from Theorem 6.5.3 that if

$$\lim_{x \to a+} \frac{f'(x)}{g'(x)} \tag{6.15}$$

exists, then so does the limit (6.14). The question is whether we can extend this observation to say that if $f'(x)/g'(x)$ exists throughout some interval $(a, a+\delta)$, where $\delta > 0$, but the limit (6.15) does not exist, then the limit (6.14) also does not exist.

(a) Show that the answer to the question posed is "no" by considering the functions defined by

$$f(x) = x^2 \sin \frac{1}{x},$$

for all $x \neq 0$, and $g(x) = \sin x$. Use the facts that

$$\lim_{x \to 0+} x \sin \frac{1}{x} = 0$$

(see Example 4.4.1),

$$\lim_{x \to 0+} \cos \frac{1}{x}$$

does not exist (Example 6.4.1) and

$$\lim_{x \to 0+} \frac{\sin x}{x} = 1$$

(Example 6.5.1) in order to demonstrate that $f'(x)/g'(x)$ exists throughout the interval $(0, \pi/2)$ and that

$$\lim_{x \to 0+} \frac{f'(x)}{g'(x)}$$

does not exist but

$$\lim_{x \to 0+} \frac{f(x)}{g(x)} = 0.$$

(b) i. Show that if $f'(x)/g'(x)$ exists throughout some interval $(a, a+\delta)$, where $\delta > 0$, then for each $x \in (a, a+\delta)$ there exists $\xi \in (a, x)$ such that

$$\frac{f(x)}{g(x)} = \frac{f'(\xi)}{g'(\xi)}.$$

ii. Why is it therefore not valid to conclude that the answer to the question posed is "yes" by using the following argument?

If the limit (6.15) does not exist, then

$$\lim_{x \to a+} \frac{f'(\xi)}{g'(\xi)}$$

does not exist since $a < \xi < x$ for each x. Because

$$\frac{f(x)}{g(x)} = \frac{f'(\xi)}{g'(\xi)},$$

it therefore follows that the limit (6.14) also does not exist.

6.6   A Discrete Version of l'Hôpital's Rule

L'Hôpital's rule is a powerful tool for calculating limits of indeterminate forms. However, in applications one may encounter an indeterminate form $f(x)/g(x)$ whose limit exists even though functions f and g are not differentiable or the limit of $f'(x)/g'(x)$ does not exist. In 1988, Huang [8] dealt with this problem by proving a discrete version of l'Hôpital's rule.

Let f be a function defined at all $x > a$ for some number a. For each $x > a$ and $h > a - x$ let us define

$$\Delta_h f(x) = f(x + h) - f(x).$$

Then

$$f'(x) = \lim_{h \to 0} \frac{\Delta_h f(x)}{h}.$$

Huang's result shows that, under certain circumstances,

$$\lim_{x \to \infty} \frac{f(x)}{g(x)} = \lim_{x \to \infty} \frac{\Delta_h f(x)}{\Delta_h g(x)}.$$

Theorem 6.6.1. *Let f and g be functions defined at all $x > a$ for some number a. Suppose that*

1.

$$\lim_{x \to \infty} f(x) = \lim_{x \to \infty} g(x) = 0,$$

and
2. for some $h > 0$,

(a) either $\Delta_h g(x) > 0$ for all $x > a$ or $\Delta_h g(x) < 0$ for all $x > a$, and
(b)

$$\lim_{x \to \infty} \frac{\Delta_h f(x)}{\Delta_h g(x)} = L,$$

where L may be any real number, ∞, or $-\infty$.

Then

$$\lim_{x \to \infty} \frac{f(x)}{g(x)} = L.$$

Proof. Let us assume that $\Delta_h g(x) > 0$ for all $x > a$; the argument in the other case is similar.

Case 1: Suppose that L is a real number. Choose $\varepsilon > 0$. By assumption there exists $N > a$ such that

$$L - \varepsilon < \frac{f(x+h) - f(x)}{g(x+h) - g(x)} < L + \varepsilon$$

for all $x \geq N$. Fix $x \geq N$. Then for all positive integers k we have $x + (k-1)h \geq N$, and so

$$L - \varepsilon < \frac{f(x+kh) - f(x+(k-1)h)}{g(x+kh) - g(x+(k-1)h)} < L + \varepsilon.$$

Since

$$g(x + kh) - g(x + (k - 1)h) > 0,$$

it follows that

$$(L - \varepsilon)(g(x + kh) - g(x + (k - 1)h))$$
$$< f(x + kh) - f(x + (k - 1)h)$$
$$< (L + \varepsilon)(g(x + kh) - g(x + (k - 1)h)).$$

Summation from $k = 1$ to $k = n$, where n is a positive integer, gives

$$(L - \varepsilon)(g(x + nh) - g(x)) < f(x + nh) - f(x) < (L + \varepsilon)(g(x + nh) - g(x)),$$

by the telescoping property. Taking limits as $n \to \infty$, we obtain

$$(L - \varepsilon)(-g(x)) \leq -f(x) \leq (L + \varepsilon)(-g(x)),$$

by Theorem 4.6.2. Our initial assumption shows that the sequence $\{g(x + nh)\}$ is increasing. Since

$$\lim_{n \to \infty} g(x + nh) = 0,$$

it follows that $g(x) < 0$. Thus

$$L - \varepsilon \leq \frac{f(x)}{g(x)} \leq L + \varepsilon,$$

from which the required result follows.

Case 2: If $L = \infty$, then for each $M > 0$ there exists $N > a$ such that

$$f(x + h) - f(x) > M(g(x + h) - g(x))$$

whenever $x \geq N$. Fix such an x. Using the method employed in the previous case, we find that

$$f(x + nh) - f(x) > M(g(x + nh) - g(x))$$

for all positive integers n. By letting $n \to \infty$, we see that

$$\frac{f(x)}{g(x)} \geq M,$$

and the result follows.

Case 3: The case where $L = -\infty$ can be dealt with by applying the argument of case 2 to the functions $-f$ and g.

$\square$

By applying the proof of this theorem to sequences, we obtain the following result.

Corollary 6.6.2. *Let $\{a_n\}$ and $\{b_n\}$ be sequences that converge to 0. Let h be a positive integer for which there exists a real number N such that the sign of $b_{n+h} - b_n$ is constant for all $n \geq N$. Suppose that*

$$\lim_{n \to \infty} \frac{a_{n+h} - a_n}{b_{n+h} - b_n} = L,$$

where L may be any real number, ∞, or $-\infty$. Then

$$\lim_{n \to \infty} \frac{a_n}{b_n} = L.$$

Example 6.6.1. For each $x > 1$ define

$$f(x) = \frac{\sin 2\pi x}{\lfloor x \rfloor}$$

and

$$g(x) = \sqrt{x+1} - \sqrt{x} = \frac{1}{\sqrt{x+1} + \sqrt{x}}.$$

Then

$$\lim_{x \to \infty} f(x) = \lim_{x \to \infty} g(x) = 0.$$

We wish to evaluate $\lim_{x \to \infty} f(x)/g(x)$. Evidently, l'Hôpital's rule is not applicable. Taking $h = 1$ in Theorem 6.6.1, we obtain

$$\Delta_1 g(x) = g(x+1) - g(x)$$
$$= \sqrt{x+2} - \sqrt{x}$$
$$= \frac{2}{\sqrt{x+2} + \sqrt{x}}$$
$$> 0$$

and

$$\Delta_1 f(x) = f(x+1) - f(x)$$
$$= \frac{\sin 2\pi(x+1)}{\lfloor x+1 \rfloor} - \frac{\sin 2\pi x}{\lfloor x \rfloor}$$
$$= \left(\frac{1}{\lfloor x \rfloor + 1} - \frac{1}{\lfloor x \rfloor} \right) \sin 2\pi x$$
$$= -\frac{\sin 2\pi x}{\lfloor x \rfloor (\lfloor x \rfloor + 1)},$$

so that

$$\frac{\Delta_1 f(x)}{\Delta_1 g(x)} = -\frac{(\sqrt{x+2} + \sqrt{x}) \sin 2\pi x}{2 \lfloor x \rfloor (\lfloor x \rfloor + 1)};$$

hence

$$\left| \frac{\Delta_1 f(x)}{\Delta_1 g(x)} \right| < \frac{2\sqrt{x+2}}{2(x-1)^2} \to 0$$

as $x \to \infty$. We conclude that

$$\lim_{x \to \infty} \frac{f(x)}{g(x)} = 0.$$

$\triangle$

We now establish another version of the theorem.

Theorem 6.6.3. *Let f and g be functions defined at all $x > a$ for some number a, and suppose that they are bounded on every finite subinterval of (a, ∞). Suppose also that*

1.

$$\lim_{x \to \infty} g(x) = \infty,$$

and

2. for some $h > 0$,

(a) either $\Delta_h g(x) > 0$ for all $x > a$ or $\Delta_h g(x) < 0$ for all $x > a$, and
(b)

$$\lim_{x \to \infty} \frac{\Delta_h f(x)}{\Delta_h g(x)} = L,$$

where L may be any real number, ∞, or $-\infty$.

Then

$$\lim_{x \to \infty} \frac{f(x)}{g(x)} = L.$$

Proof. Let us assume that $\Delta_h g(x) > 0$ for all $x > a$, as the argument in the other case is similar. Since $\lim_{x \to \infty} g(x) = \infty$, we may also assume that $g(x) > 0$ for all $x > a$.

Case 1: Suppose L is a real number. Choose $\varepsilon > 0$. Arguing as in the proof of Theorem 6.6.1, we deduce the existence of $N > a$ such that

$$\left| \frac{f(x+nh) - f(x)}{g(x+nh) - g(x)} - L \right| < \varepsilon \qquad (6.16)$$

for all $x \geq N$ and all positive integers n.
Choose $x \geq N+h$, and write $x = N + \theta h$. Let $j = \lfloor \theta \rfloor$ and $r = N + (\theta - j)h$. Thus $j > 0$, $x = r + jh$, and $r \in [N, N+h)$. Furthermore,

$$g(x) - g(r) = g(r + jh) - g(r) > 0,$$

so that

$$0 < \frac{g(r)}{g(x)} < 1;$$

hence

$$\left| 1 - \frac{g(r)}{g(x)} \right| < 1,$$

and so

$$\left| \frac{f(x)}{g(x)} - L \right| \leq \left| \frac{f(r) - Lg(r)}{g(x)} \right| + \left| \frac{f(x) - f(r)}{g(x) - g(r)} - L \right|$$

because

$$\frac{f(x)}{g(x)} - L = \frac{f(x) - Lg(x)}{g(x)}$$

$$= \frac{f(r) - Lg(r) + f(x) - f(r) - Lg(x) + Lg(r)}{g(x)}$$

$$= \frac{f(r) - Lg(r)}{g(x)} + \frac{g(x) - g(r)}{g(x)} \left(\frac{f(x) - f(r) - L(g(x) - g(r))}{g(x) - g(r)} \right)$$

$$= \frac{f(r) - Lg(r)}{g(x)} + \left(1 - \frac{g(r)}{g(x)}\right)\left(\frac{f(x) - f(r)}{g(x) - g(r)} - L\right).$$

Note that

$$\left|\frac{f(x) - f(r)}{g(x) - g(r)} - L\right| = \left|\frac{f(r + jh) - f(r)}{g(r + jh) - g(r)} - L\right| < \varepsilon$$

by inequality (6.16). Furthermore, since f and g are bounded on $[N, N + h)$ and $\lim_{x \to \infty} g(x) = \infty$, there exists $N_1 \geq N$ such that

$$\left|\frac{f(r) - Lg(r)}{g(x)}\right| < \varepsilon$$

for all $x \geq N_1$. For all such x we therefore have

$$\left|\frac{f(x)}{g(x)} - L\right| < 2\varepsilon,$$

and so

$$\lim_{x \to \infty} \frac{f(x)}{g(x)} = L.$$

Case 2: Suppose that $L = \infty$.
We first show that $\Delta_h f(x) > 0$ for sufficiently large x. Choose $M > 0$. There exists $N > a$ such that

$$\frac{f(x + h) - f(x)}{g(x + h) - g(x)} > M$$

for all $x \geq N$. Hence

$$\Delta_h f(x) = f(x + h) - f(x) > M(g(x + h) - g(x)) > 0$$

for all $x \geq N$, as claimed.

As in Theorem 6.6.1, we see that

$$\frac{f(x + nh) - f(x)}{g(x + nh) - g(x)} > M$$

for all $x \geq N$ and all positive integers n. Choose $x \geq N + h$, and write $x = r + jh$, where $r \in [N, N + h)$ and j is a positive integer. Thus

$$\frac{f(x) - f(r)}{g(x) - g(r)} = \frac{f(r + jh) - f(r)}{g(r + jh) - g(r)} > M,$$

so that

$$f(x) > M(g(x) - g(r)) + f(r).$$

Hence

$$\lim_{x \to \infty} f(x) = \infty,$$

and so we may assume also that $f(x) > 0$ for all $x > a$.
    Finally,

$$\lim_{x \to \infty} \frac{\Delta_h g(x)}{\Delta_h f(x)} = \lim_{x \to \infty} \frac{1}{\frac{\Delta_h f(x)}{\Delta_h g(x)}}$$

and

$$\lim_{x \to \infty} \frac{\Delta_h f(x)}{\Delta_h g(x)} = \infty.$$

Consequently,

$$\lim_{x \to \infty} \frac{\Delta_h g(x)}{\Delta_h f(x)} = 0.$$

It therefore follows from case 1 that

$$\lim_{x \to \infty} \frac{g(x)}{f(x)} = 0,$$

whence

$$\lim_{x \to \infty} \frac{f(x)}{g(x)} = \lim_{x \to \infty} \frac{1}{\frac{g(x)}{f(x)}} = \infty,$$

because $f(x)/g(x) > 0$ for all $x > a$.

Case 3: The case where $L = -\infty$ can be handled by the same argument with f replaced by $-f$.

$\square$

We immediately obtain the corresponding result for sequences.

Corollary 6.6.4. *Let $\{a_n\}$ and $\{b_n\}$ be sequences, where $\lim_{n \to \infty} b_n = \infty$. Let h be a positive integer for which there exists a real number N such that the sign of $b_{n+h} - b_n$ is constant for all $n \geq N$. Suppose that*

$$\lim_{n \to \infty} \frac{a_{n+h} - a_n}{b_{n+h} - b_n} = L,$$

where L may be any real number, ∞, or $-\infty$. Then

$$\lim_{n \to \infty} \frac{a_n}{b_n} = L.$$

By taking $h = 1$, we obtain the following result, which is due to Stolz.

Corollary 6.6.5. *Let $\{a_n\}$ be a sequence and $\{b_n\}$ an increasing sequence such that* $\lim_{n \to \infty} b_n = \infty$. *If*

$$\lim_{n \to \infty} \frac{a_{n+1} - a_n}{b_{n+1} - b_n} = L$$

for some real L, then

$$\lim_{n \to \infty} \frac{a_n}{b_n} = L.$$

Example 6.6.2. Let $x_1 = 1/2$ and

$$x_{n+1} = x_n(1 - x_n)$$

for all $n \geq 1$. It is easy to see by induction that $0 < x_n < 1$ for all n, and so $x_{n+1} < x_n$ for all n. Thus the sequence $\{x_n\}$ is decreasing. It therefore converges to some number L. From the recurrence relation we obtain

$$L = L(1 - L)$$

by taking limits, and we infer that $L = 0$. Thus the sequence $\{1/x_n\}$ is increasing and approaches infinity as n does so. Noting first that

$$x_n - x_{n+1} = x_n - x_n(1 - x_n) = x_n^2$$

for all n, we conclude from Stolz's theorem, with $a_n = n$ and $b_n = 1/x_n$ for all n, that

$$\lim_{n \to \infty} n x_n = \lim_{n \to \infty} \frac{n}{\frac{1}{x_n}}$$

$$= \lim_{n \to \infty} \frac{(n + 1) - n}{\frac{1}{x_{n+1}} - \frac{1}{x_n}}$$

$$= \lim_{n \to \infty} \frac{x_n x_{n+1}}{x_n - x_{n+1}}$$

$$= \lim_{n\to\infty} \frac{x_{n+1}}{x_n}$$

$$= \lim_{n\to\infty} (1 - x_n)$$

$$= 1.$$

$\triangle$

Corollary 6.6.6. *Let a be a positive number and let k be a function that is defined and positive at all $x > a$ and bounded on every finite subinterval of (a, ∞). Suppose that*

$$\lim_{x\to\infty} \frac{k(x+1)}{k(x)} = L,$$

where L may be any positive number or ∞. Then

$$\lim_{x\to\infty} (k(x))^{1/x} = L.$$

Proof. Let $f(x) = \log k(x)$ and $g(x) = x$ for all $x > a$. Then

$$\frac{f(x+1) - f(x)}{g(x+1) - g(x)} = \log k(x+1) - \log k(x) = \log \frac{k(x+1)}{k(x)} \to \log L$$

as $x \to \infty$. It therefore follows from Theorem 6.6.3 that

$$\lim_{x\to\infty} \frac{\log k(x)}{x} = \log L.$$

Hence

$$\lim_{x\to\infty} (k(x))^{1/x} = \lim_{x\to\infty} \exp \log (k(x))^{1/x} = \lim_{x\to\infty} \exp \frac{\log k(x)}{x} = \exp \log L = L.$$

$\square$

As a consequence of Corollary 6.6.6, we obtain the corresponding result for sequences.

Corollary 6.6.7. *If $\{a_n\}$ is a sequence of positive terms and*

$$\lim_{n\to\infty} \frac{a_{n+1}}{a_n} = L,$$

then

$$\lim_{n\to\infty} a_n^{1/n} = L.$$

Exercises 6.6.

1. Use the work of this section to show that

$$\lim_{n\to\infty} \left(\sum_{j=1}^{n} (2j-1)^2 \Big/ \sum_{j=1}^{n} j^2 \right) = 4.$$

2. Use the work of this section to show that

$$\lim_{n\to\infty} \frac{1}{n^{k+1}} \sum_{j=1}^{n} k^k = \frac{1}{k+1}$$

for all positive integers k.

3. Use Theorem 6.6.3 to find

$$\lim_{x\to\infty} \frac{\sin x + x}{x}.$$

4. For each $x \geq 0$ let $k(x)$ be the integer such that

$$x = y(x) + 2k(x)\pi$$

for some number $y(x) \in [0, 2\pi)$, and let

$$g(x) = \sin y(x) + k(x).$$

Show that

$$\lim_{x\to\infty} \frac{x}{g(x)} = 2\pi.$$

5. Use the work in this section to find

$$\lim_{x\to\infty} \frac{\log x}{x}$$

and

$$\lim_{x\to\infty} \frac{x}{e^x}.$$

6. If $\{a_n\}$ and $\{b_n\}$ are sequences such that

$$\lim_{n\to\infty} a_n = \lim_{n\to\infty} b_n = 0$$

and

$$\lim_{n \to \infty} \frac{a_n}{b_n} = L$$

for some number L, show that

$$\lim_{n \to \infty} \left(\sum_{j=1}^{n} a_j \Big/ \sum_{j=1}^{n} b_j \right) = L.$$

7. Prove that if

$$\lim_{n \to \infty} (x_n - x_{n+1}) = L,$$

then

$$\lim_{n \to \infty} \frac{x_n}{n} = L.$$

6.7   Differentiation of Power Series

We move on to consider derivatives of power series. Let us define

$$f(z) = \sum_{j=0}^{\infty} a_j (z - c)^j$$

for each z within the circle of convergence of the power series on the right-hand side. If we differentiate the power series term by term, we obtain a new series,

$$\sum_{j=1}^{\infty} j a_j (z - c)^{j-1},$$

which is said to be the corresponding **derived series**. It is natural to ask whether the derived series has the same circle of convergence and, if so, whether the function it defines is f'. It turns out that the answers to these questions are affirmative within the interior of the circle of convergence.

First we investigate the radius of convergence.

Theorem 6.7.1. *A power series and its derived series have the same radius of convergence.*

Proof. It suffices to consider power series with center 0. Let r_1 and r_2 be the radii of convergence of the power series

$$\sum_{j=0}^{\infty} a_j z^j$$

and its derived series, respectively. For each positive integer n we have

$$|a_n z^n| \leq |n a_n z^n| = |z| |n a_n z^{n-1}|.$$

Thus if the derived series converges, then so does the given series by the comparison test. We deduce that $r_1 \geq r_2$.

It remains to show that $r_2 \geq r_1$. This inequality certainly holds if $r_1 = 0$. Assume therefore that $r_1 > 0$. Choose z and r such that $0 < |z| < r < r_1$. Then

$$\sum_{j=0}^{\infty} |a_j r^j|$$

is convergent, and so

$$0 = \lim_{n \to \infty} |a_n| r^n = r \lim_{n \to \infty} |a_n| r^{n-1}.$$

As it therefore converges to 0, the sequence $\{|a_n| r^{n-1}\}$ is bounded above by some number M. Thus

$$|n a_n z^{n-1}| = n |a_n| r^{n-1} \left| \frac{z}{r} \right|^{n-1}$$

$$\leq M n \left| \frac{z}{r} \right|^{n-1}.$$

Now the series

$$\sum_{j=1}^{\infty} j \left| \frac{z}{r} \right|^{j-1}$$

converges by the ratio test, since

$$\frac{(n+1)|z|^n}{r^n} \cdot \frac{r^{n-1}}{n|z|^{n-1}} = \frac{n+1}{n} \cdot \frac{|z|}{r}$$

and this quantity approaches $|z|/r < 1$ as $n \to \infty$. Therefore the derived series converges at $|z|$ by the comparison test. We conclude that $r_2 \geq r_1$, as required. □

By applying the theorem k times, we see that the series $\sum_{j=0}^{\infty} a_j z^j$ has the same radius of convergence as

$$\sum_{j=k}^{\infty} j(j-1)\cdots(j-k+1)a_j z^{j-k} = \sum_{j=k}^{\infty} \frac{j!}{(j-k)!}a_j z^{j-k}.$$

The theorem cannot be applied to deduce information about convergence on the circle of convergence or, in the real case, at the end points of the interval of convergence. For instance, the real series

$$\sum_{j=1}^{\infty} \frac{(-1)^{j-1}}{j} x^j$$

is convergent when $x = 1$, by Leibniz's test. Therefore the radius of convergence, r, is at least 1. The derived series,

$$\sum_{j=0}^{\infty} (-1)^j x^j,$$

diverges when $x = 1$. Thus, $r = 1$. However, the two series exhibit different behaviour in regard to convergence at at least one endpoint of the interval of convergence.

Theorem 6.7.2. *Let*

$$f(z) = \sum_{j=0}^{\infty} a_j z^j$$

for all z such that $|z| < r$, where r is the radius of convergence of the power series and is nonzero. Then f is differentiable at all z such that $|z| < r$, and

$$f'(z) = \sum_{j=1}^{\infty} j a_j z^{j-1}$$

for all such z.

Proof. Choose w such that $|w| < r$. Also, choose t such that $|w| < t < r$, and z such that $|z| < t$ and $z \neq w$. Define

$$\Delta = \frac{f(z) - f(w)}{z - w} - \sum_{j=1}^{\infty} j a_j w^{j-1}.$$

$$= \frac{1}{z-w} \sum_{j=1}^{\infty} a_j (z^j - w^j) - \sum_{j=1}^{\infty} j a_j w^{j-1}$$

$$= \sum_{j=2}^{\infty} a_j \left(\frac{z^j - w^j}{z-w} - j w^{j-1} \right).$$

The challenge is to show that $\Delta \to 0$ as $z \to w$. It will be met by finding an upper bound for $|\Delta|$.

We begin by investigating the expression in parentheses. Noting that

$$\sum_{k=0}^{j-1} z^k w^{j-k-1} = \frac{z^j - w^j}{z-w}$$

for all $j > 0$, by Theorem 1.5.5, and that

$$\sum_{k=0}^{j-1} w^{j-1} = w^{j-1} \sum_{k=0}^{j-1} 1 = j w^{j-1}$$

for all $j > 0$, we find that

$$\frac{z^j - w^j}{z-w} - j w^{j-1} = \sum_{k=0}^{j-1} z^k w^{j-k-1} - \sum_{k=0}^{j-1} w^{j-1}$$

$$= \sum_{k=0}^{j-1} (z^k w^{j-k-1} - w^{j-1})$$

$$= \sum_{k=1}^{j-1} w^{j-k-1} (z^k - w^k)$$

$$= \sum_{k=1}^{j-1} w^{j-k-1} (z - w) \sum_{m=0}^{k-1} z^m w^{k-m-1}$$

$$= (z - w) \sum_{k=1}^{j-1} \sum_{m=0}^{k-1} z^m w^{j-m-2}.$$

Since $t > \max\{|z|, |w|\}$, for every nonnegative s and u we have

$$|z^s w^u| = |z|^s |w|^u$$

$$\leq t^s t^u$$

$$= t^{s+u}.$$

Thus

$$\left| a_j \left(\frac{z^j - w^j}{z - w} - j w^{j-1} \right) \right| = |a_j| \left| \frac{z^j - w^j}{z - w} - j w^{j-1} \right|$$

$$= |a_j| \left| (z - w) \sum_{k=1}^{j-1} \sum_{m=0}^{k-1} z^m w^{j-m-2} \right|$$

$$\leq |a_j| |z - w| \sum_{k=1}^{j-1} \sum_{m=0}^{k-1} |z^m w^{j-m-2}|$$

$$\leq |a_j| |z - w| \sum_{k=1}^{j-1} \sum_{m=0}^{k-1} t^{j-2}$$

$$= |a_j| |z - w| t^{j-2} \sum_{k=1}^{j-1} \sum_{m=0}^{k-1} 1$$

$$= |a_j| |z - w| t^{j-2} \sum_{k=1}^{j-1} k$$

$$= \frac{j(j-1)}{2} |z - w| |a_j t^{j-2}|$$

for all $j > 1$, since

$$\sum_{k=1}^{j-1} k = \frac{j(j-1)}{2}$$

by Theorem 1.5.4. As $0 < t < r$, the series $\sum_{j=0}^{\infty} |a_j t^j|$ converges, and it follows by two applications of Theorem 6.7.1 that

$$\sum_{j=2}^{\infty} |j(j-1)a_j t^{j-2}|$$

also converges. Writing

$$S = \sum_{j=2}^{\infty} |j(j-1)a_j t^{j-2}|$$

$$= \sum_{j=2}^{\infty} j(j-1)|a_j t^{j-2}|,$$

we obtain

$$|\Delta| \le \sum_{j=2}^{\infty} \left| a_j \left(\frac{z^j - w^j}{z - w} - j w^{j-1} \right) \right|$$

$$\le \frac{|z - w|}{2} \sum_{j=2}^{\infty} j(j-1)|a_j t^{j-2}|$$

$$= \frac{|z - w|S}{2}.$$

Hence $\Delta \to 0$ as $z \to w$, as required. $\qquad \square$

Corollary 6.7.3. *Let r be the radius of convergence of the series $\sum_{j=0}^{\infty} a_j z^j$. Then the series*

$$\sum_{j=0}^{\infty} a_j \frac{z^{j+1}}{j+1}$$

converges for all z such that $|z| < r$.

Proof. This result is immediate from the theorem, for the first series is the derivative of the second. $\qquad \square$

The application of Theorem 6.7.2 is sometimes referred to as term-by-term differentiation and that of Corollary 6.7.3 as term-by-term integration. We will study the general concept of integration in the next chapter.

Theorem 6.7.2 implies that the limit of a power series is continuous within the interior of the interval of convergence.

Example 6.7.1. Theorem 3.2.1 shows that

$$\sum_{j=0}^{\infty} (-1)^j z^j = \frac{1}{1+z}$$

for all z such that $|z| < 1$, since $(-1)^j z^j = (-z)^j$. For all such z differentiation gives

$$-\frac{1}{(1+z)^2} = \sum_{j=1}^{\infty} (-1)^j j z^{j-1}$$

$$= \sum_{j=0}^{\infty} (-1)^{j+1} (j+1) z^j,$$

so that

$$\frac{1}{(1+z)^2} = \sum_{j=0}^{\infty}(-1)^j(j+1)z^j$$

for all z such that $|z| < 1$.

For instance, take $z = -1/3$. Then

$$\sum_{j=0}^{\infty}\frac{j+1}{3^j} = \frac{1}{\left(\frac{2}{3}\right)^2} = \frac{9}{4}.$$

Hence

$$\sum_{j=1}^{\infty}\frac{j}{3^j} = \sum_{j=0}^{\infty}\left(\frac{j+1}{3^j} - \frac{1}{3^j}\right) = \frac{9}{4} - \frac{1}{\frac{2}{3}} = \frac{3}{4}.$$

$\triangle$

Example 6.7.2. Since

$$\exp(z) = e^z = \sum_{j=0}^{\infty}\frac{z^j}{j!}$$

for all z, it follows that

$$\exp'(z) = \sum_{j=1}^{\infty}\frac{jz^{j-1}}{j!} = \sum_{j=1}^{\infty}\frac{z^{j-1}}{(j-1)!} = \sum_{j=0}^{\infty}\frac{z^j}{j!} = e^z = \exp(z),$$

as we saw earlier. $\triangle$

It is possible to express the logarithm function as a power series. Indeed, we have the following theorem.

Theorem 6.7.4. *For all $x \in (-1, 1)$*

$$\log(1 + x) = \sum_{j=1}^{\infty}(-1)^{j+1}\frac{x^j}{j}.$$

Proof. Note first that the power series is absolutely convergent, by Example 3.15.3. Its derivative is

$$\sum_{j=1}^{\infty}(-1)^{j+1}\frac{jx^{j-1}}{j} = \sum_{j=1}^{\infty}(-x)^{j-1} = \sum_{j=0}^{\infty}(-x)^j = \frac{1}{1+x},$$

by Theorem 3.2.1. This expression is also the derivative of $\log(1 + x)$, and so there exists c such that

$$\sum_{j=1}^{\infty}(-1)^{j+1}\frac{x^j}{j} = \log(1 + x) + c.$$

Substituting $x = 0$ gives

$$0 = \log 1 + c = c$$

and the theorem follows. □

If $x = -1$, then the series above diverges since the harmonic series does so. Thus the radius of convergence is 1. If $x = 1$, then the series is alternating and converges. We would naturally expect it to converge to $\log 2$. We will confirm this fact later.

Since $|-x| = |x| < 1$ for all $x \in (-1, 1)$, we also have

$$\log(1 - x) = \sum_{j=1}^{\infty}(-1)^{j+1}\frac{(-x)^j}{j} = -\sum_{j=1}^{\infty}\frac{x^j}{j}.$$

We proceed to extend this result. For all $x \in [-1, 1)$ let

$$t(x) = \frac{1 + x}{1 - x}. \tag{6.17}$$

Note that $t(x) > 0$ for all $x \in (-1, 1)$ and that t is continuous at all $x \in [-1, 1)$. Since $t(-1) = 0$ and

$$\lim_{x\to 1^-}\frac{1 + x}{1 - x} = \infty,$$

the range of t must be the set of all nonnegative real numbers. Furthermore, for all $x \in (-1, 1)$,

$$\log t(x) = \log\frac{1 + x}{1 - x}$$
$$= \log(1 + x) - \log(1 - x)$$
$$= \sum_{j=1}^{\infty}(-1)^{j+1}\frac{x^j}{j} + \sum_{j=1}^{\infty}\frac{x^j}{j}$$
$$= 2\sum_{k=0}^{\infty}\frac{x^{2k+1}}{2k + 1}. \tag{6.18}$$

But if we write t instead of $t(x)$ for convenience, Eq. (6.17) shows that

$$t - tx = 1 + x,$$

whence

$$t - 1 = x(t + 1).$$

Since $t > 0$, it follows that

$$x = \frac{t - 1}{t + 1},$$

and so we obtain the following theorem.

Theorem 6.7.5. *For all $t > 0$*

$$\log t = 2 \sum_{k=0}^{\infty} \frac{1}{2k + 1} \left(\frac{t - 1}{t + 1} \right)^{2k+1}. \tag{6.19}$$

For instance,

$$\log 2 = 2 \sum_{k=0}^{\infty} \frac{1}{(2k + 1)3^{2k+1}}. \tag{6.20}$$

Equation (6.20) may be used to approximate $\log 2$ to a high level of accuracy. Even if we use only five terms of the series, we obtain $0.6931\ldots$, which is correct to two decimal places.

Theorem 6.7.5 can be used to find a quick way of approximating $\log t$ for each positive t. First find an integer k such that

$$2^k \le t < 2^{k+1}.$$

Thus

$$1 \le \frac{t}{2^k} < 2,$$

so that Theorem 6.7.4 can be used to approximate $\log(t/2^k)$. But

$$\log \frac{t}{2^k} = \log t - k \log 2,$$

and so $\log t$ may be calculated from the formula

$$\log t = \log \frac{t}{2^k} + k \log 2.$$

This method is particularly effective for large values of t.

The logarithm function features in the definition of an important mathematical constant. Note first that for every $j > 0$ the mean-value theorem yields the existence of a $\xi \in (j, j + 1)$ such that

$$\frac{\log(j + 1) - \log j}{j + 1 - j} = \log' \xi.$$

In other words,

$$\log(j + 1) - \log j = \frac{1}{\xi}.$$

Since $0 < j < \xi < j + 1$, it follows that

$$\frac{1}{j + 1} < \log(j + 1) - \log j < \frac{1}{j} \qquad (6.21)$$

for all $j > 0$. Thus for all $n > 1$ we have

$$\sum_{j=1}^{n-1} \frac{1}{j + 1} < \sum_{j=1}^{n-1} (\log(j + 1) - \log j) < \sum_{j=1}^{n-1} \frac{1}{j}. \qquad (6.22)$$

Now let

$$S_n = \sum_{j=1}^{n} \frac{1}{j} = \sum_{j=0}^{n-1} \frac{1}{j + 1}$$

for all $n > 0$. Then Eq. (6.22) yields

$$S_n - 1 < \log n < S_{n-1}$$

for all $n > 1$, where the telescoping property was used to evaluate the middle summation. Thus

$$\log n + \frac{1}{n} < S_{n-1} + \frac{1}{n} = S_n < 1 + \log n. \qquad (6.23)$$

Let

$$a_n = S_n - \log n$$

for all $n \in \mathbb{N}$. We shall prove the sequence $\{a_n\}$ convergent. First, since

$$S_{n+1} - S_n = \frac{1}{n+1},$$

we have

$$a_{n+1} - a_n = \frac{1}{n+1} - \log(n+1) + \log n,$$

and it follows that $a_{n+1} < a_n$ because

$$\log(n+1) > \frac{1}{n+1} + \log n$$

by Eq. (6.21). Thus $\{a_n\}$ is a decreasing sequence. It is bounded below by 0: Inequalities (6.23) give

$$a_n = S_n - \log n > \frac{1}{n} > 0.$$

Hence $\{a_n\}$ converges to some number γ, which is called **Euler's constant**. Thus

$$\gamma = \lim_{n \to \infty} a_n = \lim_{n \to \infty} \left(\sum_{j=1}^{n} \frac{1}{j} - \log n \right).$$

From inequalities (6.23) we have

$$\frac{1}{n} < S_n - \log n < 1,$$

so that $0 \le \gamma \le 1$. In fact, γ is approximately 0.577.

We can use these ideas to obtain a convenient power series expansion for $\log 2$. Define

$$\varepsilon_n = S_n - \log n - \gamma$$

for all $n > 0$. Then $\varepsilon_n \to 0$ as $n \to \infty$, and

$$S_n = \log n + \gamma + \varepsilon_n.$$

Now we put

$$T_n = \sum_{j=1}^{n} \frac{(-1)^{j+1}}{j}$$

for all $n > 0$. Thus

$$T_{2n} = \sum_{j=1}^{2n} \frac{(-1)^{j+1}}{j}$$

$$= \sum_{j=1}^{2n} \frac{1}{j} - 2\sum_{j=1}^{n} \frac{1}{2j}$$

$$= S_{2n} - S_n$$

$$= \log 2n + \gamma + \varepsilon_{2n} - \log n - \gamma - \varepsilon_n$$

$$= \log 2 + \varepsilon_{2n} - \varepsilon_n$$

$$\rightarrow \log 2$$

as $n \rightarrow \infty$, so that

$$\log 2 = \sum_{j=1}^{\infty} \frac{(-1)^{j+1}}{j}.$$

We have now extended Theorem 6.7.4 to the case $x = 1$. However, this result does not provide an efficient method for approximating $\log 2$. For instance, by taking 20 terms of the series we obtain the approximation 0.6669, rounded to four decimal places. This result is much worse than that reached by taking only five terms of the series given in Eq. (6.20).

The next theorem gives a power series expansion, due to Gregory, for $\arctan x$ that is valid for all x such that $|x| < 1$.

Theorem 6.7.6. *For all x such that $|x| < 1$,*

$$\arctan x = \sum_{j=0}^{\infty} (-1)^j \frac{x^{2j+1}}{2j+1}.$$

Proof. It follows from Example 3.15.3 that the power series is absolutely convergent. Its derivative is

$$\sum_{j=0}^{\infty} (-1)^j x^{2j} = \sum_{j=0}^{\infty} (-x^2)^j$$

$$= \frac{1}{1+x^2}$$

$$= \arctan' x$$

for all x such that $|x| < 1$. By Corollary 6.3.4 there is a constant c such that

$$\sum_{j=0}^{\infty} (-1)^j \frac{x^{2j+1}}{2j+1} = \arctan x + c$$

for all such x. Since $\arctan 0 = 0$, we have $c = 0$ and the theorem follows. $\square$

We finish this section with an example of an application of the differentiation of power series to the solution of a certain type of differential equation. Such equations arise, for instance, in wave mechanics.

Example 6.7.3. The differential equation

$$y'' - 2xy' + 2\lambda y = 0, \tag{6.24}$$

where $\lambda \geq 0$, is known as **Hermite's equation** of **order** λ. We aim to find a power series solution.

Let

$$y(x) = \sum_{j=0}^{\infty} a_j x^j$$

be a solution. From Eq. (6.24) we obtain

$$0 = \sum_{j=2}^{\infty} j(j-1)a_j x^{j-2} - 2x \sum_{j=1}^{\infty} ja_j x^{j-1} + 2\lambda \sum_{j=0}^{\infty} a_j x^j$$

$$= \sum_{j=0}^{\infty} (j+2)(j+1)a_{j+2} x^j - 2 \sum_{j=0}^{\infty} ja_j x^j + 2\lambda \sum_{j=0}^{\infty} a_j x^j$$

$$= \sum_{j=0}^{\infty} ((j+2)(j+1)a_{j+2} - 2ja_j + 2\lambda a_j)x^j$$

for all x for which the series converges. Therefore

$$(j+2)(j+1)a_{j+2} + 2(\lambda - j)a_j = 0$$

for each j, so that

$$a_{j+2} = -\frac{2(\lambda - j)}{(j+2)(j+1)} a_j.$$

Note that $y(0) = a_0$. Similarly, since

$$y'(x) = \sum_{j=1}^{\infty} j a_j x^{j-1},$$

we have $y'(0) = a_1$. These two equations are called the initial conditions for the differential equation. The series for y is uniquely determined if a_0 and a_1 are known. In fact, an inductive argument shows that

$$a_{2j} = (-1)^j \frac{2^j \lambda(\lambda - 2)(\lambda - 4) \cdots (\lambda - 2j + 2)}{(2j)!} a_0$$

and

$$a_{2j+1} = (-1)^j \frac{2^j (\lambda - 1)(\lambda - 3)(\lambda - 5) \cdots (\lambda - 2j + 1)}{(2j + 1)!} a_1$$

for all $j \geq 0$. Recalling the convention that the empty product is 1, we therefore obtain

$$y(x) = a_0 \sum_{j=0}^{\infty} (-1)^j \frac{2^j \prod_{k=0}^{j-1}(\lambda - 2k)}{(2j)!} x^{2j} + a_1 \sum_{j=0}^{\infty} (-1)^j \frac{2^j \prod_{k=0}^{j-1}(\lambda - 2k - 1)}{(2j + 1)!} x^{2j+1}.$$

In the case where λ is a positive integer, the first series is a polynomial if λ is even and the second series is a polynomial if λ is odd. The solution is therefore a polynomial of degree λ if either λ is even and $(a_0, a_1) = (1, 0)$ or λ is odd and $(a_0, a_1) = (0, 1)$. In either case it is called the **Hermite polynomial** of **degree** λ and is denoted by H_λ. For example, for all x we have

$$H_1(x) = x,$$

$$H_2(x) = 1 - 2x^2,$$

$$H_3(x) = x - \frac{2x^3}{3},$$

$$H_4(x) = 1 - 4x^2 + \frac{4x^4}{3},$$

$$H_5(x) = x - \frac{4x^3}{3} + \frac{4x^5}{15},$$

$$H_6(x) = 1 - 6x^2 + 4x^4 - \frac{8x^6}{15}.$$

There are other ways of defining the Hermite polynomials. For instance, it can be shown [16] that they satisfy the recursive relation

$$H_{n+1}(x) = 2xH_n(x) - 2nH_{n-1}(x)$$

for all $n > 1$ and all x.

It is not hard to see that the solution converges absolutely for all real x. This fact also follows from the following beautiful result, which is due to Fuchs [13]. Consider the differential equation

$$y'' + p(x)y' + q(x)y = 0,$$

where $y(0)$ and $y'(0)$ are given. Let $r > 0$. If p and q can be represented by power series on the interval $(-r, r)$, then the differential equation has a unique power series solution and this solution also converges on $(-r, r)$. In other words, the radius of convergence of the solution is at least as big as the minimum of the radii of convergence of p and q.   △

Exercises 6.7.

1. Verify the formulas for the derivatives of the sine and cosine functions by differentiating their power series.

2. Let

$$f(z) = \sum_{j=0}^{\infty} \frac{z^{3j}}{(3j)!}.$$

   (a) Find the radius of convergence of the series.
   (b) Show that

$$f(z) + f'(z) + f''(z) = e^z$$

   for all z within the circle of convergence.

3. Suppose that the radius of convergence of

$$f(z) = \sum_{j=0}^{\infty} a_j z^j$$

   is $r > 0$. Show that

$$f^{(n)}(z) = \sum_{j=0}^{\infty} \frac{(n+j)!}{n!} a_{n+j} z^j$$

   for all nonnegative integers n and all z such that $|z| < r$.

4. Suppose that

$$f(z) = 1 + \sum_{j=1}^{\infty} a_j z^j$$

is the series solution of the differential equation $f'(z) = f(z)$. Show that

$$a_{n+1} = \frac{a_n}{n+1}$$

for all $n \in \mathbb{N}$ and hence that $f(z) = e^z$ for all z.

5. Suppose that $f(x) = \sum_{j=0}^{\infty} a_j x^j$ is the series solution of $f'(x) = 1 + x^2$ and that $f(0) = 0$. Show that $f(x) = \tan x$ for all x for which $\cos x \neq 0$.

6. Suppose that $f(x) = \sum_{j=0}^{\infty} a_j z^j$ is the series solution of $f''(z) = -f(z)$ and that $f(0) = 0$ and $f'(0) = 1$. Show that $f(z) = \sin z$ for all z.

7. Suppose that $f(x) = \sum_{j=0}^{\infty} a_j z^j$ is the series solution of $f''(z) = -f(z)$ and that $f(0) = 1$ and $f'(0) = 0$. Show that $f(z) = \cos z$ for all z.

8. Is the solution of Hermite's equation satisfying the following conditions a polynomial:

$$\lambda = 3, a_0 = 1, a_1 = 0?$$

9. (Airy's equation) Suppose that $f(x) = \sum_{j=0}^{\infty} a_j x^j$ is the series solution of

$$f''(x) - xf(x) = 0.$$

First, show that $a_2 = 0$ and

$$a_{n+2} = \frac{a_{n-1}}{(n+1)(n+2)}$$

for all $n \in \mathbb{N}$. Hence for all $k \in \mathbb{N}$, show that

$$a_{3k} = \frac{a_0}{2 \cdot 3 \cdot 5 \cdot 6 \cdots (3k-1)3k},$$

$$a_{3k+1} = \frac{a_1}{3 \cdot 4 \cdot 6 \cdot 7 \cdots 3k(3k+1)},$$

$$a_{3k+2} = 0.$$

10. (Legendre's equation) Suppose that $\sum_{j=0}^{\infty} a_j x^j$ is the series solution of

$$(1 - x^2) f''(x) - 2xf'(x) + \alpha(\alpha + 1) f(x) = 0.$$

Substitute the power series for $f(x)$ into the differential equation to obtain a series $p(x)$. Show that the constant term of the latter series satisfies the equation

$$2a_2 + \alpha(\alpha + 1)a_0 = 0,$$

that the coefficient of x satisfies

$$6a_3 + (-2 + \alpha(\alpha + 1))a_1 = 0,$$

and that, for all $n > 1$, the coefficient of x^n satisfies

$$(n + 2)(n + 1)a_{n+2} + (-n(n - 1) - 2n + \alpha(\alpha + 1))a_n = 0.$$

Writing

$$p(x) = a_0 p_1(x) + a_1 p_2(x),$$

show that the series $p(x)$ converges absolutely whenever $|x| < 1$.

When n is a nonnegative integer, either p_1 or p_2 is a polynomial. Show that if $n = 0$, then $p_1(x) = 1$ and

$$p_2(x) = \frac{1}{2} \log \frac{1 + x}{1 - x}$$

and that if $n = 1$, then $p_2(x) = x$ and

$$p_1(x) = 1 - \frac{1}{2} \log \frac{1 + x}{1 - x}.$$

11. Use the fact that

$$\lim_{n \to \infty} \left(\sum_{j=1}^{n} \frac{1}{j} - \log n \right) = \gamma$$

to show that

$$\lim_{n \to \infty} \frac{1}{\log n} \sum_{j=1}^{n} \frac{1}{j} = 1.$$

Derive this result also from Stolz's theorem.

Chapter 7
The Riemann Integral

In this chapter we use the idea of an area to motivate a concept called an integral of a function, and we show that the process of finding an integral of a function is closely related to that of obtaining an antiderivative of the function, that is, a function whose derivative is the given function. Some techniques for finding integrals are derived and the use of integrals for testing the convergence of certain types of series are discussed. All functions under consideration are assumed to be real-valued functions of one or more real variables.

7.1 Area Under a Curve

Suppose we wish to approximate the area bounded by the curve $y = f(x)$ and the lines $x = a$, $x = b$, and $y = 0$ (see Fig. 7.1. For the sake of clarity, the figure is drawn for a function f such that $f(x) \geq 0$ for all $x \in [a, b]$.)

We approximate the area using small rectangular regions. Let us first introduce a few definitions and notation.

Definition 7.1.1. Let $[a, b]$ be a closed interval. By a **partition** of $[a, b]$ we mean a (finite) sequence $x_0, x_1, \ldots, x_n$ of numbers such that

$$a = x_0 < x_1 < \ldots < x_n = b.$$

(More formally, $x_0 = a$, $x_n = b$, and $x_{j+1} > x_j$ for all $j < n$.)

This partition P is denoted by $(x_0, x_1, \ldots, x_n)$. The numbers $x_0, x_1, \ldots, x_n$ are called the **division points** of P. The set of division points of P is denoted by $\hat{P}$. We say that $\hat{P}$ **determines** P. If $(x_0, x_1, \ldots, x_n)$ is a partition P of $[a, b]$, we define the **norm** or **mesh** of P to be the maximum of the quantities $x_{j+1} - x_j$, where

© Springer Science+Business Media New York 2015   333
C.H.C. Little et al., *Real Analysis via Sequences and Series*, Undergraduate
Texts in Mathematics, DOI 10.1007/978-1-4939-2651-0_7

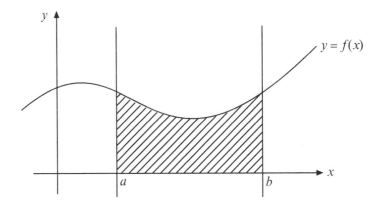

Fig. 7.1 Area under the graph of f between a and b

$0 \leq j < n$. This number is denoted by $\mu(P)$. Each interval $[x_j, x_{j+1}]$ is called a **subinterval** of P, and the elements of (x_j, x_{j+1}) are called **interior points** of $[x_j, x_{j+1}]$.

Recall that a function f mapping $[a, b]$ into $\mathbb{R}$ is bounded if there exist m and M such that $m \leq f(x) \leq M$ for all $x \in [a, b]$. If $(x_0, x_1, \ldots, x_n)$ is a partition of $[a, b]$ and f is bounded on $[a, b]$, then f is also bounded on each subinterval $[x_j, x_{j+1}]$ of $[a, b]$, where $j < n$. For each $j < n$ we set

$$M_j(f) = \sup\{f(x) \mid x \in [x_j, x_{j+1}]\}$$

and

$$m_j(f) = \inf\{f(x) \mid x \in [x_j, x_{j+1}]\}.$$

If f and g are functions that are defined and bounded on $[a, b]$ and $f(x) \leq g(x)$ for all $x \in [a, b]$, then it is clear that $M_j(f) \leq M_j(g)$ and $m_j(f) \leq m_j(g)$ for each j. For the rest of this chapter, unless we state otherwise, we assume that f is a function that is defined and bounded on a closed interval $[a, b]$. Given such a function f and a partition $(x_0, x_1, \ldots, x_n)$ of $[a, b]$, we shall also take $M_j(f)$ and $m_j(f)$ to be defined as above.

Remark. There might not exist $c \in [x_j, x_{j+1}]$ such that $f(c) = M_j(f)$, and similarly for $m_j(f)$. For example, let $f : [0, 2] \to \mathbb{R}$ be defined by

$$f(x) = \begin{cases} x - \lfloor x \rfloor & \text{if } x \in [0, 1] \\ \lfloor x \rfloor - x & \text{if } x \in (1, 2] \end{cases}$$

Fig. 7.2 Graph of $x - \lfloor x \rfloor$
for $x \in [0, 1]$ and $\lfloor x \rfloor - x$ for
$x \in (1, 2]$

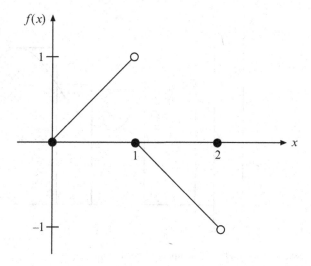

(see Fig. 7.2). Let P be the partition $(0, 1, 2)$ of $[0, 2]$. Then $M_0(f) = 1$ and $m_1(f) = -1$. It is clear that there is no $c \in [0, 1]$ such that $f(c) = 1$ and no $c \in [1, 2]$ such that $f(c) = -1$. However, if f is continuous on $[a, b]$, then there is always a number $c \in [x_j, x_{j+1}]$ such that $f(c) = M_j(f)$ (see Corollary 5.3.2). A corresponding statement holds for $m_j(f)$. The number c also exists if f is nondecreasing or nonincreasing on $[a, b]$. For instance, if f is nondecreasing on $[a, b]$, then $M_j(f) = f(x_{j+1})$ and $m_j(f) = f(x_j)$.

Let $(x_0, x_1, \ldots, x_n)$ be a partition P of some closed interval $J \subseteq [a, b]$. Then we define

$$U(P, f) = \sum_{j=0}^{n-1} M_j(f)(x_{j+1} - x_j).$$

The number $U(P, f)$ is called the **upper (Riemann) sum** of f over J relative to P. An upper sum of f over the whole of $[a, b]$ can be thought of geometrically as the shaded area in Fig. 7.3. This sum is an approximation, from above, for the area A bounded by the curve $y = f(x)$ and the lines $x = a$, $x = b$, and $y = 0$. It is clear that if $c \in (a, b)$ and P_1 and P_2 are partitions of $[a, c]$ and $[c, b]$, respectively, then

$$U(P, f) = U(P_1, f) + U(P_2, f),$$

where P is the partition of $[a, b]$ for which $\hat{P} = \hat{P}_1 \cup \hat{P}_2$.

The sum

$$L(P, f) = \sum_{j=0}^{n-1} m_j(f)(x_{j+1} - x_j)$$

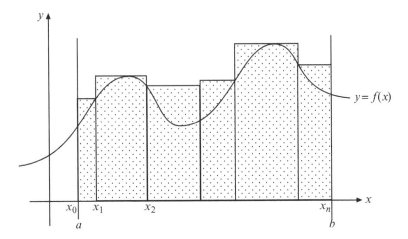

Fig. 7.3 Illustration of an upper Riemann sum

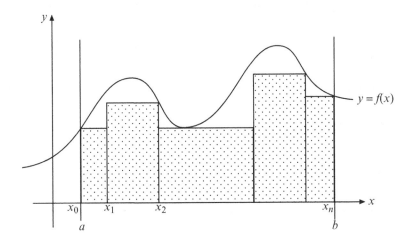

Fig. 7.4 Illustration of a lower Riemann sum

is called the **lower (Riemann) sum** of f over J relative to the partition P of J. If $J = [a, b]$, then this lower sum gives the shaded area in Fig. 7.4 and constitutes an approximation for A from below. Note also that

$$L(P, f) = L(P_1, f) + L(P_2, f),$$

where P_1 and P_2 are partitions of $[a, c]$ and $[c, b]$, respectively, for some $c \in (a, b)$, and P is the partition of $[a, b]$ for which $\hat{P} = \hat{P}_1 \cup \hat{P}_2$.

We also have

$$U(P, f) - L(P, f) = \sum_{j=0}^{n-1} M_j(f)(x_{j+1} - x_j) - \sum_{j=0}^{n-1} m_j(f)(x_{j+1} - x_j)$$

$$= \sum_{j=0}^{n-1} (M_j(f) - m_j(f))(x_{j+1} - x_j).$$

In addition, if f and g are functions that are defined and bounded on J and satisfy $f(x) \leq g(x)$ for all $x \in J$, then $L(P, f) \leq L(P, g)$ and $U(P, f) \leq U(P, g)$.

Example 7.1.1. If P is the partition (a, b) of $[a, b]$ determined by $\{a, b\}$, then

$$L(P, f) = m(b - a)$$

and

$$U(P, f) = M(b - a),$$

where $M = \sup\{f(x) \mid x \in [a, b]\}$ and $m = \inf\{f(x) \mid x \in [a, b]\}$. △

The sum

$$S(P, f) = \sum_{j=0}^{n-1} f(c_j)(x_{j+1} - x_j), \tag{7.1}$$

where $c_j \in [x_j, x_{j+1}]$ for all $j < n$, is called a **Riemann sum** of f over J relative to P and the intermediate points $c_0, c_1, \ldots, c_{n-1}$. It gives the shaded area in Fig. 7.5 if $J = [a, b]$. Observe that the notation does not indicate the dependence of the Riemann sum $S(P, f)$ on the intermediate points. Whenever we need to make this dependence explicit, then we write $S(P, f, c)$ instead of $S(P, f)$, where $c = (c_0, c_1, \ldots, c_{n-1})$.

Example 7.1.2. If $f(x) = k$ for all $x \in [a, b]$, then

$$S(P, f) = \sum_{j=0}^{n-1} k(x_{j+1} - x_j)$$

$$= k(b - a),$$

by the telescoping property. In this case we also have

$$U(P, f) = L(P, f) = S(P, f).$$

△

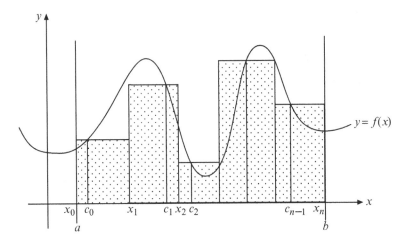

Fig. 7.5 Illustration of a Riemann sum

If f and g are functions defined and bounded on J, and k and l are constants, then

$$S(P, kf + lg) = \sum_{j=0}^{n-1} (kf(c_j) + lg(c_j))(x_{j-1} - x_j)$$

$$= k \sum_{j=0}^{n-1} f(c_j)(x_{j+1} - x_j) + l \sum_{j=0}^{n-1} g(c_j)(x_{j+1} - x_j)$$

$$= kS(P, f) + lS(P, g).$$

It is also clear that

$$L(P, f) \leq S(P, f) \leq U(P, f). \tag{7.2}$$

Intuitively, as the partition becomes "finer," we expect the Riemann sum to provide a better approximation for the required area. However, it is not always the case that if $\mu(P_1) \leq \mu(P_2)$; we necessarily obtain a better approximation for the area by using P_1 rather than P_2.

Definition 7.1.2. We say that a partition Q of $[a, b]$ is a **refinement** of a partition P if $\hat{P} \subseteq \hat{Q}$; that is, every division point of P is a division point of Q.

Let P and Q be partitions of $[a, b]$. We define $P \cup Q$ to be the partition determined by the set $\hat{P} \cup \hat{Q}$ of division points.

It is clear that if P and Q are partitions of $[a, b]$, then $P \cup Q$ is a refinement of both P and Q and that if Q is a refinement of P, then $\mu(Q) \leq \mu(P)$.

Theorem 7.1.1. *Suppose P and Q are partitions of $[a,b]$ and that Q is a refinement of P. Then for each function f defined and bounded on $[a,b]$ we have*

$$L(P, f) \leq L(Q, f)$$

and

$$U(P, f) \geq U(Q, f).$$

Proof. To establish the first inequality, it suffices to prove the result for the case where Q contains exactly one more point than P. An easy induction then gives us the general case.

Suppose $\hat{Q} = \hat{P} \cup \{y\}$, where $P = (x_0, x_1, \ldots, x_n)$ and $x_j < y < x_{j+1}$ for some integer $j < n$. Let

$$m_1' = \inf\{f(x) \mid x \in [x_j, y]\}$$

and

$$m_2' = \inf\{f(x) \mid x \in [y, x_{j+1}]\}.$$

Clearly, $m_1' \geq m_j(f)$ and $m_2' \geq m_j(f)$. Thus

$$
\begin{aligned}
L(Q, f) - L(P, f) &= m_1'(y-x_j)+m_2'(x_{j+1}-y)-m_j(f)(x_{j+1}-x_j) \\
&\geq m_j(f)(y-x_j)+m_j(f)(x_{j+1}-y)-m_j(f)(x_{j+1}-x_j) \\
&= 0.
\end{aligned}
$$

The proof of the second inequality is similar. $\square$

Corollary 7.1.2. *Let P and Q be any two partitions of $[a,b]$. Then $L(P, f) \leq U(Q, f)$.*

Proof. Since $P \cup Q$ is a refinement of both P and Q, Theorem 7.1.1 implies that

$$L(P, f) \leq L(P \cup Q, f) \leq U(P \cup Q, f) \leq U(Q, f).$$

$\square$

Exercises 7.1.

1. Prove that, for any partitions P and Q of an interval $[a,b]$,

$$m(b-a) \leq L(P, f) \leq U(Q, f) \leq M(b-a),$$

where $M = \sup\{f(x) \mid x \in [a,b]\}$ and $m = \inf\{f(x) \mid x \in [a,b]\}$.

2. Consider the function f defined by $f(x) = x$ for all $x \in [0, 1]$. For each positive integer n, the partition

$$P_n = \left(0, \frac{1}{n}, \frac{2}{n}, \dots, 1 \right)$$

divides the interval $[0, 1]$ into n subintervals of length $1/n$. Find

$$\lim_{n \to \infty} U(P_n, f)$$

and

$$\lim_{n \to \infty} L(P_n, f).$$

7.2   Upper and Lower Integrals

The collections of upper and lower sums of f over $[a, b]$ are nonempty and bounded below and above, respectively. Hence we can make the following definition.

Definition 7.2.1. The **lower integral** $\underline{\int} f$ of a bounded function f over $[a, b]$ is the least upper bound of the set of lower sums of f relative to partitions of $[a, b]$. Similarly, the **upper integral** $\overline{\int} f$ of f over $[a, b]$ is the greatest lower bound of the set of upper sums of f relative to partitions of $[a, b]$.

From Corollary 7.1.2 it is clear that

$$L(P, f) \le \underline{\int} f \le \overline{\int} f \le U(P, f) \tag{7.3}$$

for every partition P of $[a, b]$. Moreover, if f and g are bounded functions such that $f(x) \le g(x)$ for all $x \in [a, b]$, then $\underline{\int} f \le \underline{\int} g$ and $\overline{\int} f \le \overline{\int} g$. It is also clear from inequalities (7.2) that

$$\underline{\int} f \le S(P, f) \le \overline{\int} f$$

for every Riemann sum $S(P, f)$ of f over $[a, b]$ relative to P.

Example 7.2.1. Let $f : [0, 1] \to \mathbb{R}$ be defined by

$$f(x) = \begin{cases} 0 \text{ if } x \text{ is irrational,} \\ 1 \text{ if } x \text{ is rational.} \end{cases}$$

For each partition $P = (x_0, x_1, \ldots, x_n)$ of $[0, 1]$ we have $M_j(f) = 1$ and $m_j(f) = 0$ for all $j < n$. Hence

$$U(P, f) = \sum_{j=0}^{n-1} (x_{j+1} - x_j)$$

$$= x_n - x_0$$

$$= 1$$

and

$$L(P, f) = 0.$$

Thus $\underline{\int} f = 0$ and $\overline{\int} f = 1$.   △

7.3   The Riemann Integral

In this section we define the (Riemann) integral of a function and show that when the integral exists, it coincides with both the upper and lower integrals.

Definition 7.3.1. Let f be a bounded function from $[a, b]$ into $\mathbb{R}$. We say that the Riemann sums of f over $[a, b]$ **converge** to a number I if for every $\varepsilon > 0$ there is a $\delta > 0$ such that if P is a partition of $[a, b]$ with norm $\mu(P) < \delta$, then

$$|S(P, f) - I| < \varepsilon$$

for every Riemann sum $S(P, f)$ of f over $[a, b]$ relative to P.

Remark 1. As in Proposition 2.2.3, convergence will follow if this inequality can be established with ε replaced by $c\varepsilon$ for some $c > 0$.

Remark 2. If the Riemann sums of f over $[a, b]$ converge to I, then $S(P, f) < I + \varepsilon$ for every Riemann sum $S(P, f)$ of f over $[a, b]$ relative to P. Thus $\underline{\int} f < I + \varepsilon$, and since ε is arbitrary it follows that $\underline{\int} f \leq I$. Similarly, $I \leq \overline{\int} f$.

Example 7.3.1. If $f(x) = k$ for each $x \in [a, b]$, then the Riemann sums of f over $[a, b]$ converge to $k(b-a)$, the value of each such Riemann sum (cf. Example 7.1.2).   △

The proof of the following theorem is analogous to that of Theorem 2.2.2 and therefore omitted.

Theorem 7.3.1. Let f be a bounded function on $[a, b]$ whose Riemann sums converge to I_1 and I_2. Then $I_1 = I_2$.

Definition 7.3.2. Let f be a function that is bounded on $[a, b]$. If the Riemann sums converge to some number I, then we say that f is (Riemann) **integrable** over $[a, b]$. We call I the (Riemann) **integral** of f over $[a, b]$, and write

$$\int_a^b f(x)\, dx = I,$$

or simply $\int_a^b f = I$ or $\int f = I$. The function f is called the **integrand** of I.

For instance, it follows from Example 7.3.1 that

$$\int_a^b k\, dx = k(b - a).$$

In particular,

$$\int_a^b 0\, dx = 0.$$

For typographical convenience, an integral of the form $\int_a^b \frac{f(x)}{g(x)}\, dx$ is sometimes written as $\int_a^b \frac{f(x)\, dx}{g(x)}$.

Theorem 7.3.2. *Let $f : [a, b] \to \mathbb{R}$ be integrable. Then every sequence $\{S(P_n, f)\}$ of Riemann sums for f over $[a, b]$ satisfying $\lim_{n \to \infty} \mu(P_n) = 0$ must converge to $\int f$.*

Proof. Let $I = \int f$. For each $\varepsilon > 0$, there exists $\delta > 0$ such that

$$|S(P, f) - I| < \varepsilon$$

for every partition P of $[a, b]$ with $\mu(P) < \delta$. For every sequence $\{S(P_n, f)\}$ of Riemann sums for f over $[a, b]$ satisfying $\lim_{n \to \infty} \mu(P_n) = 0$, there is an N such that $\mu(P_n) < \delta$ whenever $n \geq N$. Therefore $|S(P_n, f) - I| < \varepsilon$ for each $n \geq N$, and so

$$\lim_{n \to \infty} S(P_n, f) = I. \qquad \square$$

The converse of this theorem is given as an exercise.

Definition 7.3.3. We say that $U(P, f) - L(P, f)$ **converges** to 0 over a closed interval $[a, b]$ if for every $\varepsilon > 0$ there is a $\delta > 0$ such that if P is any partition of $[a, b]$ satisfying $\mu(P) < \delta$, then

$$U(P, f) - L(P, f) < \varepsilon.$$

Remark 1. Again, to establish convergence it suffices to prove the preceding inequality with ε replaced by $c\varepsilon$ for some $c > 0$.

Remark 2. Suppose that $U(P, f) - L(P, f)$ converges to 0 over $[a, b]$. By inequalities (7.3) we see that

$$U(P, f) - L(P, f) \geq \overline{\int} f - \underline{\int} f$$

for every partition P of $[a, b]$. However, if $\overline{\int} f > \underline{\int} f$, then

$$\overline{\int} f - \underline{\int} f > 0,$$

and so we may find $\delta > 0$ such that

$$U(P, f) - L(P, f) < \overline{\int} f - \underline{\int} f$$

for every partition P of $[a, b]$ with norm less than δ. This contradiction shows that if $U(P, f) - L(P, f)$ converges to 0, then $\underline{\int} f = \overline{\int} f$.

Theorem 7.3.3. *Let f be a function that is bounded on $[a, b]$. Then the following two statements are equivalent:*

1. $U(P, f) - L(P, f)$ converges to 0 over $[a, b]$;
2. f is integrable over $[a, b]$.

Proof. Suppose (1) holds. Given $\varepsilon > 0$, there exists $\delta > 0$ such that

$$U(P, f) - L(P, f) < \varepsilon$$

for every partition P of $[a, b]$ with $\mu(P) < \delta$. Now

$$U(P, f) - L(P, f) = \left(U(P, f) - \overline{\int} f \right) + \left(\overline{\int} f - \underline{\int} f \right) + \left(\underline{\int} f - L(P, f) \right).$$

Since each term in parentheses is nonnegative,

$$\overline{\int} f - \underline{\int} f \leq U(P, f) - L(P, f) < \varepsilon$$

whenever $\mu(P) < \delta$. As ε is arbitrary, $\overline{\int} f = \underline{\int} f$. Let $I = \underline{\int} f$. It follows that $L(P, f) \leq I \leq U(P, f)$, and as we also have $L(P, f) \leq S(P, f) \leq U(P, f)$ for each Riemann sum $S(P, f)$, we deduce that

$$|S(P, f) - I| \leq U(P, f) - L(P, f) < \varepsilon,$$

as required.

Conversely, suppose that (2) holds, and let $I = \int f$. Then given any $\varepsilon > 0$, there is a $\delta > 0$ such that

$$I - \varepsilon < S(P, f) < I + \varepsilon$$

for each partition P of $[a, b]$ with norm less than δ and each Riemann sum $S(P, f)$ of f over $[a, b]$ relative to P. Fix such a $P = (x_0, x_1, \ldots, x_n)$ and choose $c_j \in [x_j, x_{j+1}]$ for each $j < n$. Since $M_j(f) - \varepsilon$ is less than the least upper bound of f on $[x_j, x_{j+1}]$, we may choose each c_j so that $f(c_j) > M_j(f) - \varepsilon$. Therefore

$$U(P, f) = \sum_{j=0}^{n-1} M_j(f)(x_{j+1} - x_j)$$

$$< \sum_{j=0}^{n-1} f(c_j)(x_{j+1} - x_j) + \varepsilon \sum_{j=0}^{n-1}(x_{j+1} - x_j)$$

$$= S(P, f) + \varepsilon(b - a)$$

$$< I + \varepsilon(b - a + 1).$$

Similarly,

$$L(P, f) > I - \varepsilon(b - a + 1),$$

and so

$$U(P, f) - L(P, f) < 2\varepsilon(b - a + 1)$$

whenever $\mu(P) < \delta$. Therefore $U(P, f) - L(P, f)$ converges to 0, as required. □

From Remarks 2 after Definitions 7.3.1 and 7.3.3 we obtain the following corollary.

Corollary 7.3.4. *If f is integrable over $[a, b]$, then $\underline{\int} f = \overline{\int} f = \int f$.*

The proof of the next result is adapted from [7].

Lemma 7.3.5. *Let $f: [a, b] \to \mathbb{R}$ be a bounded function. Then for every $\varepsilon > 0$ there is a $\delta > 0$ such that*

$$U(P, f) < \overline{\int} f + \varepsilon$$

and

$$L(P, f) > \underline{\int} f - \varepsilon$$

for every partition P of $[a, b]$ for which $\mu(P) < \delta$.

Proof. The lemma is certainly true if $f(x) = 0$ for all $x \in [a, b]$, for in that case we have $U(P, f) = \overline{\int} f = L(P, f) = \underline{\int} f = 0$. We therefore assume that $f(x) \neq 0$ for some $x \in [a, b]$.

Choose $\varepsilon > 0$. By the definition of the upper integral there exists a partition P_1 such that

$$U(P_1, f) < \overline{\int} f + \frac{\varepsilon}{2}.$$

Suppose that P_1 contains k interior points. Theorem 7.1.1 shows that P_1 may be refined if necessary so that $k > 0$. Let δ be the minimum length of any subinterval of P_1. In view of Theorem 7.1.1, we may also assume that

$$\delta < \frac{\varepsilon}{6Mk},$$

where $M = \sup_{x \in [a,b]} |f(x)| > 0$.

Now let P be any partition with $\mu(P) < \delta$, and let $Q = P \cup P_1$. Then Theorem 7.1.1 shows that

$$U(Q, f) \leq U(P_1, f) < \overline{\int} f + \frac{\varepsilon}{2}. \tag{7.4}$$

We wish to obtain an upper bound for $U(P, f) - U(Q, f)$. By the choice of δ, there is at most one point of P_1 in each subinterval of P. Let S be the set of subintervals J of P such that some interior point of J belongs to P_1. These are precisely the subintervals of P that are not subintervals of Q. Rather, each such subinterval of P is the union of just two subintervals of Q. The contributions to $U(P, f)$ and to $U(Q, f)$ of each subinterval of P not in S are equal and therefore cancel in the expression $U(P, f) - U(Q, f)$. Now choose an interval $J = [r, t]$ in S and let s be its unique interior point in P_1. Thus $J_1 = [r, s]$ and $J_2 = [s, t]$ are subintervals of Q. The contribution of J to $|U(P, f)|$ is no greater than $M(t - r) < M\delta$. Similarly, the contribution of each of J_1 and J_2 to $|U(Q, f)|$ is less than $M\delta$. Therefore the sum of these contributions to $|U(P, f)|$ and $|U(Q, f)|$ is less than $3M\delta$. Moreover since P_1 has just k interior points, we see that $|S| \leq k$. Using the triangle inequality, we therefore conclude that

$$U(P, f) - U(Q, f) \leq |U(P, f) - U(Q, f)| \leq 3kM\delta < \frac{\varepsilon}{2}. \tag{7.5}$$

Combining inequalities (7.4) and (7.5), we obtain

$$U(P, f) = U(Q, f) + U(P, f) - U(Q, f) < \overline{\int} f + \varepsilon,$$

and the first of the required inequalities is proved. The proof of the second is similar.
□

Theorem 7.3.6. *A bounded function f is integrable if and only if $\underline{\int} f = \overline{\int} f$.*

Proof. We have already observed the necessity of the equation. Suppose therefore that $\underline{\int} f = \overline{\int} f = I$, and choose $\varepsilon > 0$. By Lemma 7.3.5 there exists $\delta > 0$ such that

$$I - \varepsilon < L(P, f) \le U(P, f) < I + \varepsilon$$

for every partition P for which $\mu(P) < \delta$. Hence

$$U(P, f) - L(P, f) < \varepsilon + \varepsilon = 2\varepsilon,$$

so that f is integrable by Theorem 7.3.3.   □

Corollary 7.3.7. *A bounded function defined on $[a, b]$ is integrable if and only if for every $\varepsilon > 0$ there is a partition P of $[a, b]$ such that*

$$U(P, f) - L(P, f) < \varepsilon. \tag{7.6}$$

Proof. The necessity is clear from Theorem 7.3.3.

Given a partition P of $[a, b]$ satisfying Eq. (7.6), we have

$$L(P, f) \le \underline{\int} f \le \overline{\int} f \le U(P, f).$$

Hence

$$\overline{\int} f - \underline{\int} f \le U(P, f) - L(P, f) < \varepsilon.$$

Since ε is arbitrary, it follows that $\overline{\int} f \le \underline{\int} f$. Therefore $\overline{\int} f = \underline{\int} f$, and so the result is a consequence of Theorem 7.3.6.   □

The next corollary is obtained by translating the previous one into terms of sequences and is often useful.

Corollary 7.3.8. *Let $f : [a, b] \to \mathbb{R}$ be a bounded function. Then f is integrable if and only if there exists a sequence $\{P_n\}$ of partitions of $[a, b]$ such that*

$$\lim_{n \to \infty} (U(P_n, f) - L(P_n, f)) = 0.$$

If f is integrable, then

$$\int f = \lim_{n \to \infty} U(P_n, f) = \lim_{n \to \infty} L(P_n, f).$$

Theorem 7.3.9. *If a function f is integrable over closed intervals $[a, c]$ and $[c, b]$, then f is integrable over $[a, b]$.*

Proof. We must show that $U(P, f) - L(P, f)$ converges to 0 over $[a, b]$. Choose $\varepsilon > 0$. Since f is integrable on $[a, c]$, there exists $\delta_1 > 0$ such that

$$U(P_1, f) - L(P_1, f) < \varepsilon$$

for every partition P_1 of $[a, c]$ satisfying $\mu(P_1) < \delta_1$. Similarly, there exists $\delta_2 > 0$ such that

$$U(P_2, f) - L(P_2, f) < \varepsilon$$

for every partition P_2 of $[c, b]$ satisfying $\mu(P_2) < \delta_2$. Let $\delta = \min\{\delta_1, \delta_2\}$, and choose a partition P of $[a, b]$ such that $\mu(P) < \delta$. We may refine it if necessary to ensure that $c \in \hat{P}$. Then $\hat{P} \cap [a, c]$ is the set of division points of a partition P_1 of $[a, c]$ with norm less than δ_1. Similarly, $\hat{P} \cap [c, b]$ is the set of division points of a partition P_2 of $[c, b]$ with norm less than δ_2. Moreover

$$U(P, f) = U(P_1, f) + U(P_2, f)$$

and

$$L(P, f) = L(P_1, f) + L(P_2, f);$$

hence

$$U(P, f) - L(P, f) = U(P_1, f) - L(P_1, f) + U(P_2, f) - L(P_2, f) < \varepsilon + \varepsilon = 2\varepsilon,$$

as required. □

By means of an easy inductive argument, we obtain the following corollary.

Corollary 7.3.10. *Let $(x_0, x_1, \ldots, x_n)$ be a partition of a closed interval $[a, b]$. Let f be a function defined on $[a, b]$, and suppose that f is integrable over $[x_j, x_{j+1}]$ for each $j < n$. Then f is integrable over $[a, b]$.*

We shall show that if f is continuous on $[a, b]$, then it is integrable over $[a, b]$. In fact, we prove the following more general theorem.

Theorem 7.3.11. *If f is continuous on (a, b) and bounded on $[a, b]$, then f is integrable over $[a, b]$.*

Proof. We suppose first that f is continuous at a (but not necessarily at b). Choose ε such that $0 < \varepsilon < b - a$, and let $c = b - \varepsilon > a$. Then f is continuous on $[a, c]$, and Theorem 5.5.2 implies the existence of $\delta > 0$ such that

$$|f(x) - f(y)| < \varepsilon$$

whenever $|x - y| < \delta$ and $x, y \in [a, c]$.

Take any partition $P = (x_0, x_1, \ldots, x_n)$ of $[a, b]$ with norm $\mu(P) < \delta$, and refine it if necessary so that $c = x_r$ for some r such that $0 < r < n$. By Theorem 7.3.3 it suffices to show that

$$U(P, f) - L(P, f) < k\varepsilon$$

for some $k > 0$. Let P_1 and P_2 be the partitions of $[a, c]$ and $[c, b]$, respectively, such that $\widehat{P_1}$ and $\widehat{P_2}$ are the intersections of $\hat{P}$ with $[a, c]$ and $[c, b]$, respectively. The function f is continuous on $[x_j, x_{j+1}]$ for each j such that $0 \le j < r$, so that $M_j(f) = f(c_j)$ for some $c_j \in [x_j, x_{j+1}]$ and $m_j(f) = f(d_j)$ for some $d_j \in [x_j, x_{j+1}]$. Thus

$$U(P_1, f) - L(P_1, f) = \sum_{j=0}^{r-1} (M_j(f) - m_j(f))(x_{j+1} - x_j)$$

$$= \sum_{j=0}^{r-1} (f(c_j) - f(d_j))(x_{j+1} - x_j)$$

$$< \varepsilon \sum_{j=0}^{r-1} (x_{j+1} - x_j)$$

$$= \varepsilon(c - a).$$

Since f is bounded on $[c, b]$, there exists M such that $|f(x)| < M$ for all $x \in [c, b]$. Note also that

$$M_j(f) - m_j(f) = |M_j(f) - m_j(f)| \le |M_j(f)| + |m_j(f)| \le 2M$$

for each j such that $r \le j < n$, by the triangle inequality. It follows that

$$U(P_2, f) - L(P_2, f) = \sum_{j=r}^{n-1} (M_j(f) - m_j(f))(x_{j+1} - x_j)$$

$$\le 2M(b - c)$$

$$= 2M\varepsilon.$$

We therefore deduce that

$$U(P, f) - L(P, f) = U(P_1, f) - L(P_1, f) + U(P_2, f) - L(P_2, f)$$

$$< (c - a)\varepsilon + 2M\varepsilon$$

$$= (c - a + 2M)\varepsilon,$$

as required.

We have now proved the theorem in the case where f is continuous at a. Similarly, the theorem holds if f is continuous at b. The remaining case is dealt with by noting that f is continuous at c and applying these results and Theorem 7.3.9 to the intervals $[a, c]$ and $[c, b]$. $\square$

Corollary 7.3.12. *Let f be a function that is continuous over an interval $[a, b]$ except at a finite number of points. If f is bounded on $[a, b]$, then f is integrable over $[a, b]$.*

Proof. Let P be a partition $(x_0, x_1, \ldots, x_n)$ of $[a, b]$ such that $\hat{P}$ contains all points in $[a, b]$ where f is not continuous. By Theorem 7.3.11, f is integrable on $[x_j, x_{j+1}]$ for each $j < n$. Now apply Corollary 7.3.10. $\square$

By applying Theorems 5.3.1 and 7.3.11, we also obtain the following result.

Corollary 7.3.13. *If a function f is continuous over an interval $[a, b]$, then f is integrable over $[a, b]$.*

Lebesgue showed that a function may be discontinuous at infinitely many points yet be integrable.

Example 7.3.2. Consider the function $f : [0, 1] \to \mathbb{R}$ defined by

$$f(x) = \begin{cases} x \lfloor \frac{1}{x} \rfloor & \text{if } x \neq 0, \\ 0 & \text{if } x = 0. \end{cases}$$

Thus for each

$$x \in \left(\frac{1}{n+1}, \frac{1}{n} \right],$$

where $n \in \mathbb{N}$, we have $\lfloor 1/x \rfloor = n$. In this case it follows that $f(x) = nx$, so that

$$\frac{n}{n+1} < f(x) \leq 1.$$

Thus f is bounded on $(1/(n+1), 1/n]$ for all $n \in \mathbb{N}$. Consequently it is bounded on $[1/n, 1]$ for all $n \in \mathbb{N}$. Moreover

$$\lim_{x \to \left(\frac{1}{n+1} \right)^+} f(x) = \frac{n}{n+1} < 1,$$

whereas

$$f\left(\frac{1}{n+1} \right) = \frac{1}{n+1} \lfloor n + 1 \rfloor = 1.$$

Hence f is discontinuous at $1/(n+1)$ for each $n \in \mathbb{N}$.

Choose $\varepsilon > 0$ and $N \in \mathbb{N}$ so large that $1/N < \varepsilon$. By Corollary 7.3.12, f is integrable on $[1/N, 1]$. Let $P = (x_0, x_1, \ldots, x_n)$ be a partition of $[0, 1]$. By Theorem 7.3.3, we may refine P so that $x_k = 1/N$, for some $k < n$, and

$$\sum_{j=k}^{n-1} (M_j(f) - m_j(f))(x_{j+1} - x_j) < \varepsilon.$$

Now for each $x < 1/N$ we have $x \in (1/(M+1), 1/M]$ for some $M \in \mathbb{N}$ such that $M \geq N$, so that

$$1 \geq f(x) > \frac{M}{M+1} \geq \frac{N}{N+1}$$

since the sequence $\{n/(n+1)\}$ is increasing. [This observation is easily checked for all $n > 0$ by dividing both sides of the inequality

$$n^2 > n^2 - 1 = (n-1)(n+1)$$

by $n(n+1)$.] As

$$\frac{N}{N+1} > \frac{N-1}{N} = 1 - \frac{1}{N},$$

it follows that

$$\sum_{j=0}^{k-1} (M_j(f) - m_j(f))(x_{j+1} - x_j) < \sum_{j=0}^{k-1} \left(1 - \left(1 - \frac{1}{N} \right) \right)(x_{j+1} - x_j)$$

$$= \frac{1}{N}(x_k - x_0)$$

$$= \frac{1}{N^2}$$

$$< \varepsilon^2.$$

We conclude that

$$U(P, f) - L(P, f) < \varepsilon^2 + \varepsilon = \varepsilon(\varepsilon + 1);$$

therefore f is integrable over $[0, 1]$. $\triangle$

Two more important classes of integrable functions are the increasing and decreasing bounded functions.

Theorem 7.3.14. *A bounded function that is increasing on an interval $[a, b]$ or decreasing on $[a, b]$ is also integrable over $[a, b]$.*

Proof. Let f be a bounded function that is increasing on $[a, b]$. Given $\varepsilon > 0$, let P be a partition $(x_0, x_1, \ldots, x_n)$ of $[a, b]$ with $\mu(P) < \varepsilon$. Then

$$U(P, f) - L(P, f) = \sum_{j=0}^{n-1}(M_j(f) - m_j(f))(x_{j+1} - x_j)$$

$$< \varepsilon \sum_{j=0}^{n-1}(f(x_{j+1}) - f(x_j))$$

$$= \varepsilon(f(b) - f(a)).$$

Moreover $f(b) > f(a)$ since f is increasing on $[a, b]$. Thus $U(P, f) - L(P, f)$ converges to 0, and so f is integrable over $[a, b]$.

The argument is similar if f is decreasing on $[a, b]$. $\square$

Exercises 7.2.

1. Let $f : [a, b] \to \mathbb{R}$. Prove that the following statements are equivalent:

   (a) f is integrable over $[a, b]$.
   (b) For each $\varepsilon > 0$ there exists $\delta > 0$ such that if P and Q are partitions of $[a, b]$ with norm less than δ, then

$$|S(P, f) - S(Q, f)| < \varepsilon$$

   for all Riemann sums $S(P, f)$ and $S(Q, f)$ of f over $[a, b]$ relative to P and Q, respectively.

   This result is known as the Cauchy criterion for integrability.

2. Let $f : [a, b] \to \mathbb{R}$. Suppose that for each $\varepsilon > 0$ there exist integrable functions g and h from $[a, b]$ to $\mathbb{R}$ such that

$$\int (g - h) < \varepsilon$$

and

$$g(x) \leq f(x) \leq h(x)$$

for all $x \in [a, b]$. Show that f is integrable.

This result is called the sandwich theorem for integrals. In particular it shows that if there are integrable functions g and h mapping $[a, b]$ into $\mathbb{R}$ such that $g(x) \le f(x) \le h(x)$ for all $x \in [a, b]$ and $\int g = \int h = I$, then f is integrable and $\int f = I$.

3. Let

$$f_n(x) = \frac{nx^{n-1}}{1 + x}$$

for all $n > 1$ and $x \ne -1$. Define $A_n = \int_0^1 f_n$ for all $n > 1$, and show that

$$\lim_{n \to \infty} A_n = \frac{1}{2}.$$

[Hint: $nx^{n-1}/2 \le f_n(x) \le nx^{n-2}/2$ for all $n > 2$.]

4. Let $f : [0, 1] \to \mathbb{R}$ be defined by

$$f(x) = \begin{cases} 1 & \text{if } x = \frac{1}{n} \text{ for some } n \in \mathbb{N}, \\ 0 & \text{otherwise.} \end{cases}$$

Show that f is integrable and $\int f = 0$. [Hint: Choose $\varepsilon > 0$ and an integer $m > 1/\varepsilon$. Let

$$\left(0, \frac{1}{m}, x_2, x_3, \dots, x_{2m-1}, 1 \right)$$

be a partition P of $[0, 1]$ satisfying the inequalities

$$x_{2k} > \frac{1}{m - k + 1} > x_{2k-1}$$

and

$$x_{2k} - x_{2k-1} < \frac{\varepsilon}{m}$$

for each integer k such that $1 < k < m$. Compute $U(P, f) - L(P, f)$.]

5. Let $f : [a, b] \to \mathbb{R}$ be a function. Suppose that any sequence $\{S(P_n, f)\}$ of Riemann sums of f over $[a, b]$ is convergent if $\lim_{n \to \infty} \mu(P_n) = 0$. Prove that f is integrable.

This result is the converse of Theorem 7.3.2 and therefore completes a sequential characterization of integrability.

7.4 Basic Properties of Integrals

The next theorem summarizes the fundamental properties of integrals. We prepare for it with the following lemma.

Lemma 7.4.1. *Let f be a real-valued function defined on an interval $[a, b]$, and let $M = \sup \mathcal{R}_f$ and $m = \inf \mathcal{R}_f$. Define*

$$g(x, y) = f(y) - f(x)$$

for all $(x, y) \in \mathcal{D}_f^2$. Then

$$\sup \mathcal{R}_g = M - m.$$

Proof. For each $x, y \in [a, b]$ we have $f(y) \leq M$ and $f(x) \geq m$. Hence

$$g(x, y) \leq M - m,$$

so that $M - m$ is an upper bound of $\mathcal{R}_g$.

Now choose $\varepsilon > 0$. We must show that $M - m - \varepsilon$ is not an upper bound of $\mathcal{R}_g$. Since $M = \sup \mathcal{R}_f$, there exists $d \in [a, b]$ such that

$$f(d) > M - \frac{\varepsilon}{2}.$$

Similarly there exists $c \in [a, b]$ such that

$$f(c) < m + \frac{\varepsilon}{2}.$$

Hence

$$g(c, d) = f(d) - f(c) > M - m - \varepsilon,$$

as required. □

Theorem 7.4.2. *Let f, g be functions and $[a, b]$ an interval.*

1. *If f is integrable over $[a, b]$, then for every constant k, the function kf is also integrable over $[a, b]$ and*

$$\int_a^b kf = k \int_a^b f.$$

2. *If f and g are integrable over $[a, b]$, then so is $f + g$ and*

$$\int_a^b (f + g) = \int_a^b f + \int_a^b g.$$

3. *If f is integrable over $[a, b]$ and $c \in (a, b)$, then f is integrable over $[a, c]$ and $[c, b]$ and*

$$\int_a^b f = \int_a^c f + \int_c^b f.$$

4. *If f and g are integrable over $[a, b]$, then so is fg.*

Proof. 1. The result is clear if $k = 0$. Suppose therefore that $k \neq 0$.

Let $(x_0, x_1, \ldots, x_n)$ be a partition P of $[a, b]$. Choose $c_j \in [x_j, x_{j+1}]$ for each $j < n$, and let $c = (c_0, c_1, \ldots, c_{n-1})$. It is immediate from Eq. (7.1) that

$$S(P, kf, c) = kS(P, f, c).$$

The desired result now follows from the observation that if

$$|S(P, f, c) - I| < \varepsilon,$$

then

$$|S(P, kf, c) - kI| = |k||S(P, f, c) - I| < |k|\varepsilon.$$

2. Let $I_1 = \int_a^b f$ and $I_2 = \int_a^b g$. Choose $\varepsilon > 0$. There is a $\delta_1 > 0$ such that if P is a partition of $[a, b]$ with norm less than δ_1, then

$$|S(P, f) - I_1| < \varepsilon$$

for every Riemann sum $S(P, f)$ of f over $[a, b]$ relative to P. Similarly, there is a $\delta_2 > 0$ such that

$$|S(P, g) - I_2| < \varepsilon$$

for every partition P of $[a, b]$ with $\mu(P) < \delta_2$ and every Riemann sum $S(P, g)$ of g over $[a, b]$ relative to P. Let $\delta = \min\{\delta_1, \delta_2\}$, and choose a partition P of $[a, b]$ with norm less than δ. Let $P = (x_0, x_1, \ldots, x_n)$, choose $c_j \in [x_j, x_{j+1}]$ for each $j < n$, and let $c = (c_0, c_1, \ldots, c_{n-1})$. Since $\mu(P) < \delta_1$ and $\mu(P) < \delta_2$, we have

$$|S(P, f + g, c) - (I_1 + I_2)| = |S(P, f, c) + S(P, g, c) - (I_1 + I_2)|$$
$$\leq |S(P, f, c) - I_1| + |S(P, g, c) - I_2|$$
$$< 2\varepsilon,$$

as required.

3. Choose $\varepsilon > 0$. Then

$$\sum_{j=0}^{n-1}(M_j(f) - m_j(f))(y_{j+1} - y_j) < \varepsilon$$

for every partition $Q = (y_0, y_1, \ldots, y_n)$ of $[a, b]$ with small enough norm. Choose such a partition Q with $y_l = c$ for some l. Then we have both

$$\sum_{j=0}^{l-1}(M_j(f) - m_j(f))(y_{j+1} - y_j) < \varepsilon$$

and

$$\sum_{j=l}^{n-1}(M_j(f) - m_j(f))(y_{j+1} - y_j) < \varepsilon,$$

and we deduce that f is integrable over both $[a, c]$ and $[c, b]$.

Let $\int_a^b f = I$, $\int_a^c f = I_1$, and $\int_c^b f = I_2$. There exists $\delta_0 > 0$ such that

$$|S(P, f) - I| < \varepsilon$$

for every Riemann sum $S(P, f)$ of f over $[a, b]$ relative to any partition P of $[a, b]$ satisfying $\mu(P) < \delta_0$. Similarly, there exist $\delta_1 > 0$ such that

$$|S(P, f) - I_1| < \varepsilon$$

for every Riemann sum $S(P, f)$ of f over $[a, c]$ relative to any partition P of $[a, c]$ satisfying $\mu(P) < \delta_1$ and $\delta_2 > 0$ such that

$$|S(P, f) - I_2| < \varepsilon$$

for every Riemann sum $S(P, f)$ of f over $[c, b]$ relative to any partition P of $[c, b]$ satisfying $\mu(P) < \delta_2$. Let $\delta = \min\{\delta_0, \delta_1, \delta_2\}$, and let P_1 and P_2 be partitions of $[a, c]$ and $[c, b]$, respectively, with norm less than δ. Let P be the partition of $[a, b]$ for which $\hat{P} = \widehat{P_1 \cup P_2}$; then $\mu(P) < \delta$. Let $P_1 = (x_0, x_1, \ldots, x_k)$ and $P_2 = (x_k, x_{k+1}, \ldots, x_n)$. Choose $c_j \in [x_j, x_{j+1}]$ for each $j < n$, and let $c = (c_0, c_1, \ldots, c_{n-1})$, $c' = (c_0, c_1, \ldots, c_{k-1})$ and $c'' = (c_k, c_{k+1}, \ldots, c_{n-1})$. Then

$$S(P, f, c) = S(P_1, f, c') + S(P_2, f, c'').$$

Hence

$$|I - (I_1 + I_2)|$$
$$= |I - S(P, f, c) + S(P_1, f, c') + S(P_2, f, c'') - (I_1 + I_2)|$$
$$\leq |I - S(P, f, c)| + |S(P_1, f, c') - I_1| + |S(P_2, f, c'') - I_2|$$
$$< 3\varepsilon.$$

As ε is arbitrary, we conclude that $I = I_1 + I_2$, as required.

4. First we prove the result for the special case where $f = g$. Choose $\varepsilon > 0$. There exists $\delta > 0$ such that

$$U(P, f) - L(P, f) = \sum_{j=0}^{n-1}(M_j(f) - m_j(f))(x_{j+1} - x_j) < \varepsilon$$

for all partitions $P = (x_0, x_1, \ldots, x_n)$ of $[a, b]$ with norm less than δ. Choose such a partition P and let

$$M = \sup\{|f(x)| \mid x \in [a, b]\}.$$

We may assume that $M > 0$, the required result being clear if $f(x) = 0$ for all $x \in [a, b]$. Note that

$$M_j(f^2) - m_j(f^2) = \sup\{f^2(c_j) - f^2(d_j) \mid \{c_j, d_j\} \subset [x_j, x_{j+1}]\}$$

for each $j < n$, by Lemma 7.4.1. Moreover

$$f^2(c_j) - f^2(d_j) = (f(c_j) + f(d_j))(f(c_j) - f(d_j))$$
$$\leq (|f(c_j)| + |f(d_j)|)|f(c_j) - f(d_j)|$$
$$\leq 2M|f(c_j) - f(d_j)|,$$

and so

$$M_j(f^2) - m_j(f^2) \leq 2M \cdot \sup\{|f(c_j) - f(d_j)| \mid \{c_j, d_j\} \subset [x_j, x_{j+1}]\}$$
$$= 2M \cdot \sup\{f(c_j) - f(d_j) \mid \{c_j, d_j\} \subset [x_j, x_{j+1}]\}$$
$$= 2M(M_j(f) - m_j(f)),$$

again by Lemma 7.4.1. Hence

$$U(P, f^2) - L(P, f^2) \leq 2M(U(P, f) - L(P, f))$$
$$< 2M\varepsilon$$

whenever $\mu(P) < \delta$. Thus $U(P, f^2) - L(P, f^2)$ converges to 0 and so f^2 is integrable over $[a, b]$.

The general case follows from this special case, the equation

$$fg = \frac{(f+g)^2 - (f-g)^2}{4}$$

and parts (1) and (2).

$\square$

We now introduce the conventions that

$$\int_a^a f = 0$$

for every a, and

$$\int_a^b f = -\int_b^a f$$

whenever $b < a$. With these conventions we can extend Theorem 7.4.2(3) as follows. In this theorem no assumption is made about the sizes of a, b, and c.

Theorem 7.4.3. *If a, b, c are any real numbers and f is integrable over a closed interval containing a, b, c, then*

$$\int_a^b f = \int_a^c f + \int_c^b f.$$

Proof. The definitions above imply the theorem immediately if $a = b$, $a = c$, or $b = c$. If $c < a < b$, then

$$\int_c^b f = \int_c^a f + \int_a^b f,$$

so that

$$\int_a^b f = \int_c^b f - \int_c^a f = \int_a^c f + \int_c^b f.$$

The argument is similar if $a < b < c$. The case where $a < c < b$ is covered by Theorem 7.4.2(3). If $b < a$, then

$$\int_a^b f = -\int_b^a f = -\int_b^c f - \int_c^a f = \int_a^c f + \int_c^b f.$$

$\square$

Theorem 7.4.4. *If* $m \leq f(x) \leq M$ *for all* $x \in [a,b]$ *and* f *is integrable over* $[a,b]$, *then*

$$m(b-a) \leq \int_a^b f \leq M(b-a).$$

Proof. Let P be a partition $(x_0, x_1, \ldots, x_n)$ of $[a,b]$. Since

$$U(P, f) = \sum_{j=0}^{n-1} M_j(f)(x_{j+1} - x_j)$$

$$\leq M \sum_{j=0}^{n-1} (x_{j+1} - x_j)$$

$$= M(b-a),$$

it follows that

$$\underline{\int} f \leq \overline{\int} f \leq M(b-a).$$

The remaining inequality is proved similarly.   □

The following corollary is immediate.

Corollary 7.4.5. *If* $f(x) \geq 0$ *for all* $x \in [a,b]$ *and* f *is integrable over* $[a,b]$, *then*

$$\int_a^b f \geq 0.$$

Corollary 7.4.6. *If* $f(x) \geq g(x)$ *for all* $x \in [a,b]$ *and* f *and* g *are integrable over* $[a,b]$, *then*

$$\int_a^b f \geq \int_a^b g.$$

Proof. This result is immediate from the fact that

$$0 \leq \int_a^b (f - g) = \int_a^b f - \int_a^b g.$$ □

Corollary 7.4.7. *Suppose that f is continuous on $[a,b]$. If $f(x) \geq 0$ for all $x \in [a,b]$ and $f(x) > 0$ for some $x \in [a,b]$, then*

$$\int_a^b f > 0.$$

Proof. The continuous function f is integrable on $[a,b]$. Choose $c \in [a,b]$ such that $f(c) > 0$, and without loss of generality suppose that $c \neq b$. By Theorem 4.4.6 and the continuity of f there exists a number $\delta \in (0, b-c)$ such that $f(x) \geq f(c)/2$ for all $x \in [c,d]$, where $d = c + \delta$. Hence

$$\int_c^d f \geq \frac{f(c)}{2}(d - c) > 0,$$

and so

$$\int_a^b f = \int_a^c f + \int_c^d f + \int_d^b f > 0,$$

since $\int_a^c f \geq 0$ and $\int_d^b f \geq 0$. $\qquad \square$

If $f : [a,b] \to \mathbb{R}$ is a function, then we define $|f|$ to be the function given by

$$|f|(x) = |f(x)|$$

for all $x \in [a,b]$.

Theorem 7.4.8. *If a function f is integrable over $[a,b]$, then so is $|f|$. Moreover*

$$\left| \int_a^b f \right| \leq \int_a^b |f|.$$

Proof. Choose $\varepsilon > 0$. There exists a partition $P = (x_0, x_1, \ldots, x_n)$ of $[a,b]$ such that

$$U(P, f) - L(P, f) < \varepsilon.$$

For all $j < n$ and all $t, t' \in [x_j, x_{j+1}]$ we have

$$\big| |f(t)| - |f(t')| \big| \leq |f(t) - f(t')|$$
$$\leq \sup\{ f(x) - f(x') \mid \{x, x'\} \subset [x_j, x_{j+1}] \}$$
$$= M_j(f) - m_j(f),$$

by Lemma 7.4.1. Hence

$$M_j(|f|) - m_j(|f|) = \sup\{|f(x)| - |f(x')| \mid \{x, x'\} \subset [x_j, x_{j+1}]\}$$
$$\leq M_j(f) - m_j(f),$$

so that

$$U(P, |f|) - L(P, |f|) \leq U(P, f) - L(P, f) < \varepsilon.$$

We conclude that $|f|$ is integrable over $[a, b]$. The desired inequality follows from Corollary 7.4.6, since $|\int_a^b f| = \pm \int_a^b f$. □

Theorem 7.4.9 (Mean-Value Theorem for Integrals). *Suppose that f and g are continuous on $[a, b]$ and that $g(x) \geq 0$ for all $x \in [a, b]$. Then there exists $\xi \in [a, b]$ such that*

$$\int_a^b fg = f(\xi) \int_a^b g.$$

Proof. Since f is continuous on $[a, b]$, $f(x)$ attains a minimum m and a maximum M on $[a, b]$. As $g(x) \geq 0$ for all $x \in [a, b]$, it follows that

$$m \int g \leq \int fg \leq M \int g.$$

Therefore the theorem holds for every $\xi \in [a, b]$ if $\int g = 0$. In the remaining case we have

$$m \leq \frac{\int fg}{\int g} \leq M.$$

By the intermediate-value theorem there exists $\xi \in [a, b]$ such that

$$f(\xi) = \frac{\int fg}{\int g},$$

and the result follows. □

By taking $g(x) = 1$ for all x, we obtain the following corollary.

Corollary 7.4.10. *If f is continuous on $[a, b]$, then there exists $\xi \in [a, b]$ such that*

$$\int_a^b f = f(\xi)(b - a).$$

Fig. 7.6 Illustration for
Corollary 7.4.10

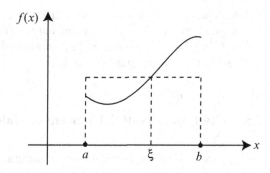

Remark. This corollary asserts that, for a function f that is continuous on $[a, b]$, the area bounded by the curve $y = f(x)$ and the lines $x = a$, $x = b$, and $y = 0$ is equal to the area of some rectangle with side $[a, b]$ (see Fig. 7.6).

Exercises 7.3.

1. For all x in a closed interval $[a, b]$, define

$$f^+(x) = \max\{f(x), 0\}$$

and

$$f^-(x) = \min\{f(x), 0\}.$$

Show that f is integrable over $[a, b]$ if both f^+ and f^- are.

2. Prove that if both f and g are integrable over a closed interval $[a, b]$, then so are $\max\{f, g\}$ and $\min\{f, g\}$. [Hint: Note that

$$\max\{f, g\} = \frac{1}{2}(f + g + |f - g|).$$

Find a similar formula for $\min\{f, g\}$.]

3. Let $f: [a, b] \to \mathbb{R}$ be continuous. Show that

$$\lim_{p \to \infty} \left(\int_0^1 |f(x)|^p \right)^{1/p} = \max_{x \in [0,1]} |f(x)|$$

and

$$\lim_{p \to -\infty} \left(\int_0^1 |f(x)|^p \right)^{1/p} = \min_{x \in [0,1]} |f(x)|.$$

4. Let $f : [a, b] \to \mathbb{R}$ be a continuous function, and suppose that $f(x) \le 0$ for all $x \in [a, b]$. Show that $\int f = 0$ if and only if $f(x) = 0$ for all $x \in [a, b]$.
5. Let $f : [a, b] \to \mathbb{R}$ be an integrable function such that $f(x) \ge 0$ for all $x \in [a, b]$. Show that $\sqrt{f}$ is integrable on $[a, b]$.

7.5 The Fundamental Theorem of Calculus

This section develops some techniques of integration. First we establish a relationship between the concepts of an integral and an antiderivative.

Theorem 7.5.1 (Fundamental Theorem of Calculus). *Let f be continuous on $[a, b]$, and for each $x \in [a, b]$ define*

$$F(x) = \int_a^x f.$$

Then $F'(x) = f(x)$ for all $x \in [a, b]$. Moreover, if G is any function on $[a, b]$ such that $G' = f$, then

$$\int_a^b f = G(b) - G(a).$$

Proof. Since f is continuous on $[a, b]$, the function F indeed exists. We show that $F'(t) = f(t)$ for all $t \in [a, b]$. By definition,

$$
\begin{aligned}
F'(t) &= \lim_{x \to t} \frac{F(x) - F(t)}{x - t} \\
&= \lim_{x \to t} \frac{1}{x - t} \left(\int_a^x f - \int_a^t f \right) \\
&= \lim_{x \to t} \frac{1}{x - t} \int_t^x f \\
&= \lim_{x \to t} \frac{f(\xi)(x - t)}{x - t} \\
&= \lim_{x \to t} f(\xi)
\end{aligned}
$$

for some number ξ between x and t. (In the penultimate step of the above calculation we used the mean-value theorem for integrals.)

In order to prove that $F'(t) = f(t)$, it therefore suffices to show that

$$\lim_{x \to t} f(\xi) = f(t).$$

Choose $\varepsilon > 0$. Since f is continuous at t, there exists $\delta > 0$ such that $|f(x) - f(t)| < \varepsilon$ whenever $|x - t| < \delta$. Choose x such that $0 < |x - t| < \delta$. Since ξ lies between x and t, we have $|\xi - t| \leq |x - t| < \delta$, and so $|f(\xi) - f(t)| < \varepsilon$, as required.

Now suppose G is another antiderivative of f. Then $G'(x) = F'(x) = f(x)$ for all $x \in [a, b]$, so that

$$(F - G)'(x) = 0$$

for each such x. Hence $F(x) - G(x)$ is a constant. Thus

$$G(b) - G(a) = F(b) - F(a) = \int_a^b f - \int_a^a f = \int_a^b f.$$

$\square$

We sometimes write $F(b) - F(a)$ as $F(x)|_a^b$.

Example 7.5.1. Since $F'(x) = 1/x$ if $F(x) = \log x$ for all $x > 0$, we have

$$\int_1^x \frac{dt}{t} = F(x) - F(1) = \log x$$

for all $x > 0$. This integral is sometimes used as a definition of the logarithm function. $\triangle$

Example 7.5.2. Let

$$f(x) = \frac{x^{n+1}}{n + 1}$$

for all $x \in [a, b]$, where $n \neq -1$. Then

$$f'(x) = \frac{1}{n + 1} \cdot (n + 1)x^n = x^n$$

for all such n. The fundamental theorem therefore implies that

$$\int_a^b x^n \, dx = \frac{x^{n+1}}{n + 1}\Big|_a^b$$

for all $n \neq -1$. Similar arguments show that

$$\int_a^b \sin x \, dx = -\cos x|_a^b,$$

$$\int_a^b \cos x \, dx = \sin x|_a^b,$$

and

$$\int_a^b e^x \, dx = e^x \big|_a^b.$$

△

The fundamental theorem of calculus asserts that if we differentiate the integral of a continuous function, then we recover the original function. It is pertinent to ask the following question: Is it always true that

$$\int_a^b f' = f(b) - f(a)?$$

The answer is no, for f' might not be integrable.

Example 7.5.3. Let f be defined by

$$f(x) = \begin{cases} x^2 \sin \frac{1}{x^2} & \text{if } x \neq 0, \\ 0 & \text{if } x = 0. \end{cases}$$

Then

$$f'(x) = 2x \sin \frac{1}{x^2} - \frac{2}{x} \cos \frac{1}{x^2}$$

for all $x \neq 0$. It is easy to see that f' is not integrable on any interval that contains 0 and has positive length as it is not bounded on any such interval.   △

However, we do have the following theorem.

Theorem 7.5.2. *If f is differentiable on $[a,b]$ and f' is integrable over $[a,b]$, then*

$$\int_a^b f' = f(b) - f(a).$$

Proof. Take any partition $P = (x_0, x_1, \ldots, x_n)$ of $[a, b]$. By the mean-value theorem, there exist numbers $\xi_0, \xi_1, \ldots, \xi_{n-1}$ such that $\xi_j \in (x_j, x_{j+1})$ and

$$f(x_{j+1}) - f(x_j) = f'(\xi_j)(x_{j+1} - x_j)$$

for each $j < n$. Therefore

$$L(P, f') = \sum_{j=0}^{n-1} m_j(f')(x_{j+1} - x_j)$$

$$\leq \sum_{j=0}^{n-1} f'(\xi_j)(x_{j+1} - x_j)$$

$$= \sum_{j=0}^{n-1} (f(x_{j+1}) - f(x_j))$$

$$= f(b) - f(a),$$

by the telescoping property. Similarly,

$$U(P, f') \geq f(b) - f(a).$$

Hence

$$\underline{\int} f' \leq f(b) - f(a) \leq \overline{\int} f'.$$

As f' is integrable over $[a, b]$, it follows that

$$\int_a^b f' = f(b) - f(a).$$

$\square$

Theorem 7.5.3. *Suppose that*

$$f(x) = \sum_{j=0}^{\infty} a_j x^j$$

for all x such that $|x| < r$, where r is the radius of convergence of the power series and is nonzero. Then f is integrable over every closed subinterval of $(-r, r)$, and for each $x \in (-r, r)$ we have

$$\int_0^x f = \sum_{j=0}^{\infty} \frac{a_j}{j+1} x^{j+1}. \tag{7.7}$$

Proof. Being differentiable at all $x \in (-r, r)$ by Theorem 6.7.2, f is continuous, and therefore integrable, on every closed subinterval of $(-r, r)$.

By Theorem 6.7.1, r is also the radius of convergence for the series on the right-hand side of Eq. (7.7). We may therefore define

$$G(x) = \sum_{j=0}^{\infty} \frac{a_j}{j+1} x^{j+1}$$

for each $x \in (-r, r)$. It follows from Theorem 6.7.2 that G is differentiable on $(-r, r)$ and $G'(x) = f(x)$ for each $x \in (-r, r)$. We conclude that G' is integrable over every closed subinterval of $(-r, r)$. Therefore Theorem 7.5.2 shows that

$$\int_0^x f = G(x) - G(0) = G(x). \qquad \square$$

The technique implicit in the following theorem is called **integration by substitution**.

Theorem 7.5.4. *Let g be a function that is differentiable on an interval $[c, d]$ and assume that g' is continuous on $[c, d]$. Let f be a function that is continuous on the range of g. Suppose that $g(c) = a$ and $g(d) = b$. Then*

$$\int_a^b f = \int_c^d (f \circ g)g'.$$

Proof. Note that since $f \circ g$ and g' are continuous, $(f \circ g)g'$ is also continuous, and therefore integrable, over every interval on which it is defined.

For each $x \in \mathcal{R}_g$ define

$$H(x) = \int_a^x f$$

and for each $x \in [c, d]$ let

$$G(x) = \int_c^x (f \circ g)g'.$$

Then, by the chain rule and the fundamental theorem of calculus,

$$\begin{aligned}
(H \circ g)'(x) &= H'(g(x))g'(x) \\
&= f(g(x))g'(x) \\
&= G'(x).
\end{aligned}$$

Hence the functions $H \circ g$ and G differ by a constant, and so

$$G(d) - G(c) = H(g(d)) - H(g(c)) = H(b) - H(a).$$

The required result now follows since $H(a) = G(c) = 0.$ \qquad $\square$

Example 7.5.4. In order to evaluate

$$\int_c^d \frac{dx}{x \log x},$$

where $c > 1$ and $d > 1$, set $f(x) = 1/x$ for all $x \neq 0$ and $g(x) = \log x$ for all $x > 0$. Then $g'(x) = 1/x$ for all $x > 0$. For all such x it follows that

$$\frac{1}{x \log x} = \frac{g'(x)}{g(x)} = f(g(x))g'(x).$$

Hence

$$\int_c^d \frac{dx}{x \log x} = \int_{\log c}^{\log d} \frac{dx}{x}$$

$$= \log x \big|_{\log c}^{\log d}$$

$$= \log \log d - \log \log c. \qquad \triangle$$

The use of the next theorem is a technique called **integration by parts**.

Theorem 7.5.5. *If f and g are both differentiable over an interval $[a,b]$ and f' and g' are integrable over $[a,b]$, then*

$$\int_a^b fg' = f(b)g(b) - f(a)g(a) - \int_a^b f'g.$$

Proof. It is clear that both the integrals concerned exist. Let

$$F(x) = f(x)g(x)$$

for all $x \in [a, b]$. Then

$$f(b)g(b) - f(a)g(a) = F(b) - F(a)$$

$$= \int F'$$

$$= \int (fg' + f'g)$$

$$= \int fg' + \int f'g,$$

and the result follows. $\qquad \square$

Example 7.5.5. For each nonnegative integer n, define

$$I_n = \int_0^{\pi/2} \sin^n x \, dx.$$

For instance, $I_0 = \pi/2$ and

$$I_1 = \int_0^{\pi/2} \sin x \, dx = -\cos x |_0^{\pi/2} = 1.$$

For the case where $n > 1$, put $f(x) = \sin^{n-1} x$ and $g(x) = -\cos x$ for all x. Thus

$$I_n = -\sin^{n-1} x \cos x |_0^{\pi/2} + (n-1) \int_0^{\pi/2} \sin^{n-2} x \cos^2 x \, dx$$

$$= (n-1) \int_0^{\pi/2} \sin^{n-2}(x)(1 - \sin^2 x) \, dx$$

$$= (n-1)(I_{n-2} - I_n).$$

Hence

$$n I_n = (n-1) I_{n-2},$$

so that

$$I_n = \frac{n-1}{n} I_{n-2}.$$

Easy inductions now show that

$$I_{2n} = \frac{\pi}{2} \prod_{m=1}^{n} \frac{2m-1}{2m}$$

and

$$I_{2n+1} = \prod_{m=1}^{n} \frac{2m}{2m+1}$$

for each positive integer n. For example,

$$\int_0^{\pi/2} \sin^2 x \, dx = \frac{\pi}{4}$$

and

$$\int_0^{\pi/2} \sin^3 x \, dx = \frac{2}{3}.$$

Note also that

$$\int_0^{\pi/2} \cos^n x \, dx = I_n$$

for every nonnegative integer n. This formula can be verified either similarly or by replacing x with $\pi/2 - x$.   △

Integration by parts has a number of interesting applications. We illustrate this point by using the preceding example to derive a famous formula due to Wallis:

$$\lim_{n \to \infty} \frac{2^{2n}(n!)^2}{(2n)!\sqrt{2n}} = \sqrt{\frac{\pi}{2}}. \tag{7.8}$$

When $0 \le x \le \pi/2$ and $n \in \mathbb{N}$, we have

$$0 \le \sin^{2n+1} x \le \sin^{2n} x \le \sin^{2n-1} x.$$

Using the notation of Example 7.5.5, we therefore deduce that

$$0 \le I_{2n+1} \le I_{2n} \le I_{2n-1},$$

and so

$$\prod_{m=1}^{n} \frac{2m}{2m+1} \le \frac{\pi}{2} \prod_{m=1}^{n} \frac{2m-1}{2m} \le \prod_{m=1}^{n-1} \frac{2m}{2m+1} = \frac{2n+1}{2n} \prod_{m=1}^{n} \frac{2m}{2m+1}.$$

Applying the result

$$\prod_{m=1}^{n} (2m-1)(2m+1) = 1 \cdot 3 \cdot 3 \cdot 5 \cdot \ldots \cdot (2n-1)(2n+1)$$

$$= (2n+1) \prod_{m=1}^{n} (2m-1)^2,$$

we find that

$$\frac{\pi}{2} \ge \prod_{m=1}^{n} \frac{2m \cdot 2m}{(2m-1)(2m+1)},$$

$$= \frac{1}{2n+1} \prod_{m=1}^{n} \frac{(2m)^2}{(2m-1)^2}$$

$$= \frac{P_n}{2n+1},$$

where

$$P_n = \left(\prod_{m=1}^{n} \frac{2m}{2m-1} \right)^2$$

for all positive integers n. Similarly,

$$\frac{\pi}{2} \le \frac{2n+1}{2n} \cdot \frac{P_n}{2n+1} = \frac{P_n}{2n}.$$

In other words,

$$a_n \le \frac{\pi}{2} \le b_n,$$

where $a_n = P_n/(2n+1)$ and $b_n = P_n/(2n)$ for each positive integer n. Therefore

$$b_n - a_n = \frac{P_n}{2n} - \frac{P_n}{2n+1} = \frac{P_n}{2n(2n+1)} = \frac{a_n}{2n} \le \frac{\pi}{4n},$$

so that $b_n - a_n \to 0$ as $n \to \infty$ by the sandwich theorem. As

$$0 \le \frac{\pi}{2} - a_n \le b_n - a_n,$$

the sandwich theorem also shows that

$$\lim_{n \to \infty} a_n = \lim_{n \to \infty} b_n = \frac{\pi}{2}.$$

Since

$$\prod_{m=1}^{n} 2m = 2 \cdot 4 \cdot \ldots \cdot (2n) = 2^n n!$$

and

$$\prod_{m=1}^{n} (2m-1) = 1 \cdot 3 \cdot \ldots \cdot (2n-1) = \frac{(2n)!}{2 \cdot 4 \cdot \ldots \cdot (2n)} = \frac{(2n)!}{2^n n!},$$

we find that

$$b_n = \frac{1}{2n} \left(\prod_{m=1}^{n} \frac{2m}{2m-1} \right)^2$$

$$= \frac{1}{2n} \left(\frac{2^{2n}(n!)^2}{(2n)!} \right)^2.$$

Therefore

$$\sqrt{\frac{\pi}{2}} = \lim_{n \to \infty} \sqrt{b_n}$$

$$= \lim_{n \to \infty} \frac{2^{2n}(n!)^2}{\sqrt{2n}(2n)!},$$

as required.

We can use Wallis's formula to establish a formula, attributed to Stirling, for approximating factorials:

$$n! \sim \frac{\sqrt{2\pi} \, n^{\frac{2n+1}{2}}}{e^n}.$$

If we write

$$c_n = \frac{n!e^n}{n^{\frac{2n+1}{2}}}$$

for all positive integers n, then our goal is to show that

$$\lim_{n \to \infty} c_n = \sqrt{2\pi}. \tag{7.9}$$

Note that

$$\frac{c_n}{c_{n+1}} = \frac{n!e^n}{n^{\frac{2n+1}{2}}} \cdot \frac{(n+1)^{\frac{2n+3}{2}}}{(n+1)!e^{n+1}}$$

$$= \frac{1}{e} \left(\frac{n+1}{n} \right)^{\frac{2n+1}{2}}.$$

Now define $d_n = \log c_n$ for all $n \in \mathbb{N}$. Then

$$d_n - d_{n+1} = \log c_n - \log c_{n+1}$$

$$= \log \frac{c_n}{c_{n+1}}$$

$$= \frac{2n+1}{2} \log \frac{n+1}{n} - 1.$$

Next we introduce a change of variable. Let

$$t = \frac{1}{2n+1}.$$

Noting that $0 < t < 1$, we have

$$2n+1 = \frac{1}{t},$$

so that

$$n = \frac{1}{2}\left(\frac{1}{t} - 1\right) = \frac{1-t}{2t}$$

and

$$n+1 = \frac{1-t}{2t} + 1 = \frac{1+t}{2t};$$

hence

$$\frac{n+1}{n} = \frac{1+t}{1-t}.$$

Using Eq. (6.18), we therefore find that

$$d_n - d_{n+1} = \frac{1}{2t}\log\frac{1+t}{1-t} - 1$$

$$= \frac{1}{2t}\cdot 2\sum_{j=0}^{\infty}\frac{t^{2j+1}}{2j+1} - 1$$

$$= \sum_{j=1}^{\infty}\frac{t^{2j}}{2j+1}$$

$$> 0.$$

Therefore the sequence $\{d_n\}$ is decreasing. In order to show that it is bounded below, we resume the calculation above:

$$d_n - d_{n+1} = \sum_{j=1}^{\infty}\frac{t^{2j}}{2j+1}$$

$$< t^2 \sum_{j=1}^{\infty} t^{2j-2}$$

$$= t^2 \sum_{j=0}^{\infty} t^{2j}$$

$$= \frac{t^2}{1 - t^2}$$

$$= \frac{1}{(2n + 1)^2 \left(1 - \frac{1}{(2n+1)^2}\right)}$$

$$= \frac{1}{(2n + 1)^2 - 1}$$

$$= \frac{1}{4n(n + 1)}$$

$$= \frac{1}{4n} - \frac{1}{4(n + 1)}.$$

Hence

$$d_n - \frac{1}{4n} - d_{n+1} + \frac{1}{4(n + 1)} < 0$$

and we conclude that the sequence $\{d_n - 1/(4n)\}$ is increasing. Therefore

$$d_n > d_n - \frac{1}{4n} \geq d_1 - \frac{1}{4}$$

for all positive integers n.

Decreasing but bounded below, the sequence $\{d_n\}$ converges. Let C be its limit. Then

$$\lim_{n \to \infty} c_n = \lim_{n \to \infty} e^{d_n} = e^C,$$

and it remains only to show that $e^C = \sqrt{2\pi}$. We have already proved that

$$n! \sim n^{\frac{2n+1}{2}} e^{C-n}.$$

Using Wallis's formula, we therefore find that

$$\sqrt{\frac{\pi}{2}} \sim \frac{2^{2n} n^{2n+1} e^{2C-2n}}{(2n)^{\frac{4n+1}{2}} e^{C-2n} \sqrt{2n}} = \frac{e^C}{2}.$$

Consequently $e^C \sim \sqrt{2\pi}$, and the required equation follows because C is constant. The proof of the next theorem also uses integration by parts.

Theorem 7.5.6. *Let I be an open interval and let $a \in I$. Let $f : I \to \mathbb{R}$ be a function. Let n be a nonnegative integer and suppose that $f^{(j)}$ exists and is continuous for each nonnegative integer $j \le n + 1$. Then for each $x \in I$ we have*

$$f(x) = P_n(x) + R_n(x),$$

where

$$P_n(x) = \sum_{j=0}^{n} \frac{f^{(j)}(a)}{j!} (x - a)^j,$$

$$R_n(x) = \frac{1}{n!} \int_a^x (x - t)^n f^{(n+1)}(t)\, dt$$

and $0^0 = 1$.

Proof. For every $j \in \{0, 1, \ldots, n + 1\}$, the function $f^{(j)}$ is continuous, and therefore integrable, on the closed interval with ends a and x. Consequently $R_n(x)$ exists, by Theorem 7.4.2(4).

We proceed by induction on n. First, for all $x \in I$ we have $P_0(x) = f(a)$ and

$$R_0(x) = \int_a^x f'(t)\, dt = f(x) - f(a),$$

by Theorem 7.5.2. Therefore the theorem holds for $n = 0$.

Suppose therefore that $n > 0$ and the theorem holds for $n - 1$. Integration by parts yields

$$R_n(x) = \frac{1}{n!} \left((x - t)^n f^{(n)}(t) \Big|_a^x + n \int_a^x (x - t)^{n-1} f^{(n)}(t)\, dt \right)$$

$$= -\frac{f^{(n)}(a)}{n!} (x - a)^n + \frac{1}{(n-1)!} \int_a^x (x - t)^{n-1} f^{(n)}(t)\, dt$$

for all $x \in I$. By the inductive hypothesis, it follows that

$$f(x) = P_{n-1}(x) + R_{n-1}(x)$$

$$= \sum_{j=0}^{n-1} \frac{f^{(j)}(a)}{j!} (x - a)^j + \frac{1}{(n-1)!} \int_a^x (x - t)^{n-1} f^{(n)}(t)\, dt$$

$$= \sum_{j=0}^{n-1} \frac{f^{(j)}(a)}{j!} (x - a)^j + \frac{f^{(n)}(a)}{n!} (x - a)^n + R_n(x)$$

$$= \sum_{j=0}^{n} \frac{f^{(j)}(a)}{j!}(x-a)^j + \frac{1}{n!} \int_a^x (x-t)^n f^{(n+1)}(t)\, dt,$$

as required.   □

In 1947, Niven proved that π is irrational. In 1986, Parks proved a more general result from which the irrationality of both π and e follows. We present this result now. Its proof uses integration by parts.

Theorem 7.5.7. *Let c be a positive number and let $f : [0, c] \to \mathbb{R}$ be a differentiable function such that $f(x) > 0$ for all $x \in (0, c)$. Suppose there exist functions $f_1, f_2, \ldots$ that are differentiable on $[0, c]$ and satisfy the conditions that $f_1' = f$, $f_k' = f_{k-1}$ for all $k > 1$ and $f_k(0)$ and $f_k(c)$ are integers for all k. Then c is irrational.*

Proof. Let S be the set of all real polynomials p such that $p^{(k)}(0)$ and $p^{(k)}(c)$ are integers for all nonnegative integers k. In other words, p and all its derivatives yield integers when evaluated at 0 and at c. The set S is closed under multiplication: If p and q are polynomials in S, then so is pq, for it is easily seen by induction that $(pq)^{(k)}$ is a sum of products of derivatives of p and q. (We consider p and q to be zeroth derivatives of themselves.) Note also that $p \in S$ if $p(0)$ and $p(c)$ are integers and $p' \in S$.

We show next that if $p \in S$, then $\int_0^c fp$ is an integer. To this end, we first claim that it is an integer if and only if $\int_0^c f_1 p'$ is. Indeed, using integration by parts, we find that

$$\int_0^c fp = f_1(c)p(c) - f_1(0)p(0) - \int_0^c f_1 p'.$$

From the hypotheses and the assumption that $p \in S$ we see that the first two terms on the right-hand side are integers, and our claim follows. Proceeding inductively, we find that $\int_0^c fp$ is an integer if and only if $\int_0^c f_\Delta p^{(\Delta)}$ is an integer, where Δ is the degree of p. Note that $p^{(\Delta)}$ is a constant function, and the constant is an integer since $p \in S$. Moreover the differentiable function $f_\Delta = f_{\Delta+1}'$ is continuous, and hence integrable, on $[0, c]$. Therefore, by the fundamental theorem of calculus, we have

$$\int_0^c f_\Delta p^{(\Delta)} = p^{(\Delta)}(c) \int_0^c f_{\Delta+1}'$$

$$= p^{(\Delta)}(c)(f_{\Delta+1}(c) - f_{\Delta+1}(0)),$$

an integer. We conclude that $\int_0^c fp$ is an integer.

Assume now that c is rational. Then there are positive integers m and n such that $c = m/n$. Define

$$P_k(x) = \frac{x^k(m-nx)^k}{k!}$$

for all $x \in [0, c]$ and all nonnegative integers k, where $0^0 = 1$.

We shall show that $P_k \in S$ for all k. First, it is clear that the constant polynomial P_0 is in S. Suppose therefore that $k > 0$ and that $P_{k-1} \in S$. Observe first that $P_k(0) = 0$, and that $P_k(c) = 0$ since $m = nc$. For all $x \in [0, c]$ we have

$$\begin{aligned} P_k'(x) &= \frac{kx^{k-1}(m-nx)^k - knx^k(m-nx)^{k-1}}{k!} \\ &= \frac{kx^{k-1}(m-nx)^{k-1}(m-2nx)}{k!} \\ &= P_{k-1}(x)(m-2nx). \end{aligned}$$

The first factor is a polynomial in S by the inductive hypothesis. The second is in S since m, $m - 2nc = -m$, and $-2n$ are all integers. Since S is closed under multiplication, we conclude that $P_k' \in S$ and therefore that $P_k \in S$ for all k as $P_k(0)$ and $P_k(c)$ are integers.

Next observe that $P_k(x) > 0$ for all $x \in (0, c)$ and all k, a property shared by $f(x)$. Thus $\int_0^c fP_k > 0$. But this integral must be an integer, since $P_k \in S$. Hence

$$\int_0^c f(x)P_k(x)\,dx \geq 1 \tag{7.10}$$

for all nonnegative integers k. However, setting

$$M = \max\{x(m-nx) \mid x \in [0, c]\}$$

and

$$L = \max\{f(x) \mid x \in [0, c]\},$$

we obtain

$$\begin{aligned} \int_0^c f(x)P_k(x)\,dx &\leq \int_0^c L\frac{M^k}{k!}\,dx \\ &= \frac{cLM^k}{k!}. \end{aligned}$$

This result gives a contradiction, for

$$\lim_{k\to\infty} \frac{M^k}{k!} = 0$$

by Example 2.5.6.   □

Corollary 7.5.8. *If $0 < |\theta| \le \pi$ and $\cos\theta$ and $\sin\theta$ are rational, then θ is irrational.*

Proof. Since $\cos\theta$ and $\sin\theta$ are rational, so are $\cos|\theta|$ and $\sin|\theta|$. Hence we can find a positive integer n such that $n\cos|\theta|$ and $n\sin|\theta|$ are integers. Now apply the theorem with $c = |\theta|$ and $f(x) = n\sin x$ for all $x \in [0, |\theta|]$ to conclude that $|\theta|$, and hence θ, is irrational. $\qquad\square$

Corollary 7.5.9. *For every positive rational number $a \ne 1$, $\log a$ is irrational.*

Proof. Suppose first that $a > 1$, so that $\log a > 0$. Write $a = m/n$ for positive integers m and n, and apply the theorem with $c = \log a$ and $f(x) = ne^x$ for all $x \in [0, \log a]$, noting that $ne^c = ne^{\log a} = na = m$.

If $0 < a < 1$, then $1/a > 1$ and we conclude that $\log 1/a$ is irrational by the previous case. That $\log a$ is irrational now follows from the equation

$$\log \frac{1}{a} = -\log a. \qquad\square$$

Remark 1. By taking $\theta = \pi$ in Corollary 7.5.8, we find that π is irrational.

Remark 2. Let a, b, c be positive rational numbers and suppose that $a^2 + b^2 = c^2$. There is a number θ such that $0 < \theta < \pi/2$, $\sin\theta = a/c$, and $\cos\theta = b/c$. Then both $\sin\theta$ and $\cos\theta$ are rational, so that θ is irrational. In other words, $\arctan(a/b)$ is irrational.

Remark 3. Corollary 7.5.9 confirms the irrationality of e, since $\log e = 1$. In fact, e^a is irrational for every nonzero rational number a.

In Sect. 6.4 we introduced the number π. We now give it a geometric interpretation by showing that the circumference of a unit circle is 2π. For this purpose we need the notion of arc length.

Let f be a continuous real-valued function defined on a closed interval I with partition $(x_0, x_1, \ldots, x_n)$. For each j let P_j be the point $(x_j, f(x_j))$. For each $j < n$ the distance between P_{j+1} and P_j is $|P_{j+1} - P_j|$. The sum of these distances gives an approximation for the length of the graph of f. If

$$\lim_{n\to\infty} \sum_{j=0}^{n-1} |P_{j+1} - P_j|$$

exists and is finite, then we define it to be the **(arc) length** of the graph of f over I. It is evaluated in the following theorem.

Theorem 7.5.10. *Let f be a differentiable function from an interval $[a, b]$ into $\mathbb{R}$. Suppose that f' is continuous on $[a, b]$. Then the length of the graph of f over $[a, b]$ is*

$$\int_a^b \sqrt{1 + (f')^2}.$$

Proof. By applying the mean-value theorem to f over $[x_j, x_{j+1}]$, where $j < n$, we discover a $\xi_j \in (x_j, x_{j+1})$ for which

$$f(x_{j+1}) - f(x_j) = f'(\xi_j)(x_{j+1} - x_j).$$

Therefore

$$\sum_{j=0}^{n-1} |P_{j+1} - P_j| = \sum_{j=0}^{n-1} \sqrt{(x_{j+1} - x_j)^2 + (f(x_{j+1}) - f(x_j))^2}$$

$$= \sum_{j=0}^{n-1} \sqrt{(x_{j+1} - x_j)^2 + (f')^2(\xi_j)(x_{j+1} - x_j)^2}$$

$$= \sum_{j=0}^{n-1} (x_{j+1} - x_j)\sqrt{1 + (f')^2(\xi_j)},$$

which is a Riemann sum for the function $\sqrt{1 + (f')^2}$. This function is continuous, and therefore integrable, on $[a, b]$. Its Riemann sums converge to the required arc length, and the result follows.   □

Example 7.5.6. Since $\arccos(1/\sqrt{2}) = \pi/4$, the length of the graph of the function

$$f(x) = \sqrt{1 - x^2},$$

where $x \in [0, 1/\sqrt{2}]$, is an eighth of the circumference of the unit circle. In order to evaluate it, we first use the chain rule to compute

$$f'(x) = \frac{1}{2\sqrt{1 - x^2}} \cdot (-2x)$$

$$= -\frac{x}{\sqrt{1 - x^2}}.$$

This function is continuous at all $x \neq \pm 1$. Therefore

$$\int_0^{1/\sqrt{2}} \sqrt{1 + (f')^2} = \int_0^{1/\sqrt{2}} \sqrt{1 + \frac{x^2}{1 - x^2}}\, dx$$

$$= \int_0^{1/\sqrt{2}} \frac{dx}{\sqrt{1 - x^2}}$$

$$= \arcsin x \,|_0^{1/\sqrt{2}}$$

$$= \frac{\pi}{4}.$$

Consequently the circumference of the unit circle is $8 \cdot \pi/4 = 2\pi$.   △

Exercises 7.4.

1. Use the result of Example 7.5.1 to prove the following:

   (a) If $F(x) = \log x$ for all $x > 0$ and $cx > 0$ for some constant c, then

   $$F'(cx) = \frac{1}{x}$$

   for all $x > 0$.

   (b) If $x > 0$ and $y > 0$, then

   $$\log xy = \log x + \log y.$$

   (c) If $x > 0$ and p is a rational number, then

   $$\log x^p = p \log x.$$

   (d)

   $$\lim_{x \to \infty} \log x = \infty$$

   and

   $$\lim_{x \to 0+} \log x = -\infty.$$

2. Let f be a function that is integrable over $[a, b]$ and let $F(x) = \int_a^x f$ for all $x \in [a, b]$.

   (a) Show that for all $x, y \in [a, b]$ satisfying $x < y$ we have

   $$|F(x) - F(y)| < M(y - x),$$

   where M is the least upper bound of $|f|$ on $[a, b]$.

   (b) Hence show that F is continuous on $[a, b]$. (Note that f need not be continuous, and therefore the fundamental theorem of calculus cannot be used.)

3. By choosing a particular type of partition of the interval $[1, 2]$ and suitable Riemann sums for

   $$\int_1^2 \frac{dx}{x^2},$$

evaluate

$$\lim_{n \to \infty} n \sum_{j=1}^{n} \frac{1}{(n+j)^2}.$$

(The evaluation of this limit is due to Darboux.)

4. By choosing a particular type of partition of the interval $[0, 1]$ and suitable Riemann sums for $\int_0^1 e^x \, dx$, evaluate

$$\lim_{n \to \infty} \frac{1}{n} \sum_{j=1}^{n} e^{\frac{j-1}{n}}.$$

5. By choosing a particular type of partition of the interval $[0, 1]$ and suitable Riemann sums for $\int_0^1 x^m \, dx$, evaluate

$$\lim_{n \to \infty} \frac{1}{n^{m+1}} \sum_{j=1}^{n} j^m.$$

6. Let $f : [a, b] \to \mathbb{R}$ be integrable. Show that if f is continuous at a point $c \in (a, b)$, then $\int_a^x f$ is differentiable at c. (Note that this result implies the fundamental theorem of calculus.)

7. Let $f : [a, b] \to \mathbb{R}$ be a continuous function such that $f(x) > 0$ for all $x \in [a, b]$. Show that $\int_a^x f$ is increasing on $[a, b]$.

8. Suppose that f'' is continuous on $[a, b]$. Show that

$$\int_a^b x f''(x) \, dx = b f'(b) - f(b) - (a f'(a) - f(a)).$$

9. Let f be a continuous real-valued function and let u and v be differentiable functions. Define

$$F(x) = \int_{u(x)}^{v(x)} f(t) \, dt.$$

Show that

$$F'(x) = f(v(x))v'(x) - f(u(x))u'(x).$$

10. Let f and g be functions from $[a, b]$ to $\mathbb{R}$ such that g is continuous, f monotonic, and f' integrable. Show that there exists $c \in [a, b]$ for which

$$\int_a^b fg = f(a) \int_a^c g + f(b) \int_c^b g.$$

This result is known as the second mean-value theorem. [Hint: Let

$$G(x) = \int_a^x g(t)\, dt$$

for each $x \in [a, b]$. Apply integration by parts to

$$\int_a^b fg = \int_a^b fG'$$

and then use the mean-value theorem.]

11. For each $m, n \in \mathbb{N}$ evaluate the following integrals:

(a) $\int_{-\pi}^{\pi} \sin mx \sin nx\, dx$;

(b) $\int_{-\pi}^{\pi} \cos mx \cos nx\, dx$;

(c) $\int_{-\pi}^{\pi} \sin mx \cos nx\, dx$.

7.6 The Cauchy–Schwarz Inequality

Another theorem about integrals is known as the Cauchy–Schwarz inequality.

Theorem 7.6.1 (Cauchy–Schwarz Inequality). *If f and g are integrable on a closed interval, then*

$$\left(\int fg \right)^2 \le \int f^2 \int g^2.$$

Proof. Note that the functions fg, f^2, and g^2 are integrable by Theorem 7.4.2(4). For all real x we have

$$\int (xf + g)^2 = x^2 \int f^2 + 2x \int fg + \int g^2.$$

The polynomial on the right-hand side of this equation is therefore nonnegative for all real x. Thus its discriminant cannot be positive, by the results of Example 6.3.4. In other words,

$$4 \left(\int fg \right)^2 - 4 \int f^2 \int g^2 \le 0.$$

The required inequality follows.   □

Let $f : [a, b] \to \mathbb{R}$ be integrable. We define the norm $\|f\|$ of f by the equation

$$\|f\| = \left(\int f^2 \right)^{\frac{1}{2}}.$$

With this notation we may rewrite the Cauchy–Schwarz inequality as

$$\int fg \leq \|f\|\|g\|.$$

The norm of a function satisfies the triangle inequality.

Corollary 7.6.2. *Let* $f, g: [a, b] \to \mathbb{R}$ *be integrable. Then*

$$\|f + g\| \leq \|f\| + \|g\|.$$

Proof. Note first that $\|f\| \geq 0$ for every integrable function f. Now

$$\|f + g\|^2 = \int (f + g)^2$$

$$= \int f^2 + 2 \int fg + \int g^2$$

$$\leq \|f\|^2 + 2\|f\|\|g\| + \|g\|^2$$

$$= (\|f\| + \|g\|)^2,$$

and the result follows by taking square roots of both sides.   □

A similar argument to that used to prove Theorem 7.6.1, combined with Theorem 3.12.5, yields the following corresponding result for series. The details of the proof are left as an exercise.

Theorem 7.6.3. *Let* $\{x_n\}$ *and* $\{y_n\}$ *be sequences such that* $\sum_{j=0}^{\infty} x_j^2$ *and* $\sum_{j=0}^{\infty} y_j^2$ *converge. Then* $\sum_{j=0}^{\infty} |x_j y_j|$ *converges and*

$$\left(\sum_{j=0}^{\infty} |x_j y_j| \right)^2 \leq \sum_{j=0}^{\infty} x_j^2 \sum_{j=0}^{\infty} y_j^2.$$

Exercises 7.5.

1. If f and g are integrable functions, prove that

$$\|f\| - \|g\| \leq \|f - g\|$$

and

$$\|\alpha f\| = |\alpha| \|f\|$$

for every number α.

2. Let f and g be integrable over an interval $[a, b]$. Prove that

$$\int_a^b \int_a^b \left| \begin{matrix} f(x) & g(x) \\ f(y) & g(y) \end{matrix} \right|^2 dy\, dx = 2 \left(\|f\|^2 \|g\|^2 - \left(\int_a^b f(x)g(x)\, dx \right)^2 \right).$$

(Note that the integrand on the left-hand side is the square of a determinant.) Deduce the Cauchy–Schwarz inequality from this equation.

3. Let f and g be integrable over an interval $[a, b]$.

(a) Prove that

$$\int_a^b \int_a^b (f(y) - f(x))(g(y) - g(x))\, dy\, dx$$

$$= 2 \left((b-a) \int_a^b f(x)g(x)\, dx - \int_a^b f(x)\, dx \int_a^b g(x)\, dx \right).$$

(b) Suppose that f and g are both increasing or both decreasing. Deduce that

$$\int_a^b f \int_a^b g \le (b-a) \int_a^b fg.$$

4. Prove Theorem 7.6.3.

7.7 Numerical Integration

The fundamental theorem of calculus is a very useful tool for evaluating integrals. However, it is not always applicable, as there are integrable functions such as e^{-x^2} and $\sin \sin x$ whose antiderivatives cannot be expressed explicitly in terms of such elementary functions as polynomials, the trigonometric functions, and the exponential and logarithm functions. In this section we therefore investigate the question of approximating a definite integral by means of a Riemann sum. We also wish to estimate the accuracy of the resulting approximations.

We begin with a lemma.

Lemma 7.7.1. *Let f be a function with continuous derivative on an interval $[a, b]$. For every $c \in [a, b]$ let*

$$\varepsilon = \int_a^b f - f(c)(b-a).$$

Then

$$|\varepsilon| \le \frac{(b-a)^2}{2} \sup_{x \in [a,b]} |f'(x)|.$$

Proof. Since f' is continuous on the closed interval $[a, b]$, we may define

$$M = \sup_{x \in [a,b]} |f'(x)|.$$

By the mean-value theorem (for differentiation), for every $x \in [a, b] - \{c\}$ there exists ξ between x and c such that

$$|f(x) - f(c)| = |f'(\xi)(x - c)|$$
$$\leq M|x - c|.$$

Hence

$$|\varepsilon| = \left| \int_a^b (f(x) - f(c)) \, dx \right|$$

$$\leq \int_a^b |f(x) - f(c)| \, dx$$

$$\leq M \int_a^b |x - c| \, dx$$

$$= M \left(-\int_a^c (x - c) \, dx + \int_c^b (x - c) \, dx \right)$$

$$= M \left(-\frac{(x - c)^2}{2} \Big|_a^c + \frac{(x - c)^2}{2} \Big|_c^b \right)$$

$$= M \left(\frac{(a - c)^2}{2} + \frac{(b - c)^2}{2} \right)$$

$$\leq \frac{M}{2}(b - a)^2$$

since

$$(b - a)^2 = ((b - c) + (c - a))^2$$
$$= (b - c)^2 + 2(b - c)(c - a) + (c - a)^2$$
$$\geq (b - c)^2 + (c - a)^2.$$

$\square$

Theorem 7.7.2. *Let f be a function with continuous derivative on an interval $[a, b]$ with partition $P = (x_0, x_1, \ldots, x_n)$. Then every Riemann sum $S(P, f)$ of f over $[a, b]$ relative to P satisfies the inequality*

$$\left| \int_a^b f - S(P, f) \right| \leq \frac{b-a}{2} \sup_{x \in [a,b]} |f'(x)| \mu(P).$$

Proof. As in the proof of the lemma, let

$$M = \sup_{x \in [a,b]} |f'(x)|.$$

For any Riemann sum $S(P, f, c)$ of f over $[a, b]$ relative to P and intermediate points $c_0, c_1, \ldots, c_{n-1}$, we have

$$
\left| \int_a^b f(x)\, dx - S(P, f, c) \right| = \left| \sum_{j=0}^{n-1} \int_{x_j}^{x_{j+1}} f(x)\, dx - \sum_{j=0}^{n-1} f(c_j)(x_{j+1} - x_j) \right|
$$

$$
\leq \sum_{j=0}^{n-1} \left| \int_{x_j}^{x_{j+1}} f(x)\, dx - f(c_j)(x_{j+1} - x_j) \right|
$$

$$
\leq \sum_{j=0}^{n-1} \frac{M}{2}(x_{j+1} - x_j)^2
$$

$$
\leq \frac{M}{2} \sum_{j=0}^{n-1} \mu(P)(x_{j+1} - x_j)
$$

$$
= \frac{1}{2} M \mu(P)(b - a). \qquad \square
$$

Although the upper bound this theorem provides for the error may seem crude, the following example shows that it cannot be improved.

Example 7.7.1. Let $f(x) = x$ for all $x \in [0, 1]$. Then $\int_0^1 f = 1/2$. For each $n \in \mathbb{N}$ let P_n be the partition

$$\left(0, \frac{1}{n}, \frac{2}{n}, \ldots, 1 \right),$$

so that $\mu(P_n) = 1/n$. By choosing as intermediate points the leftmost end of each subinterval of P_n or the rightmost end of each such subinterval, we obtain the lower sum $L(P_n, f)$ or the upper sum $U(P_n, f)$, respectively, of f over $[0, 1]$ relative to P_n. Now for each j the jth subinterval of P_n is $[(j-1)/n, j/n]$, so that

$$L(P_n, f) = \frac{1}{n} \sum_{j=0}^{n-1} \frac{j}{n} = \frac{n(n-1)}{2n^2} = \frac{1}{2} - \frac{1}{2n}$$

and

$$U(P_n, f) = \frac{1}{n} \sum_{j=1}^{n} \frac{j}{n} = \frac{n(n+1)}{2n^2} = \frac{1}{2} + \frac{1}{2n}.$$

Hence

$$\left| \int_0^1 f(x)\,dx - L(P_n, f) \right| = \left| \int_0^1 f(x)\,dx - U(P_n, f) \right| = \frac{1}{2n},$$

and this number is the upper bound provided by the theorem for the error.   △

We now consider the case where a Riemann sum is obtained by using the midpoints of each subinterval of a partition.

Lemma 7.7.3. *Let f be a function with continuous second derivative on an interval $[a, b]$. Let $c = (a + b)/2$ and define*

$$\varepsilon = \int_a^b f - f(c)(b - a).$$

Then

$$|\varepsilon| \le \frac{(b-a)^3}{24} \sup_{x \in [a,b]} |f''(x)|.$$

Proof. Using Theorem 7.5.6 with $n = 1$, we obtain

$$f(x) - f(c) = f'(c)(x - c) + \int_c^x (x - t) f''(t)\,dt$$

for all $x \in (a, b)$. Thus

$$\varepsilon = \int_a^b (f(x) - f(c))\,dx$$

$$= \int_a^b f'(c)(x - c)\,dx + \int_a^b \int_c^x (x - t) f''(t)\,dt\,dx.$$

But

$$\int_a^b f'(c)(x - c)\,dx = f'(c) \int_a^b \left(x - \frac{a+b}{2} \right) dx$$

$$= f'(c) \left(\int_a^b x\,dx - \frac{(a+b)(b-a)}{2} \right)$$

$$= f'(c) \left(\frac{b^2 - a^2}{2} - \frac{b^2 - a^2}{2} \right)$$

$$= 0.$$

Setting

$$M = \sup_{x \in [a,b]} |f''(x)|,$$

we therefore conclude that

$$|\varepsilon| \le \int_a^b \left| \int_c^x (x - t) f''(t) \, dt \right| dx$$

$$\le \int_a^b \int_c^x (x - t) M \, dt \, dx$$

$$= -M \int_a^b \left. \frac{(t - x)^2}{2} \right|_c^x dx$$

$$= \frac{M}{2} \int_a^b (x - c)^2 \, dx$$

$$= \frac{M}{6} \left((b - c)^3 - (a - c)^3 \right)$$

$$= \frac{M}{6} \left(\left(b - \frac{a + b}{2} \right)^3 - \left(a - \frac{a + b}{2} \right)^3 \right)$$

$$= \frac{M}{6} \cdot \frac{1}{8} \left((b - a)^3 - (a - b)^3 \right)$$

$$= \frac{M}{24} (b - a)^3. \qquad \square$$

Theorem 7.7.4. *Let f be a function such that f'' exists and is continuous on an interval $[a, b]$ with partition $(x_0, x_1, \ldots, x_n)$. Let $c = (c_0, c_1, \ldots, c_{n-1})$, where*

$$c_j = \frac{x_{j+1} + x_j}{2}$$

for all $j < n$. Then

$$\left| \int_a^b f - S(P, f, c) \right| \le \frac{b - a}{24} \sup_{x \in [a,b]} |f''(x)| (\mu(P))^2.$$

Proof. Letting

$$M = \sup_{x\in[a,b]} |f''(x)|$$

and using Lemma 7.7.3, we obtain

$$\left| \int_a^b f(x)\,dx - S(P, F, c) \right| \leq \sum_{j=0}^{n-1} \left| \int_{x_j}^{x_{j+1}} f(x)\,dx - f(c_j)(x_{j+1} - x_j) \right|$$

$$\leq \sum_{j=0}^{n-1} \frac{M}{24}(x_{j+1} - x_j)^3$$

$$\leq \frac{M}{24}(\mu(P))^2 \sum_{j=0}^{n-1}(x_{j+1} - x_j)$$

$$= \frac{1}{24}M(\mu(P))^2(b - a). \qquad \square$$

Again, the bound is sharp.

Example 7.7.2. Let $f(x) = x^2$ for all $x \in [0, 1]$, so that $\int_0^1 f = 1/3$. For each $n \in \mathbb{N}$ let P_n be the partition used in Example 7.7.1. Defining $c_0, c_1, \ldots, c_{n-1}, c$ as in the theorem, we find that

$$c_j = \frac{1}{2}\left(\frac{j}{n} + \frac{j+1}{n}\right) = \frac{j + \frac{1}{2}}{n}$$

for each $j < n$. Therefore

$$S(P_n, f, c) = \frac{1}{n}\sum_{j=0}^{n-1} \frac{1}{n^2}\left(j + \frac{1}{2}\right)^2$$

$$= \frac{1}{n^3}\sum_{j=0}^{n-1}\left(j^2 + j + \frac{1}{4}\right)$$

$$= \frac{1}{n^3}\left(\frac{n(n-1)(2n-1)}{6} + \frac{n(n-1)}{2} + \frac{n}{4}\right)$$

$$= \frac{4n^2 - 1}{12n^2}$$

$$= \frac{1}{3} - \frac{1}{12n^2}.$$

Hence

$$\left| \int_0^1 f(x)\,dx - S(P_n, f, c) \right| = \frac{1}{12n^2},$$

the bound given by the theorem. △

The reader is referred to [3] for more on numerical integration.

Exercises 7.6.

1. Let $f : [a, b] \to \mathbb{R}$ be monotonic. For each $n \in \mathbb{N}$ let $h = (b - a)/n$ and

$$P_n = (a, a + h, a + 2h, \ldots, b).$$

For each Riemann sum $S(P_n, f)$ of f over $[a, b]$ relative to P_n, show that

$$\left| \int_a^b f(x)\,dx - S(P_n, f) \right| \le \frac{|b - a|}{n} |f(b) - f(a)|.$$

2. Let $b > 0$ and

$$S = \frac{f(0) + f(b)}{2} b$$

for some function f. If $|f''(x)| \le M$ for all $x \in [0, b]$, show that

$$\left| \int_0^b f(x)\,dx - S \right| \le \frac{1}{12} M(b - a)^3.$$

[Hint: Apply integration by parts twice to evaluate $\int_0^b t(b - t) f''(t)\,dt$.]

3. Using four subintervals and choosing their midpoints as the intermediate points, approximate $\log 2 = \int_1^2 \frac{dx}{x}$ by means of a Riemann sum and estimate the error.

7.8 Improper Integrals

Although there are several types of improper integrals, here we remove only the condition that the domain of the function is a finite interval. Let f be a function that is integrable over $[a, n]$, for every integer $n \ge a$. We define

$$\int_a^\infty f = \lim_{n \to \infty} \int_a^n f,$$

provided the limit exists. In this case we say that the integral **converges**; otherwise it **diverges**. If $a \leq b \leq n$, then

$$\int_a^\infty f = \lim_{n \to \infty} \int_a^n f$$

$$= \lim_{n \to \infty} \left(\int_a^b f + \int_b^n f \right)$$

$$= \int_a^b f + \int_b^\infty f.$$

Example 7.8.1. We show that the integral

$$\int_1^\infty x^{-p} \, dx$$

converges if and only if $p > 1$.

If $p \neq 1$, then

$$\int_1^\infty x^{-p} \, dx = \lim_{n \to \infty} \int_1^n x^{-p} \, dx$$

$$= \lim_{n \to \infty} \frac{x^{1-p}}{1 - p} \Big|_1^n$$

$$= \lim_{n \to \infty} \frac{n^{1-p} - 1}{1 - p}$$

$$= \begin{cases} \frac{1}{p-1} & \text{if } p > 1 \\ \infty & \text{if } p < 1. \end{cases}$$

If $p = 1$, then the integral becomes

$$\int_1^\infty \frac{dx}{x} = \lim_{n \to \infty} \int_1^n \frac{dx}{x}$$

$$= \lim_{n \to \infty} \log x \Big|_1^n$$

$$= \lim_{n \to \infty} \log n$$

$$= \infty. \qquad\qquad \triangle$$

One of the basic tools for establishing convergence of an improper integral is the comparison test. It is analogous to the comparison test for convergence of a series.

Theorem 7.8.1 (Comparison Test). *Let* $a \in \mathbb{R}$. *Suppose that* f *and* g *are integrable functions over* $[a, b]$ *for all* $b > a$ *and that* $|f(x)| \le g(x)$ *for all* $x \ge a$. *If* $\int_a^\infty g$ *converges, then so does* $\int_a^\infty f$. *Moreover*

$$\left| \int_a^\infty f \right| \le \int_a^\infty g.$$

Proof. First note that $|f|$ is also integrable over $[a, b]$ for all $b > a$, by Theorem 7.4.8.

Suppose that $f(x) \ge 0$ for all $x \ge a$. Then the sequence

$$\left\{ \int_a^n f \right\}$$

is nondecreasing and bounded above by $\int_a^\infty g$, according to Corollary 7.4.6. In this case the result follows from Theorem 2.7.1.

For the general case, let

$$f_1 = \frac{|f| + f}{2}$$

and

$$f_2 = \frac{|f| - f}{2}.$$

Then f_1 and f_2 are integrable over $[a, b]$ for all $b > a$. Furthermore, for each $x \ge a$ and $j \in \{1, 2\}$ we have

$$0 \le f_j(x) \le |f(x)| \le g(x).$$

Hence $\int_a^\infty f_1$ and $\int_a^\infty f_2$ both converge. As $f = f_1 - f_2$, $\int_a^\infty f$ also converges.
Finally, by Theorem 7.4.8 and Corollary 7.4.6,

$$\left| \int_a^b f \right| \le \int_a^b |f| \le \int_a^b g.$$

The proof is completed by taking the limit as $b \to \infty$.   □

A corresponding result holds for integrals of the form $\int_{-\infty}^a f$.
The theorem immediately implies the following corollary.

Corollary 7.8.2. *Suppose* f *is an integrable function over* $[a, b]$ *for all* $b > a$. *If* $\int_a^\infty |f|$ *converges, then so does* $\int_a^\infty f$.

Example 7.8.2. We show that the integral

$$\int_1^\infty t^{x-1} e^{-t}\, dt$$

converges for all $x > 0$. We first locate the maximum of the function f defined by

$$f(t) = t^{x+1} e^{-t}$$

for all $t > 1$. Note that

$$f'(t) = (x+1)t^x e^{-t} - t^{x+1} e^{-t} = t^x e^{-t}(x + 1 - t)$$

for all such t. Hence $f'(t) = 0$ if and only if $t = x + 1$. Since $f'(t) > 0$ when $t < x + 1$ and $f'(t) < 0$ when $t > x + 1$, the maximum of f occurs at $x + 1$. Let $M = f(x + 1)$. Then

$$\int_1^n t^{x-1} e^{-t}\, dt = \int_1^n f(t) t^{-2}\, dt \le M \int_1^n t^{-2}\, dt$$

for all $n \in \mathbb{N}$. By Example 7.8.1, $\int_1^\infty t^{-2}\, dt$ converges. Hence

$$\int_1^\infty t^{x-1} e^{-t}\, dt$$

converges by the comparison test.   △

The improper integral $\int_a^\infty f$ is said to be **absolutely convergent** if $\int_a^\infty |f|$ is convergent.

The following example, which is due to Dirichlet, shows that a convergent improper integral need not be absolutely convergent.

Example 7.8.3. We show that the integral

$$\int_1^\infty \frac{\sin x}{x}\, dx$$

is convergent but not absolutely.

Using integration by parts (Theorem 7.5.5), we obtain

$$\int_1^n \frac{\sin x}{x}\, dx = -\frac{\cos x}{x}\Big|_1^n - \int_1^n \frac{\cos x}{x^2}\, dx$$

for all $n \in \mathbb{N}$. Now

$$-\frac{\cos x}{x}\Big|_1^n = -\frac{\cos n}{n} + \cos 1 \to \cos 1$$

as $n \to \infty$, and since

$$\left| \frac{\cos x}{x^2} \right| \le \frac{1}{x^2},$$

the integral

$$\int_1^\infty \frac{\cos x}{x^2}\, dx$$

converges by the comparison test. Hence the integral in question is convergent.
    We show next that the integral is not absolutely convergent. For all

$$x \in \left[\frac{\pi}{4} + j\pi, \frac{3\pi}{4} + j\pi \right],$$

where $j \in \mathbb{N}$, we have

$$|\sin x| \ge \sin \frac{\pi}{4} = \frac{1}{\sqrt{2}}$$

and $x < (j+1)\pi$. Hence

$$
\begin{aligned}
\int_1^\infty \frac{|\sin x|}{x}\, dx &> \sum_{j=1}^\infty \int_{(j+\frac14)\pi}^{(j+\frac34)\pi} \frac{|\sin x|}{x}\, dx \\
&\ge \frac{1}{\sqrt{2}} \sum_{j=1}^\infty \int_{(j+\frac14)\pi}^{(j+\frac34)\pi} \frac{dx}{x} \\
&> \frac{1}{\sqrt{2}} \sum_{j=1}^\infty \left(\frac{\pi}{2} \cdot \frac{1}{(j+1)\pi} \right) \\
&= \frac{1}{2\sqrt{2}} \sum_{j=2}^\infty \frac{1}{j}.
\end{aligned}
$$

As $\sum_{j=2}^\infty 1/j$ diverges, we deduce that

$$\int_1^\infty \frac{|\sin x|}{x}\, dx$$

also diverges. $\triangle$

Exercises 7.7.

1. Test the following integrals for convergence:

   (a) $\int_0^\infty \frac{dx}{xe^x}$;

   (b) $\int_{-\infty}^0 e^{ax}\, dx$;

   (c) $\int_0^\infty \sin x^2\, dx$.

2. Evaluate

$$\int_1^\infty \frac{\log x}{x^2}\, dx.$$

3. (a) Suppose that $f(x) \geq 0$ and $g(x) \geq 0$ for all $x \geq a$, and let

$$\lim_{x\to\infty} \frac{f(x)}{g(x)} = L.$$

   Prove that

   i. If $0 < L < \infty$, then $\int_a^\infty f$ and $\int_a^\infty g$ both converge or both diverge.
   ii. If $L = 0$, then $\int_a^\infty f$ converges if $\int_a^\infty g$ converges.
   iii. If $L = \infty$, then $\int_a^\infty g$ converges if $\int_a^\infty f$ converges.

   (b) The result of part (a) is called the limit comparison test for integrals. Use it to prove that

$$\int_1^\infty \frac{dx}{a^2 + x^2}$$

   is convergent.

7.9   Integral Test for Convergence of a Series

Since an integral is defined in terms of Riemann sums, there are many similarities between the theories of integrals and series. One of the most important connections between the two theories is the integral test for the convergence of a series. It involves improper integrals.

Theorem 7.9.1 (Euler, Maclaurin). *Let f be a nonincreasing continuous function on the interval $[1, \infty)$, and suppose that $f(x) \geq 0$ for all $x \geq 1$. Let*

$$d_n = \sum_{j=1}^n f(j) - \int_1^n f$$

for all $n \in \mathbb{N}$. Then the sequence $\{d_n\}$ converges.

Proof. It suffices to show that $\{d_n\}$ is nonincreasing and bounded below. Since f is nonincreasing, for every positive integer j we have

$$f(j + 1) = \min\{f(x) \mid x \in [j, j + 1]\}$$

and

$$f(j) = \max\{f(x) \mid x \in [j, j + 1]\}.$$

Hence

$$f(j + 1) \le \int_j^{j+1} f \le f(j),$$

so that

$$d_{n+1} - d_n = \left(\sum_{j=1}^{n+1} f(j) - \int_1^{n+1} f\right) - \left(\sum_{j=1}^{n} f(j) - \int_1^{n} f\right)$$

$$= f(n + 1) - \int_n^{n+1} f$$

$$\le f(n + 1) - f(n + 1)$$

$$= 0$$

for each $n \in \mathbb{N}$.

We have now proved that $\{d_n\}$ is nonincreasing. It is also bounded below by 0, since

$$d_n = \sum_{j=1}^{n} f(j) - \sum_{j=1}^{n-1} \int_j^{j+1} f$$

$$\ge \sum_{j=1}^{n} f(j) - \sum_{j=1}^{n-1} f(j)$$

$$= f(n)$$

$$\ge 0$$

for each $n \in \mathbb{N}$. $\qquad\square$

Corollary 7.9.2. *Under the hypotheses of Theorem 7.9.1 we have*

$$\int_1^{n+1} f \le \sum_{j=1}^n f(j) \le f(1) + \int_1^n f$$

for all $n \in \mathbb{N}$.

Proof. From the proof of the theorem we have

$$f(j+1) \le \int_j^{j+1} f \le f(j)$$

for all $j \in \mathbb{N}$. From the second inequality we obtain

$$\int_1^{n+1} f = \sum_{j=1}^n \int_j^{j+1} f \le \sum_{j=1}^n f(j).$$

The first inequality gives

$$\sum_{j=1}^{n-1} f(j+1) \le \sum_{j=1}^{n-1} \int_j^{j+1} f = \int_1^n f.$$

Consequently

$$\sum_{j=1}^n f(j) = f(1) + \sum_{j=2}^n f(j) = f(1) + \sum_{j=1}^{n-1} f(j+1) \le f(1) + \int_1^n f. \qquad \square$$

Corollary 7.9.3 (Integral Test). *Let f be a nonincreasing continuous function on the interval $[1, \infty]$, and suppose that $f(x) \ge 0$ for all $x \ge 1$. Then the series $\sum_{j=1}^\infty f(j)$ and the integral $\int_1^\infty f$ both converge or both diverge.*

Remark. The Euler–Maclaurin theorem and the integral test are clearly true if there exists a fixed number N such that the function satisfies all the conditions for each $x \ge N$.

Example 7.9.1. Use the integral test to show that $\sum_{j=1}^\infty 1/j^p$ is convergent if and only if $p > 1$.

Solution. Let

$$f(x) = \frac{1}{x^p} > 0$$

for all $x > 0$.

If $p \leq 0$, we have

$$\lim_{n \to \infty} \frac{1}{n^p} \neq 0$$

and the series is divergent by the nth term test.

Suppose $p > 0$. Then f is clearly nonincreasing and continuous at all $x \geq 1$ and the integral test can be applied. Therefore, by the integral test and the result of Example 7.8.1, the series is convergent if and only if $p > 1$. $\triangle$

Example 7.9.2. Use the integral test to confirm the divergence of Abel's series

$$\sum_{j=2}^{\infty} \frac{1}{j \log j}.$$

Solution. Let

$$f(x) = \frac{1}{x \log x} > 0$$

for all $x > 1$. Then as $x \log x$ is increasing at all $x \geq 2$, it follows that f is decreasing at all $x \geq 2$. Moreover f is continuous at all $x \geq 2$. Hence the integral test can be applied.

Using the result of Example 7.5.4, we have

$$\int_2^{\infty} \frac{dx}{x \log x} = \lim_{n \to \infty} (\log \log n - \log \log 2) = \infty.$$

Therefore the series is divergent by the integral test. $\triangle$

Series for which the kth term involves $\log k$ are often suitable for the use of the integral test. One problem with the integral test is that the required integral may not be easy to evaluate. A comparison test is often used to modify the series so that the corresponding integral may be found easily.

Example 7.9.3. Test the convergence of

$$\sum_{j=1}^{\infty} \frac{1}{j \log(j^2 + 3j + 1)}.$$

Solution. Let

$$a_n = \frac{1}{n \log(n^2 + 3n + 1)}$$

and

$$b_n = \frac{1}{n \log n}$$

for all integers $n > 1$. Then, using l'Hôpital's rule,

$$\lim_{n \to \infty} \frac{a_n}{b_n} = \lim_{n \to \infty} \frac{n \log n}{n \log(n^2 + 3n + 1)}$$

$$= \lim_{n \to \infty} \frac{\log n}{\log(n^2 + 3n + 1)}$$

$$= \lim_{n \to \infty} \frac{\frac{1}{n}}{\frac{2n+3}{n^2+3n+1}}$$

$$= \lim_{n \to \infty} \frac{n^2 + 3n + 1}{2n^2 + 3n}$$

$$= \frac{1}{2}.$$

Since $\sum_{j=2}^{\infty} b_j$ is divergent by the previous example, $\sum_{j=1}^{\infty} a_j$ is also divergent by the limit comparison test. $\triangle$

Exercises 7.8.

1. Test the convergence of the following series:

   (a) $\sum_{j=2}^{\infty} \frac{1}{j \log^2 j}$.   (c) $\sum_{j=2}^{\infty} \frac{1}{j \log j \log \log j}$.

   (b) $\sum_{j=2}^{\infty} \frac{1}{(j+5) \log^2 j}$.

2. Does the result

$$\int_1^{\infty} \frac{dx}{x^2} = 1 \neq \sum_{j=1}^{\infty} \frac{1}{j^2}$$

   contradict the argument used in the proof of the integral test?

3. Apply the integral test to test the convergence of

$$\sum_{j=2}^{\infty} \frac{1}{j \log^p j}.$$

Chapter 8
Taylor Polynomials and Taylor Series

8.1 Taylor's Theorem

Because polynomials are so easy to study, it would be very convenient if they were the only functions with which we had to deal. Although this is not the case, it turns out that many functions that are not polynomials can be approximated as accurately as we please by polynomials. In this section we present a theorem that gives conditions under which such an approximation is possible.

Given a function f that is not a polynomial, we are interested in finding a polynomial, in a variable x, that gives a good approximation for $f(x)$ at values of x near some number a. It is convenient to write the polynomial in powers of $x - a$ rather than x. If we take the polynomial to be of degree $n \geq 0$ and denote it by P_n, then we can write

$$P_n(x) = \sum_{j=0}^{n} a_j (x - a)^j \tag{8.1}$$

for all x, where $a_0, a_1, \ldots, a_n$ are constants and we take $0^0 = 1$. We also assume that the graph of f is smooth at a in the sense that $f(a), f'(a), \ldots, f^{(n)}(a)$ exist and are known. This may seem to be rather a strong requirement, but in fact it is satisfied for every $n \geq 0$ by many important functions, such as e^x and $\sin x$ at 0 and $\log x$ at 1.

It is reasonable to expect that at a the approximation is exact and the derivatives of P_n and f are equal. In fact, we shall require that

$$P_n^{(k)}(a) = f^{(k)}(a)$$

for all integers k such that $0 \leq k \leq n$.

© Springer Science+Business Media New York 2015
C.H.C. Little et al., *Real Analysis via Sequences and Series*, Undergraduate
Texts in Mathematics, DOI 10.1007/978-1-4939-2651-0_8

We begin by deriving a formula for the coefficients of P_n. By differentiating the sides of Eq. (8.1) k times, where $0 \leq k \leq n$, we obtain

$$P_n^{(k)}(x) = \sum_{j=0}^{n} a_j \, j(j-1) \cdots (j-k+1)(x-a)^{j-k}$$

$$= \sum_{j=k}^{n} a_j \frac{j!}{(j-k)!}(x-a)^{j-k}$$

$$= a_k k! + \sum_{j=k+1}^{n} a_j \frac{j!}{(j-k)!}(x-a)^{j-k}$$

for all x. Therefore

$$f^{(k)}(a) = P_n^{(k)}(a) = a_k k!,$$

and so

$$a_k = \frac{f^{(k)}(a)}{k!}.$$

Substitution into Eq. (8.1) yields

$$P_n(x) = \sum_{j=0}^{n} \frac{f^{(j)}(a)}{j!}(x-a)^j$$

for all x. This polynomial is called the **Taylor polynomial of order n for f about** a.

Example 8.1.1. Find the Taylor polynomial of order 5 for $\cos x$ about 0.

Solution. Taking $f(x) = \cos x$ for all x and $a = 0$, we have $f(0) = 1$. Moreover

$$f'(x) = -\sin x$$

for all x, and so $f'(0) = 0$. Similarly, we find that $f^{(2)}(0) = -1$, $f^{(4)}(0) = 1$, and $f^{(3)}(0) = f^{(5)}(0) = 0$. Therefore

$$P_5(x) = 1 - \frac{x^2}{2} + \frac{x^4}{24}.$$

Notice that in this case the Taylor polynomial of order 5 is actually of degree 4 because the term in x^5 has a coefficient of 0. $\triangle$

We now prove Taylor's theorem, which gives a means of estimating the error incurred in using a Taylor polynomial as an approximation for the value of a function at some number.

Theorem 8.1.1 (Taylor). *Let a be a real number, n a nonnegative integer, and f a function such that $f^{(n+1)}(x)$ exists for all x in some open interval I containing a. Let P_n be the Taylor polynomial of order n for f about a. Then for each $x \in I - \{a\}$ there exists a number ξ between a and x such that*

$$f(x) - P_n(x) = \frac{f^{(n+1)}(\xi)}{(n+1)!}(x-a)^{n+1}.$$

Proof. Fix $x \in I - \{a\}$, and for all $t \in I$ define

$$F(t) = f(x) - \sum_{j=0}^{n} \frac{f^{(j)}(t)}{j!}(x-t)^j$$

$$= f(x) - f(t) - \sum_{j=1}^{n} \frac{f^{(j)}(t)}{j!}(x-t)^j.$$

Since $f^{(n+1)}(t)$ exists for all $t \in I$, we infer that $f^{(j)}$ must be continuous on I for all j such that $0 \leq j \leq n$. Hence F is continuous on the closed interval with ends a and x and differentiable on the corresponding open interval. Note that

$$F(x) = f(x) - f(x) = 0$$

and

$$F(a) = f(x) - \sum_{j=0}^{n} \frac{f^{(j)}(a)}{j!}(x-a)^j = f(x) - P_n(x).$$

Thus we need to show that

$$F(a) = \frac{f^{(n+1)}(\xi)}{(n+1)!}(x-a)^{n+1}$$

for some ξ between a and x.

Using the product rule, the chain rule, and the telescoping property, we obtain

$$F'(t) = -f'(t) - \sum_{j=1}^{n} \left(\frac{f^{(j+1)}(t)}{j!}(x-t)^j + \frac{f^{(j)}(t)}{j!} j (x-t)^{j-1}(-1) \right)$$

$$= -f'(t) - \sum_{j=1}^{n} \left(\frac{f^{(j+1)}(t)}{j!}(x-t)^j - \frac{f^{(j)}(t)}{(j-1)!}(x-t)^{j-1} \right)$$

$$= -f'(t) - \left(\frac{f^{(n+1)}(t)}{n!}(x-t)^n - f'(t) \right)$$

$$= -\frac{(x-t)^n}{n!} f^{(n+1)}(t).$$

Now let

$$G(t) = (x-t)^{n+1}$$

for all $t \in I$. Then

$$G'(t) = -(n+1)(x-t)^n$$

for all such t. Note also that $G(x) = 0$ and that $G'(t) \neq 0$ for all $t \neq x$. By Cauchy's mean-value formula there exists ξ between a and x such that

$$\frac{F'(\xi)}{G'(\xi)} = \frac{F(a) - F(x)}{G(a) - G(x)} = \frac{F(a)}{G(a)}.$$

Hence

$$F(a) = \frac{F'(\xi)}{G'(\xi)} G(a)$$

$$= \frac{-\frac{(x-\xi)^n}{n!} f^{(n+1)}(\xi)(x-a)^{n+1}}{-(n+1)(x-\xi)^n}$$

$$= \frac{f^{(n+1)}(\xi)}{(n+1)!}(x-a)^{n+1},$$

as required.   □

As a consequence of this theorem we have

$$f(x) = P_n(x) + \frac{f^{(n+1)}(\xi)}{(n+1)!}(x-a)^{n+1}$$

$$= \sum_{j=0}^{n} \frac{f^{(j)}(a)}{j!}(x-a)^j + \frac{f^{(n+1)}(\xi)}{(n+1)!}(x-a)^{n+1}$$

for all $x \in I - \{a\}$. Note also that the value of ξ depends on the choice of x. For each $x \in I - \{a\}$ we define

$$R_n(x) = \frac{f^{(n+1)}(\xi)}{(n+1)!}(x-a)^{n+1}. \tag{8.2}$$

Thus

$$R_n(x) = f(x) - P_n(x)$$

for each such x. The function R_n is called the **Taylor remainder** of degree n.

Example 8.1.2. Use the Taylor polynomial of order 5 for $\cos x$ about 0 to obtain an approximation for $\cos 0.1$ and use Taylor's theorem to show that the approximation gives the first eight digits in the decimal expansion of $\cos 0.1$ correctly.

Solution. From Example 8.1.1 we have

$$P_5(x) = 1 - \frac{x^2}{2} + \frac{x^4}{24}$$

for all x. The required approximation for $\cos 0.1$ is therefore

$$P_5(0.1) = 0.9950041666\ldots.$$

Since the sixth derivative of $\cos x$ is $-\cos x$, Taylor's theorem shows that there exists a number ξ between 0 and 0.1 such that

$$\cos 0.1 - P_5(0.1) = \frac{(0.1)^6}{6!}(-\cos \xi) = -0.0000000013888\ldots (\cos \xi).$$

Since $0 < \cos \xi < 1$, we have

$$-0.0000000013888\ldots < \cos 0.1 - P_5(0.1) < 0.$$

Hence

$$-0.0000000014 < \cos 0.1 - P_5(0.1) < 0,$$

and so

$$P_5(0.1) - 0.0000000014 < \cos 0.1 < P_5(0.1).$$

We conclude that

$$0.99500416526\ldots < \cos 0.1 < 0.9950041666\ldots.$$

Thus $P_5(0.1)$ gives the first eight digits in the decimal expansion of $\cos 0.1$ correctly as 0.99500416. $\triangle$

A function f is said to be **smooth** over an interval I if $f^{(k)}(x)$ exists for all $x \in I$ and all nonnegative integers k. We write $C^\infty(I)$ for the set of functions that are smooth over I.

Example 8.1.3. Let $f \in C^\infty(I)$, where I is an open interval containing 1, and suppose that $f(1) = 0$, $f'(1) = 1$, $f''(1) = 2$, and

$$f'''(x) = \frac{e^{x^2}}{1 + x}$$

for all $x \in I - \{-1\}$. Write down the Taylor polynomial of order 3 for f about 1. Suppose that this polynomial is used to approximate $f(0.5)$. Find an upper bound for the error and write down the approximate value of $f(0.5)$.

Solution. The Taylor polynomial is

$$P_3(x) = f(1) + (x - 1)f'(1) + \frac{(x - 1)^2}{2!}f''(1) + \frac{(x - 1)^3}{3!}f'''(1)$$

$$= (x - 1) + (x - 1)^2 + \frac{e(x - 1)^3}{12}.$$

The error incurred by the approximation is given by $|R_3(0.5)|$. Now,

$$f^{(4)}(\xi) = \frac{e^{\xi^2}\left(2\xi^2 + 2\xi - 1\right)}{(1 + \xi)^2},$$

where $0.5 < \xi < 1$, and since $2\xi^2 + 2\xi - 1$ is increasing at all $\xi > 0.5$, it follows that

$$|f^{(4)}(\xi)| < \frac{3e}{(1.5)^2} = \frac{4}{3}e < 4.$$

We thus have

$$|R_3(0.5)| = \frac{|f^{(4)}(\xi)||0.5 - 1|^4}{4!} < \frac{4}{4!2^4} < 0.011.$$

The approximation given by $P_3(0.5)$ is

$$P_3(0.5) = -0.5 + 0.25 - \frac{(0.125)e}{12} \approx -0.28$$

with error less than 0.011.   △

A common problem in the approximation of numbers or functions is to make the approximation correct to within a specified error tolerance. If we use the Taylor polynomial of order n about a point a to approximate a function f at a point x near a, then the error depends on n, the distance $|x - a|$, and the absolute values of the derivatives of f between x and a. The next example illustrates the problem and a solution strategy.

Example 8.1.4. Let $f(x) = \log(x + 1)$ for all $x > -1$. What degree Taylor polynomial about 0 is required to approximate $\log 1.1$ with a maximum error of 0.001? What degree Taylor polynomial about 0 is required to approximate $\log 1.5$ with the same maximum error?

Solution. For all $n \geq 1$ and all $\xi > 0$ we have

$$|f^{(n)}(\xi)| = \frac{(n-1)!}{(1+\xi)^n}$$
$$< (n-1)!.$$

To approximate $\log 1.1$, we note that $x = 0.1$ and

$$|R_n(0.1)| = \frac{|f^{(n+1)}(\xi)|(0.1)^{n+1}}{(n+1)!}$$
$$< \frac{n!}{(n+1)!10^{n+1}}$$
$$= \frac{1}{(n+1)10^{n+1}}.$$

We thus seek a value of n such that

$$(n+1)10^{n+1} \geq 1000.$$

Evidently the preceding inequality is satisfied for $n = 2$ but not for $n = 1$. We conclude that a Taylor polynomial of degree 2 suffices for the approximation.

To approximate $\log 1.5$, we have

$$|R_n(0.5)| < \frac{1}{(n+1)2^{n+1}}.$$

We seek a value of n such that

$$(n+1)2^{n+1} \geq 1000.$$

For $n = 6$,

$$7(2^7) = 896,$$

and for $n = 7$,

$$8(2^8) = 2048.$$

A Taylor polynomial of degree 7 thus suffices to make this approximation. $\triangle$

A function $f \in C^\infty(I)$ is said to have a **zero of order** k at a number $a \in I$ if $f^{(j)}(a) = 0$ for $j = 0, 1, \ldots, k - 1$ and $f^{(k)}(a) \neq 0$. If f has a zero of order $k > 0$ at a, then Taylor's theorem implies that

$$f(x) = \frac{f^{(k)}(a)}{k!}(x - a)^k + \frac{f^{(k+1)}(\xi)}{(k + 1)!}(x - a)^{k+1}$$

$$= (x - a)^k F(x) \qquad\qquad (8.3)$$

for all $x \in I - \{a\}$, where ξ is a number between a and x and

$$F(x) = \frac{f^{(k)}(a)}{k!} + \frac{f^{(k+1)}(\xi)}{(k + 1)!}(x - a)$$

for all $x \in I$. Thus

$$F(a) = \frac{f^{(k)}(a)}{k!} \neq 0.$$

Conversely, if $f(x)$ is given by Eq. (8.3) for all $x \in I - \{a\}$ and $F(a) \neq 0$, then f has a zero of order k at a.

Example 8.1.5. Let $f(x) = \sin x$ for all x and let $a = 0$. Then $f(0) = \sin 0 = 0$ and $f'(0) = \cos 0 = 1$. The function f thus has a zero of order 1 at 0. From the definition of $\sin x$ we know that

$$\sin x = \sum_{j=0}^{\infty} \frac{(-1)^j x^{2j+1}}{(2j + 1)!}$$

$$= x \sum_{j=0}^{\infty} \frac{(-1)^j x^{2j}}{(2j + 1)!}$$

$$= x F(x),$$

where

$$F(x) = \sum_{j=0}^{\infty} \frac{(-1)^j x^{2j}}{(2j + 1)!}$$

for all x. Note that $F(0) = 1 \neq 0$. $\triangle$

Taylor's theorem provides an alternative perspective on l'Hôpital's rule. Suppose that f and g are functions that are smooth over an interval I and have zeros of order $m > 0$ and $n > 0$, respectively, at a number $a \in I$. Let

$$h(x) = \frac{f(x)}{g(x)}$$

for all $x \in I - \{a\}$ for which $g(x) \neq 0$. Now, $\lim_{x \to a} h(x)$ is of the indeterminate form 0/0, and l'Hôpital's rule can be invoked to evaluate the limit, if it exists. Equation (8.3), however, shows that the functions f and g can be expressed as

$$f(x) = (x-a)^m F(x)$$

and

$$g(x) = (x-a)^n G(x),$$

where F and G are continuous functions such that

$$F(a) = \frac{f^{(m)}(a)}{m!} \neq 0$$

and

$$G(a) = \frac{g^{(n)}(a)}{n!} \neq 0.$$

Thus

$$h(x) = (x-a)^{m-n} \frac{F(x)}{G(x)}.$$

Evidently

$$\lim_{x \to a} \frac{F(x)}{G(x)} = \frac{F(a)}{G(a)} = \frac{n! f^{(m)}(a)}{m! g^{(n)}(a)} \neq 0.$$

We thus have

$$\lim_{x \to a} h(x) = \begin{cases} 0 & \text{if } m > n, \\ \frac{f^{(m)}(a)}{g^{(m)}(a)} & \text{if } m = n. \end{cases}$$

If $n > m$, it is clear that $|h(x)| \to \infty$ as $x \to a$.

In practice, this approach to evaluating limits is not notably shorter than using l'Hôpital's rule. It does, however, indicate how many applications of l'Hôpital's rule are needed, and in the case where the Taylor polynomials are known, it can prove a shorter calculation.

In determining the order of a zero it is useful to observe that if f and g are smooth functions that have zeros of order $m > 0$ and $n > 0$, respectively, at a then the function $p = fg$ can be expressed as

$$p(x) = (x-a)^{m+n} F(x) G(x),$$

where

$$F(a)G(a) = \frac{f^{(m)}(a)}{m!} \cdot \frac{g^{(n)}(a)}{n!} \neq 0;$$

consequently, p has a zero of order $m + n$ at a.

Example 8.1.6. Evaluate

$$\lim_{x \to 0} \frac{x \sin^2 x}{1 - \cos x}.$$

Solution. Let $f(x) = x \sin^2 x$ for all x and let $g(x) = 1 - \cos x$ for all x such that $\cos x \neq 1$. We know from Example 8.1.5 that $\sin x$ has a zero of order 1 at 0; consequently $\sin^2 x$ has a zero of order 2 at 0 and therefore f has a zero of order 3 at 0. Now $g(0) = 1 - \cos 0 = 0$, $g'(0) = \sin 0 = 0$, and $g''(0) = \cos 0 = 1$. The function g thus has a zero of order 2 at 0. Since $3 > 2$,

$$\lim_{x \to 0} \frac{x \sin^2 x}{1 - \cos x} = 0.$$

$\triangle$

Example 8.1.7. Evaluate

$$\lim_{x \to 0} \frac{x(1 - e^x)}{1 - \cos x}.$$

Solution. Let $f(x) = x(1 - e^x)$ for all x and let $g(x) = 1 - \cos x$ for all x for which $\cos x \neq 1$. It can be readily verified that $1 - e^x$ has a zero of order 1 at 0, and therefore f must have a zero of order 2 at 0. We know from Example 8.1.6 that g has a zero of order 2 at 0. In this case $m = n = 2$, so that the limit exists and is nonzero. We also know that two applications of l'Hôpital's rule are required to evaluate this limit. The limit can also be found by using Taylor polynomials. In detail,

$$f(x) = x \left(1 - \sum_{j=0}^{\infty} \frac{x^j}{j!} \right)$$

$$= x \left(1 - 1 - x - x \sum_{j=2}^{\infty} \frac{x^{j-1}}{j!} \right)$$

$$= x^2 (-1 + s(x))$$

and

$$g(x) = 1 - \sum_{j=0}^{\infty} \frac{(-1)^j x^{2j}}{(2j)!}$$

$$= 1 - 1 + \frac{x^2}{2!} - x^2 \sum_{j=2}^{\infty} \frac{(-1)^j x^{2j-2}}{(2j)!}$$

$$= x^2 \left(\frac{1}{2} + t(x) \right),$$

where s and t are functions such that $\lim_{x \to 0} s(x) = \lim_{x \to 0} t(x) = 0$. Consequently

$$\lim_{x \to 0} \frac{x(1 - e^x)}{1 - \cos x} = \lim_{x \to 0} \frac{x^2 (-1 + s(x))}{x^2 \left(\frac{1}{2} + t(x) \right)} = -2.$$

$\triangle$

If f has a zero of order $k > 0$ at a, then Eq. (8.3) implies that, near a, the function f behaves like $(x - a)^k F(a)$. For example, if $k = 3$ and $a = 0$, then the graph of f near 0 would be almost the same as that of $x^3 F(0)$. Thus in a neighborhood of 0 the graph would look like the cubic curve $y = Cx^3$, where C is a nonzero constant. The local shape of a curve near a critical point provides a key insight into the nature of the critical point. Recall that if a smooth function f has a relative extremum at a, then $f'(a) = 0$. The nature of the critical point depends on the higher-order derivatives at a. It is the first nonzero derivative at a that determines whether the critical point corresponds to a relative extremum. Briefly, suppose g has a critical point at a. Then a is a zero of $f(x) = g(x) - g(a)$ of order $k > 1$. The graph of f near a is nearly the same as that of $C(x - a)^k$: If k is even, the critical point will yield a local extremum; if k is odd, then the critical point cannot correspond to a relative extremum. This simple observation is formalized in the next result.

Theorem 8.1.2. *Let I be an open interval containing some number a, and let $n \in \mathbb{N}$. Let f be a function such that $f^{(k)}$ is defined and continuous on I for each positive integer $k \le n$. Suppose that $f^{(k)}(a) = 0$ for all positive integers $k < n$ but $f^{(n)}(a) \ne 0$.*

1. If n is even and $f^{(n)}(a) > 0$, then $f(a)$ is a relative minimum.
2. If n is even and $f^{(n)}(a) < 0$, then $f(a)$ is a relative maximum.
3. If n is odd, then $f(a)$ is not a relative extremum.

Proof. Since $f^{(n)}(a) \ne 0$ and $f^{(n)}$ is continuous, we may choose I so that

$$f^{(n)}(x) f^{(n)}(a) > 0$$

for all $x \in I$. Taylor's theorem shows that for each fixed $x \in I - \{a\}$ there exists ξ between x and a such that

$$f(x) = P_{n-1}(x) + R_{n-1}(x)$$
$$= f(a) + \frac{f^{(n)}(\xi)}{n!}(x - a)^n.$$

Thus

$$f^{(n)}(\xi) f^{(n)}(a) > 0.$$

1. In this case $f^{(n)}(\xi) > 0$. Hence $R_{n-1}(x) > 0$, so that $f(x) > f(a)$.
2. The proof is similar in this case.
3. Since n is odd, $(x - a)^n$ has the same sign as $x - a$. Thus the sign of $R_{n-1}(x)$ when $x > a$ is different from its sign when $x < a$. Consequently $f(a)$ is not a relative extremum.

$\square$

Example 8.1.8. Let

$$g(x) = 1 + x(1 - \cos x)$$

for all x. It is readily verified that g has a critical point at 0. Let

$$f(x) = g(x) - g(0) = x(1 - \cos x).$$

Since $1 - \cos x$ has a zero of order 2 at 0, f has a zero of order 3 at 0. We conclude that f, and therefore g, does not have a relative extremum at 0. $\triangle$

Example 8.1.9. Let

$$g(x) = 1 + x \sin x - x^2 - x^5$$

for all x. Then g has a critical point at 0. Here $f(x) = x \sin x - x^2 - x^5$ for all x, and the definition of $\sin x$ gives

$$f(x) = -\frac{x^4}{3!} - x^5 + x^6 L(x),$$

where L is some smooth function. This expression shows that f has a zero of order 4 at 0 and so the critical point yields a relative extremum. Since $g^{(4)}(0) = -4 < 0$, this extremum is a relative maximum. $\triangle$

Exercises 8.1.

1. Let $f(x) = e^x / x^e$ for all $x > 0$.

   (a) Show that the only relative minimum of f is at e.
   (b) Deduce that $e^\pi > \pi^e$.

2. Use the Taylor polynomial of order 6 for $\sin x$ about 0 to obtain an approximate value for $\sin 0.3$ and then use Taylor's theorem to prove that this approximation gives the first seven digits in the decimal expansion of $\sin 0.3$ correctly.

3. Use the Taylor polynomial of order 3 for $\log x$ about 1 to obtain an approximate value for $\log 1.06$ and then use Taylor's theorem to prove that this approximation gives the first four digits in the decimal expansion of $\log 1.06$ correctly.

4. Suppose f is a function that satisfies $f(0) = 1$, $f'(0) = 0$, and

$$f''(x) + xf(x) = 0$$

for all $x \in \mathbb{R}$. Find the Taylor polynomial of order 9 for f about 0. (The function f is called an Airy function.)

5. Let f be a function such that $f(1) = 0$, $f'(1) = 1$, $f''(1) = 2$, and

$$f'''(x) = \frac{2^x \log(x + 2)}{x + 1}$$

for all $x > -2$ satisfying $x \neq -1$.

(a) Write down the Taylor polynomial of order 3 for f about 1.

(b) Find upper bounds for the error if the Taylor polynomial of order 2 for f about 1 is used to approximate $f(1.2)$ and $f(0.5)$.

8.2 Taylor Series

Let f be a function, in a real variable x, given by

$$f(x) = \sum_{j=0}^{\infty} a_j (x - a)^j \tag{8.4}$$

for all x in the interior of the interval of convergence. Thus

$$a_j = \frac{f^{(j)}(a)}{j!}$$

for all j, as in Sect. 8.1, so that

$$f(x) = \sum_{j=0}^{\infty} \frac{f^{(j)}(a)}{j!} (x - a)^j. \tag{8.5}$$

The series on the right-hand side is called the **Taylor series** for f about a. When $a = 0$, it is also called the **Maclaurin series** for f. Note that the Taylor series about a is the only possible representation of f as a power series with center a.

One application of a Taylor series gives a proof of the binomial theorem. Recall that

$$\binom{n}{j} = \frac{n!}{j!(n-j)!},$$

where j and n are integers such that $0 \le j \le n$.

Theorem 8.2.1. *For every nonnegative integer n and every x and y,*

$$(x+y)^n = \sum_{j=0}^{n} \binom{n}{j} x^j y^{n-j}.$$

Proof. For $y = 0$ we have

$$\sum_{j=0}^{n} \binom{n}{j} x^j y^{n-j} = \sum_{j=0}^{n-1} \binom{n}{j} x^j y^{n-j} + x^n$$
$$= x^n,$$

as required.

Suppose $y = 1$, and let

$$f(x) = (x+1)^n$$

for all x. Then f is a polynomial of degree n, and so we may write

$$(x+1)^n = \sum_{j=0}^{\infty} a_j x^j$$

for some integers $a_0, a_1, \ldots$, where $a_j = 0$ for all $j > n$. Therefore

$$a_j = \frac{f^{(j)}(0)}{j!}$$

for each j. But for all $j \le n$ and all x we have

$$f^{(j)}(x) = \frac{n!}{(n-j)!}(x+1)^{n-j},$$

so that

$$f^{(j)}(0) = \frac{n!}{(n-j)!}.$$

Hence

$$a_j = \frac{n!}{j!(n-j)!} = \binom{n}{j},$$

and substitution into the summation yields

$$(x+1)^n = \sum_{j=0}^{n} \binom{n}{j} x^j,$$

as required.

In the general case with $y \neq 0$ we therefore have

$$(x+y)^n = y^n \left(\frac{x}{y} + 1\right)^n$$

$$= y^n \sum_{j=0}^{n} \binom{n}{j} \left(\frac{x}{y}\right)^j$$

$$= \sum_{j=0}^{n} \binom{n}{j} x^j y^{n-j}.$$

$\square$

Taking $x = y = 1$ yields the following corollary.

Corollary 8.2.2. *For every nonnegative integer n,*

$$\sum_{j=0}^{n} \binom{n}{j} = 2^n.$$

We now give an example of an infinitely differentiable function f whose Taylor series converges but to a sum different from f.

Example 8.2.1. Let

$$f(x) = e^{-1/x^2}$$

for all $x \neq 0$, and let $f(0) = 0$. This function is known as **Cauchy's function**. We show that its Maclaurin series does not converge to $f(x)$ for any $x \neq 0$.

First we prove that if n is a positive integer, then

$$\lim_{x \to 0} \frac{e^{-1/x^2}}{x^n} = 0. \tag{8.6}$$

We start with the case where n is even. Suppose therefore that $n = 2k$, where k is a positive integer. Then

$$\frac{e^{-1/x^2}}{x^n} = \frac{\frac{1}{x^{2k}}}{e^{1/x^2}}$$

and

$$\lim_{x \to 0} \frac{1}{x^{2k}} = \lim_{x \to 0} e^{1/x^2} = \infty.$$

Therefore we may apply l'Hôpital's rule.

At this point we apply induction. For $k = 1$ we have

$$\lim_{x \to 0} \frac{\frac{1}{x^2}}{e^{1/x^2}} = \lim_{x \to 0} \frac{\frac{-2}{x^3}}{\frac{-2}{x^3}e^{1/x^2}} = \lim_{x \to 0} \frac{1}{e^{1/x^2}} = 0.$$

Now assume that $k > 1$ and

$$\lim_{x \to 0} \frac{\frac{1}{x^{2(k-1)}}}{e^{1/x^2}} = 0.$$

Then

$$\lim_{x \to 0} \frac{\frac{1}{x^{2k}}}{e^{1/x^2}} = \lim_{x \to 0} \frac{\frac{-2k}{x^{2k+1}}}{\frac{-2}{x^3}e^{1/x^2}} = k \lim_{x \to 0} \frac{\frac{1}{x^{2(k-1)}}}{e^{1/x^2}} = k \cdot 0 = 0,$$

as required.

If n is odd, we may write $n = 2k + 1$, where k this time is a nonnegative integer. We then have

$$\lim_{x \to 0} \frac{e^{-1/x^2}}{x^{2k+1}} = \lim_{x \to 0} \left(x \cdot \frac{e^{-1/x^2}}{x^{2k+2}} \right) = 0 \cdot 0 = 0$$

since $2k + 2$ is even and positive. The proof of Eq. (8.6) is now complete.

Next we show by induction that for every $x \ne 0$ and every positive integer n we have

$$f^{(n)}(x) = e^{-1/x^2} g_n(x), \tag{8.7}$$

where

$$g_n(x) = \sum_{j=1}^{m} \frac{A_j}{x^j}$$

for some positive integer m and constants $A_1, A_2, \ldots, A_m$. For every $x \ne 0$ we have

$$f'(x) = e^{-1/x^2} \cdot \frac{2}{x^3}.$$

Thus $f'(x)$ is of the required form, with $m = 3$, $A_1 = A_2 = 0$, and $A_3 = 2$. Suppose that the desired result holds for n. Note that

$$f^{(n+1)}(x) = \frac{2}{x^3}e^{-1/x^2}g_n(x) + e^{-1/x^2}g_n'(x)$$

$$= e^{-1/x^2}\left(\frac{2}{x^3}g_n(x) + g_n'(x)\right).$$

Since

$$g_n'(x) = -\sum_{j=1}^{m}\frac{jA_j}{x^{j+1}},$$

it follows that

$$\frac{2}{x^3}g_n(x) + g_n'(x)$$

$$= \sum_{j=1}^{m}\frac{2A_j}{x^{j+3}} - \sum_{j=1}^{m}\frac{jA_j}{x^{j+1}}$$

$$= \sum_{j=4}^{m+3}\frac{2A_{j-3}}{x^j} - \sum_{j=2}^{m+1}\frac{(j-1)A_{j-1}}{x^j}$$

$$= -\frac{A_1}{x^2} - \frac{2A_2}{x^3} + \sum_{j=4}^{m+1}\left(\frac{2A_{j-3}}{x^j} - \frac{(j-1)A_{j-1}}{x^j}\right) + \frac{2A_{m-1}}{x^{m+2}} + \frac{2A_m}{x^{m+3}}.$$

Hence

$$f^{(n+1)}(x) = e^{-1/x^2}\sum_{j=1}^{r}\frac{B_j}{x^j},$$

where $r = m + 3$, $B_1 = 0$, $B_2 = -A_1$, $B_3 = -2A_2$, $B_{m+2} = 2A_{m-1}$, $B_{m+3} = 2A_m$, and

$$B_j = 2A_{j-3} - (j-1)A_{j-1}$$

for all integers j such that $4 \leq j \leq m + 1$. The proof that $f^{(n)}(x)$ is of the desired form is now complete.

Now we prove by induction that $f^{(n)}(0) = 0$ for all positive integers n. First

$$f'(0) = \lim_{x\to 0}\frac{f(x) - f(0)}{x - 0} = \lim_{x\to 0}\frac{f(x)}{x} = \lim_{x\to 0}\frac{e^{-1/x^2}}{x} = 0$$

by Eq. (8.6). Suppose that $f^{(r)}(0) = 0$ for some positive integer r. Then, using Eq. (8.7), for some positive integer m and constants $A_1, A_2, \ldots, A_m$, we have

$$f^{(r+1)}(0) = \lim_{x \to 0} \frac{f^{(r)}(x)}{x}$$

$$= \lim_{x \to 0} e^{-1/x^2} \sum_{j=1}^{m} \frac{A_j}{x^{j+1}}$$

$$= 0,$$

by Eq. (8.6).

We have now completed the proof that $f^{(n)}(0) = 0$ for all positive integers n. We deduce that the Maclaurin series for f converges to 0 regardless of the value of x. Since $f(x) \neq 0$ for all $x \neq 0$, it follows that the Maclaurin series for f converges to $f(x)$ only when $x = 0$.

It is shown in complex analysis that functions that can be represented by a Taylor series about a given point are precisely those that are differentiable throughout some neighborhood of that point. If we consider the function f of this example as a function in the complex plane, then it is not differentiable at 0. Indeed it is not even continuous at 0: If we take

$$z_n = \frac{1}{n\sqrt{i}}$$

for all $n > 0$, then $\{z_n\}$ converges to 0, but the sequence $\{f(z_n)\}$ does not converge to $f(0) = 0$ since $|e^{-1/z^2}| = |e^{-in^2}| = 1$. $\triangle$

Exercises 8.2.

1. Let f be a function such that $f^{(n)}(x)$ exists for all nonnegative integers n and for all x in an open interval I containing a number a. Show that if there is a number M such that $|f^{(n)}(x)| \leq M$ for all n and for all $x \in I$, then f is representable by a Taylor series about a.
2. Find the Maclaurin series for $(1 + x)^\alpha$ for each $\alpha \in \mathbb{R}$ and give its radius of convergence.

8.3   Some Shortcuts for Computing Taylor Series

If a function f has a Taylor series about a number a, then the coefficients of the series can be determined by evaluating the derivatives of f at a. Finding these derivatives, however, can prove tedious and awkward. In this section we look at some shortcuts when the functions involved are closely related to known Taylor series.

If f has a Taylor series about a, then the series is unique. If we can establish by any means that a power series of the correct form represents f in a neighborhood of a, then the power series must be the Taylor series. Often simple substitutions, algebraic identities, differentiation, or integration can be used to facilitate the computation of a Taylor series. The examples that follow illustrate how this is done.

Example 8.3.1. Find the Maclaurin series for $\cos x^4$.

Solution. For each $w \in \mathbb{R}$ the definition of $\cos w$ gives

$$\cos w = 1 - \frac{w^2}{2!} + \frac{w^4}{4!} - \frac{w^6}{6!} + \cdots .$$

Let $w = x^4$. Then

$$\cos x^4 = 1 - \frac{(x^4)^2}{2!} + \frac{(x^4)^4}{4!} - \frac{(x^4)^6}{6!} + \cdots$$

$$= 1 - \frac{x^8}{2!} + \frac{x^{16}}{4!} - \frac{x^{24}}{6!} + \cdots$$

$$= \sum_{j=0}^{\infty} (-1)^j \frac{x^{8j}}{(2j)!}. \qquad \triangle$$

Example 8.3.2. Find the Taylor series for e^x about 1.

Solution. For all $w \in \mathbb{R}$,

$$e^w = 1 + w + \frac{w^2}{2!} + \frac{w^3}{3!} + \cdots .$$

Let $w = x - 1$. Then

$$e^{x-1} = 1 + (x-1) + \frac{(x-1)^2}{2!} + \frac{(x-1)^3}{3!} + \cdots ,$$

and hence

$$e^x = e\left(1 + (x-1) + \frac{(x-1)^2}{2!} + \frac{(x-1)^3}{3!} + \cdots\right)$$

$$= e \sum_{j=0}^{\infty} \frac{(x-1)^j}{j!}. \qquad \triangle$$

Example 8.3.3. Find the Taylor series for $1/x$ about 2.

Solution. For all $x \neq 0$ we have

$$\frac{1}{x} = \frac{1}{2 + x - 2} = \frac{1}{2(1 - \frac{-(x-2)}{2})}.$$

Now for all w such that $|w| < 1$,

$$\frac{1}{1 - w} = 1 + w + w^2 + w^3 + \cdots;$$

hence for all x such that

$$\frac{|x - 2|}{2} < 1,$$

we have

$$\frac{1}{x} = \frac{1}{2}\left(1 - \frac{x - 2}{2} + \frac{(x - 2)^2}{2^2} - \frac{(x - 2)^3}{2^3} + \cdots\right)$$

$$= \sum_{j=0}^{\infty} \frac{(-1)^j}{2^{j+1}}(x - 2)^j.$$

$\triangle$

Example 8.3.4. Find the Maclaurin series for

$$\frac{x}{(1 + x)^2}.$$

Solution. For all w such that $|w| < 1$ we have

$$\frac{1}{1 - w} = \sum_{j=0}^{\infty} w^j,$$

and since power series can be differentiated term by term,

$$\sum_{j=1}^{\infty} jw^{j-1} = \frac{1}{(1 - w)^2}.$$

Let $w = -x$. Then

$$\frac{1}{(1 + x)^2} = \sum_{j=1}^{\infty} (-1)^{j-1} jx^{j-1},$$

so that

$$\frac{x}{(1+x)^2} = \sum_{j=1}^{\infty} (-1)^{j-1} j x^j.$$

$\triangle$

Example 8.3.5. Find the Maclaurin series for

$$\log(1+x^2).$$

Solution. The Maclaurin series can be readily obtained from the Maclaurin series for $\log(1+w)$. We derive the series from the geometric series to illustrate the use of integration.

Integration by substitution yields

$$\int_0^x \frac{2t}{1+t^2}\, dt = \log(1+x^2)$$

and for each x such that $|x| < 1$, a simple substitution in the geometric series gives

$$\frac{1}{1+x^2} = \sum_{j=0}^{\infty} (-1)^j x^{2j},$$

so that

$$\frac{2x}{1+x^2} = 2\sum_{j=0}^{\infty} (-1)^j x^{2j+1}.$$

We know that power series can be integrated term by term within the interval of convergence. We thus have

$$\begin{aligned}
\log(1+x^2) &= \int_0^x \frac{2t}{1+t^2}\, dt \\
&= 2\sum_{j=0}^{\infty} \int_0^x (-1)^j t^{2j+1}\, dt \\
&= 2\sum_{j=0}^{\infty} \frac{(-1)^j}{2j+2} x^{2j+2} \\
&= \sum_{j=0}^{\infty} \frac{(-1)^j}{j+1} x^{2j+2}.
\end{aligned}$$

$\triangle$

Example 8.3.6. Find the Maclaurin series for $\cos^2 x$.

Solution. We know the Maclaurin series for $\cos x$, but it is clear that simply squaring this series involves a multiplication of infinite series that in itself can prove awkward. Instead, we exploit the trigonometric identity

$$\cos^2 x = \frac{1 + \cos 2x}{2}.$$

The Maclaurin series for $\cos 2x$ can be derived by a simple substitution; hence

$$\cos 2x = \sum_{j=0}^{\infty} \frac{(-1)^j 2^{2j}}{(2j)!} x^{2j}.$$

We thus have

$$\cos^2 x = \frac{1}{2} + \sum_{j=0}^{\infty} \frac{(-1)^j 2^{2j-1}}{(2j!)} x^{2j} = 1 + \sum_{j=1}^{\infty} \frac{(-1)^j 2^{2j-1}}{(2j)!} x^{2j}.$$

$\triangle$

Exercises 8.3.

1. Use the geometric series to determine the Taylor series for the following functions about the indicated point a:

   (a) $\frac{1}{2+x}$, $a = -1$;   (c) $\arctan x$, $a = 0$;

   (b) $\frac{1}{x(x+1)}$, $a = 1$;   (d) $\frac{1}{(2+x)^3}$, $a = 0$.

2. Use the Maclaurin series for $\sin x$ and $\cos x$ along with trigonometric identities to determine the Taylor series for the following functions about a:

   (a) $\sin x \sin 2x$, $a = 0$;   (c) $\cos x$, $a = \frac{\pi}{2}$;

   (b) $\frac{1-\cos x}{x}$, $a = 0$;   (d) $\cos x \sin x$, $a = 0$.

3. Show that $x^{1/3}$ is representable by a Taylor series about 1 in the interval $(0, 2)$.

4. Show that $\cos x$ is representable by a Taylor series about any real number.

5. Let $f(x) = (\sin x)/x$ for all $x \neq 0$ and let $f(0) = 1$. Estimate $\int_0^1 f$ by using five terms of the Maclaurin series for f, and find an upper bound for the error.

6. For the following functions f, find the Maclaurin series for $\int_0^x f(t)\, dt$ and give the intervals of convergence:

   (a) $\sin t^2$;

   (b) e^{-t^2}.

7. Find the Maclaurin series for $\sinh x$ and $\cosh x$ and give the intervals of convergence.

8. Find the Maclaurin series for $\sin^4 x$ by expressing the function in terms of $\sin kx$ and $\cos kx$, where $k \in \mathbb{Z}$.

9. Let

$$F(x) = \int_a^x t^t \, dt.$$

Show that

$$F''(x) = (1 + \log x)F'(x).$$

Use this result to obtain the Taylor expansion

$$(x - 1) + \frac{1}{2}(x - 1)^2 + \frac{1}{3}(x - 1)^3 + \frac{1}{8}(x - 1)^4 + \dots$$

for F about 1.

Chapter 9
The Fixed-Point Problem

9.1 The Fixed-Point Problem

The theory of sequences finds many applications in the study of numerical techniques for solving equations, since many numerical methods involve the construction of a sequence of successively better approximations for the desired solution. In this chapter we show how analysis can help in the study of a numerical approximation technique for solving nonlinear equations. The work in this chapter is to a large extent based on the lecture notes of Michael Carter.

Suppose we are given a nonlinear equation, such as

$$3x^3 - 3x + 1 = 0. \tag{9.1}$$

Such an equation can always be written in the form

$$x = g(x)$$

for some function g. For example, Eq. (9.1) can be rewritten as

$$x = 3x^3 - 2x + 1 \tag{9.2}$$

or

$$x = \frac{3x^3 + 1}{3},$$

and there are many other possibilities.

A number x satisfying the equation $x = g(x)$ is called a **fixed point** of the function g because an application of g to x leaves x unchanged. For instance, the

© Springer Science+Business Media New York 2015
C.H.C. Little et al., *Real Analysis via Sequences and Series*, Undergraduate
Texts in Mathematics, DOI 10.1007/978-1-4939-2651-0_9

function given by x^2 for all x has the two fixed points 0 and 1. Evidently if we can find a good technique for determining the fixed points of a function, then we will have a good technique for solving equations, for if an equation is written in the form $x = g(x)$, then its solutions are precisely the fixed points of g.

9.2 Existence of Fixed Points

The first question to be considered is whether a given function has any fixed points at all. Once we have that information, we can consider how the fixed points may be found. We can gain considerable insight by looking at the problem graphically, because a fixed point of a function g is simply a value of the argument x at which the graph of g intersects the line $y = x$ (see Fig. 9.1).

Clearly, a function need not have a fixed point. The exponential function, whose graph is drawn in Fig. 9.2, is an example of a function with no fixed point.

The next theorem establishes a useful set of conditions under which we can be sure that a given function has a fixed point.

Theorem 9.2.1. *Let a and b be real numbers with $a < b$. Let g be a function such that $g(a) \geq a$ and $g(b) \leq b$ and g is continuous on $[a, b]$. Then g has a fixed point in $[a, b]$.*

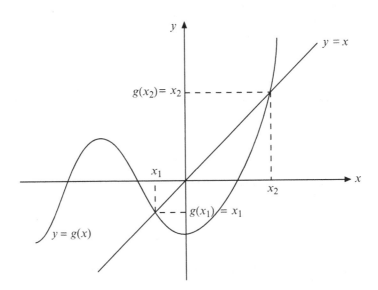

Fig. 9.1 Graph of a function g with two fixed points

Fig. 9.2 The exponential function has no fixed points

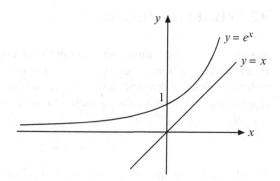

Proof. Define

$$h(x) = x - g(x)$$

for all $x \in [a,b]$. If either $h(a) = 0$ or $h(b) = 0$, then a or b, respectively, is a fixed point of g. We may suppose, therefore, that $g(a) > a$ and $g(b) < b$ and consequently that $h(a) < 0 < h(b)$. Then the intermediate-value theorem shows that there is a number $\xi \in (a,b)$ such that $h(\xi) = 0$. Thus ξ is a fixed point of g. $\quad\square$

Under the conditions of Theorem 9.2.1 the function g may have many fixed points, for its graph may touch the line $y = x$ several times. It is often important to know that g has only one fixed point in a particular interval. Geometric intuition suggests that this will be the case if $g(x)$ does not vary too rapidly as x changes, for then the graph of g will not oscillate rapidly enough to cross the line $y = x$ more than once. It therefore seems that one way of guaranteeing uniqueness of the fixed point in a particular interval is to restrict the size of g'. The next theorem makes this idea precise.

Theorem 9.2.2. *If a function g satisfies the hypotheses of Theorem 9.2.1 and in addition $|g'(x)| < 1$ for all $x \in (a,b)$, then g has a unique fixed point in $[a,b]$.*

Proof. The existence of a fixed point of g in $[a,b]$ follows from Theorem 9.2.1. Suppose that x_1 and x_2 are fixed points of g in $[a,b]$ such that $x_1 < x_2$. By the mean-value theorem applied to the interval $[x_1, x_2]$, there exists $\xi \in (x_1, x_2)$ such that

$$g'(\xi) = \frac{g(x_2) - g(x_1)}{x_2 - x_1}.$$

As x_1 and x_2 are fixed points, it follows that $g'(\xi) = 1$, despite the hypothesis that $|g'(x)| < 1$ for all $x \in (a,b)$. This contradiction shows that the fixed point in question is indeed unique. $\quad\square$

9.3 Fixed-Point Iteration

We suppose now that we are given a function g and we wish to find, as accurately as may be required, the fixed points of g. The method we shall use is called **fixed-point iteration**. It involves making an initial guess, x_0 say, and then constructing a sequence $\{x_n\}$ of successive approximations for the desired fixed point, using the formula

$$x_{n+1} = g(x_n)$$

for all nonnegative integers n. The process is illustrated graphically in Fig. 9.3.

If the sequence $\{x_n\}$ so constructed converges to some number s and g is continuous, then it is easily seen that s is a fixed point of g. Indeed, since $x_n \to s$ as $n \to \infty$ and g is continuous, it follows that $g(x_n) \to g(s)$. Therefore, taking limits on both sides of the equation $x_{n+1} = g(x_n)$ yields $s = g(s)$, as required. However, the sequence $\{x_n\}$ need not converge. The following example illustrates this and other difficulties.

Example 9.3.1. Suppose we wish to solve Eq. (9.1) given at the beginning of this chapter. First we sketch the graph of the function on the left-hand side of the equation. For instance, if we draw the graph of $3x^3 - 3x$ and move it one unit up, then Fig. 9.4 makes it appear that there are solutions near -1.1, 0.4, and 0.7.

Now let us try to solve the equation by writing it in the form (9.2) and looking for the fixed points of the corresponding function given by

$$g(x) = 3x^3 - 2x + 1$$

for all x. We show graphically in Fig. 9.5 the fates of seven iterations that start with initial guesses near one of the three values obtained above.

The following facts should be clear from the figure.

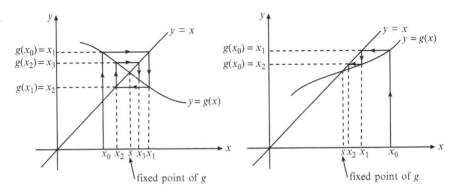

Fig. 9.3 Illustration of fixed-point iteration

Fig. 9.4 Graph of
$3x^3 - 3x + 1$

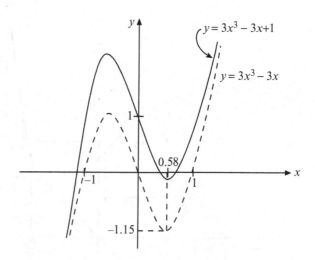

1. No iteration can converge to the fixed points located near -1.1 and 0.7.
2. Iterations such as 3–5, which start close enough to the remaining fixed point, will converge to that fixed point. The value obtained for the fixed point by this means is 0.3949 (correct to four decimal places). A closer study of the graph shows that in order for an iteration to converge to this fixed point, the initial guess must lie in the interval $[0.14, 0.74]$, approximately. Even then, the convergence is so slow that it takes many steps of an iteration to get a good approximation for the fixed point.
3. Some iterations that start near -1.1, such as iteration 7, which might be expected to diverge like iteration 2, actually converge to $0.3949\ldots$ because the third term in the iteration falls in the interval $[0.14, 0.74]$ referred to in (2), instead of overshooting or undershooting it. The figure shows how this happens.   △

It is clear from this example that trying to solve Eq. (9.1) by applying fixed-point iteration to version (9.2) is an approach whose value is very limited. Only one of the three solutions can be found this way, and that at the cost of considerable calculation because of the slow convergence of the iterations. Furthermore, the interval within which the initial guess must be placed in order for us to be sure the iterations will converge is not large. It would be helpful to know what sorts of conditions the function g must satisfy if the fixed-point iterations are to converge and to have a better understanding of what governs the rate of convergence of the iterations. The next theorem deals with these questions. As well as giving conditions under which a fixed-point iteration will converge to a fixed point of g, it also gives a bound on the error incurred by stopping the iterations after a given number of steps.

Theorem 9.3.1. *Suppose the function g satisfies the following conditions:*

1. g is continuous on the closed interval $[a, b]$;
2. $a \le g(x) \le b$ for all $x \in [a, b]$; and
3. there is a number $L < 1$ such that $|g'(x)| \le L$ for all $x \in (a, b)$.

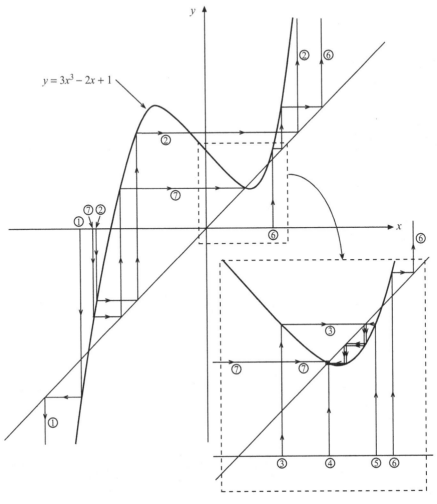

Fig. 9.5 Iterations for fixed points of $3x^3 - 2x + 1$

Then

1. *for any initial value $x_0 \in [a,b]$, the sequence $\{x_n\}$ defined by $x_{n+1} = g(x_n)$ for each $n \geq 0$ converges to a number s, which is the unique fixed point of g in $[a,b]$; and*
2. *the error $e_n = s - x_n$ satisfies*

$$|e_n| \leq \frac{L^n}{1-L}|x_1 - x_0|$$

for each $n \in \mathbb{N}$.

Proof.

1. Taking $x = a$ and $x = b$ in hypothesis (2) gives $g(a) \geq a$ and $g(b) \leq b$, respectively, so that g satisfies the hypotheses of Theorem 9.2.1. By hypothesis (3), g also satisfies the conditions of Theorem 9.2.2, and so we know that g has exactly one fixed point s in $[a, b]$. Choose any $x_0 \in [a, b]$, and let the sequence $\{x_n\}$ be defined by

$$x_{n+1} = g(x_n)$$

for each $n \geq 0$. Then, by hypothesis (2), $x_n \in [a, b]$ for all n. We may also assume that $x_n \neq s$ for all n. The mean-value theorem shows that for every positive integer n there is a number ξ_n between x_{n-1} and s such that

$$g'(\xi_n) = \frac{g(x_{n-1}) - g(s)}{x_{n-1} - s}.$$

Thus

$$x_n - s = g'(\xi_n)(x_{n-1} - s),$$

so that

$$|x_n - s| \leq L|x_{n-1} - s|$$

by hypothesis (3). In particular,

$$|x_1 - s| \leq L|x_0 - s|,$$

and if

$$|x_k - s| \leq L^k|x_0 - s|$$

for some $k > 0$, then

$$|x_{k+1} - s| \leq L|x_k - s| \leq L^{k+1}|x_0 - s|.$$

Consequently

$$|x_n - s| \leq L^n|x_0 - s|$$

for all $n \in \mathbb{N}$, by induction. Since $0 \leq L < 1$, we have

$$\lim_{n \to \infty} L^n = 0,$$

and so

$$\lim_{n\to\infty} |x_n - s| = 0.$$

It follows that

$$\lim_{n\to\infty} x_n = s,$$

as required.

2. We may assume that $x_n \neq x_{n-1}$ for all $n \in \mathbb{N}$, as $x_n = s$ otherwise. By the mean-value theorem, for every positive integer n there is a number ξ_n between x_{n-1} and x_n such that

$$g'(\xi_n) = \frac{g(x_n) - g(x_{n-1})}{x_n - x_{n-1}}.$$

Arguing as in the proof of (1), we obtain

$$|x_{n+1} - x_n| = |g'(\xi_n)(x_n - x_{n-1})| \leq L|x_n - x_{n-1}|.$$

We conclude by induction that

$$|x_{n+1} - x_n| \leq L^n |x_1 - x_0|$$

for all $n \in \mathbb{N}$.

Now fix n and let m be an integer such that $m > n$. Using the telescoping property, we have

$$
\begin{aligned}
|x_m - x_n| &= \left| \sum_{j=n}^{m-1} (x_{j+1} - x_j) \right| \\
&\leq \sum_{j=n}^{m-1} |x_{j+1} - x_j| \\
&\leq |x_1 - x_0| \sum_{j=n}^{m-1} L^j \\
&= |x_1 - x_0| \left(\sum_{j=0}^{m-1} L^j - \sum_{j=0}^{n-1} L^j \right) \\
&= |x_1 - x_0| \left(\frac{1 - L^m}{1 - L} - \frac{1 - L^n}{1 - L} \right) \\
&= |x_1 - x_0| \frac{L^n - L^m}{1 - L},
\end{aligned}
$$

since $0 \le L < 1$. Now $\lim_{m \to \infty} x_m = s$, as proved in part (1), and so

$$|e_n| = |s - x_n| \le \frac{L^n}{1 - L}|x_1 - x_0|,$$

as required. □

Example 9.3.2. Let us return to Eq. (9.1) discussed in the previous example, but this time we shall write it in the form

$$x = \left(\frac{3x - 1}{3}\right)^{1/3} = \left(x - \frac{1}{3}\right)^{1/3}$$

and look for fixed points of the function g given by

$$g(x) = \left(x - \frac{1}{3}\right)^{1/3}$$

for all x, using the information provided by Theorem 9.3.1. In order to apply the theorem we need to know for what values of x we have $|g'(x)| < 1$, and so we first sketch the graph of

$$g'(x) = \frac{1}{3}\left(x - \frac{1}{3}\right)^{-2/3}.$$

This graph is given in Fig. 9.6.

We can see from the graph that $|g'(x)| < 1$ if $x < 0.1409 \ldots$ or $x > 0.5258 \ldots$. Now we must look at the graph of g to find suitable intervals $[a, b]$ satisfying the condition that $a \le g(x) \le b$ for all $x \in [a, b]$ as well as the condition that $|g'(x)| \le L$ for some $L < 1$ and all $x \in [a, b]$. We can recognize intervals $[a, b]$

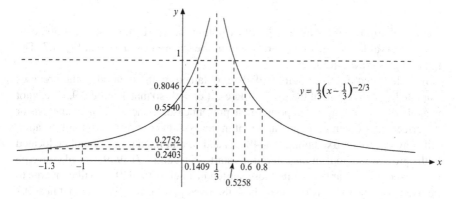

Fig. 9.6 Graph of $(1/3)(x - 1/3)^{-2/3}$

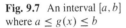

Fig. 9.7 An interval $[a, b]$
where $a \le g(x) \le b$

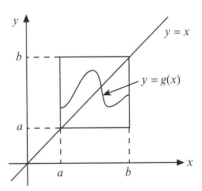

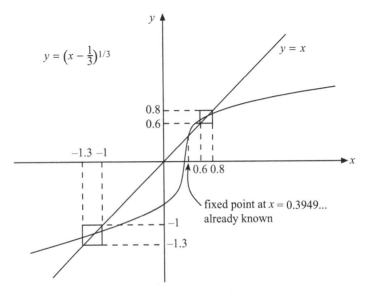

Fig. 9.8 Graph of $(x - 1/3)^{1/3}$

where $a \le g(x) \le b$ for all $x \in [a, b]$ by noting that in such a case the graph of g
on the interval $[a, b]$ must lie entirely within the square box shown in Fig. 9.7. The
graph of g is drawn in Fig. 9.8.

It is clear that the fixed point $0.3949 \ldots$ obtained by the method of the previous
example lies in a region of the graph where $g'(x) > 1$, so that Theorem 9.3.1 cannot
be applied to locate that fixed point in the present case. In fact, graphical analysis or
numerical experimentation indicates that in the present case no fixed-point iteration
will converge to the fixed point $0.3949 \ldots$, and so that fixed point cannot be located
by the present method. In fact, iterations started just above $0.3949 \ldots$ will converge
to the fixed point near 0.7, while those started just below $0.3949 \ldots$ will converge to
the fixed point near -1.1. However, both the fixed point near -1.1 and that near 0.7
can be enclosed in intervals on which Theorem 9.3.1 may be applied, for example,

the intervals $[-1.3, -1]$ and $[0.6, 0.8]$, as shown. Analysis of the graph of g' shows that $|g'(x)| \le 0.276$ for all $x \in [-1.3, -1]$ and $|g'(x)| \le 0.805$ for all $x \in [0.6, 0.8]$, so that we may apply Theorem 9.3.1 on these intervals with $L = 0.276$ and $L = 0.805$, respectively. This method therefore locates the two solutions of Eq. (9.1) not found by the method of Example 9.3.1. By combining the two methods, we can say that, correct to four decimal places, the solutions of Eq. (9.1) are $-1.1372, 0.3949$, and 0.7422. △

Considering the amount of work involved in obtaining the solutions in this example, one cannot help feeling that there must be an easier way. One point that will be noticed is the relatively rapid convergence of the iterations in the interval $[-1.3, -1]$. Note that $|g'(x)|$ is smaller in this interval, so that L is smaller and therefore L^n tends to 0 relatively fast as $n \to \infty$. Thus the error bound given by Theorem 9.3.1 will also tend to 0 relatively quickly. This observation suggests more rapid convergence of the iterations than in the case of iterations in the interval $[0.6, 0.8]$. Thus one way of speeding up the process might be to seek functions g for which $|g'(x)|$ is very small near a fixed point of g. This idea is the reason for the success of the following method.

Suppose we are given the nonlinear equation $f(x) = 0$ for all x, and we wish to write it in the form $x = g(x)$ in order to apply fixed-point iteration. One approach is to write $f(x) = 0$ in the equivalent form

$$f(x)h(x) = 0,$$

where h is any differentiable function with the property that the equation $h(x) = 0$ has no real solutions, so that $f(x) = 0$ if and only if $f(x)h(x) = 0$. Then we can rewrite $f(x) = 0$ as $x = x + f(x)h(x)$, so that

$$g(x) = x + f(x)h(x)$$

for all x.

Now let s be a fixed point of g, that is, a solution of $f(x) = 0$. In view of our previous discussion, we should like $|g'(x)|$ to be small when x is near s. If, in fact, $g'(s) = 0$, then as long as g' is continuous, we can be sure that $|g'(x)|$ will be small if x is close enough to s. Now

$$g'(x) = 1 + f(x)h'(x) + h(x)f'(x)$$

for all x, and since $f(s) = 0$ this equation gives

$$g'(s) = 1 + h(s)f'(s).$$

Therefore $g'(s) = 0$ if $f'(s) \ne 0$ and $h(s) = -1/f'(s)$. Thus we define

$$h(x) = -\frac{1}{f'(x)}$$

for all x such that $f'(x) \neq 0$, for then the value of s does not need to be known. In other words, we write the equation $f(x) = 0$ in the equivalent form

$$x = x - \frac{f(x)}{f'(x)}$$

and then use fixed-point iteration. This procedure is called **Newton's method**. If the fixed-point iteration starts close enough to a solution of $f(x) = 0$, then L in Theorem 9.3.1 will be small and so the iteration will converge rapidly to the desired solution. Difficulties arise in certain cases, most obviously if $f'(s) = 0$, but Newton's method is very useful in cases where $f'(x)$ is easily calculated and a reasonably good initial guess at the solutions can be made.

Example 9.3.3. Newton's method applied to Eq. (9.1) requires the equation to be rewritten in the form

$$x = x - \frac{3x^3 - 3x + 1}{9x^2 - 3}$$

for all x such that $9x^2 \neq 3$. Applying fixed-point iteration with starting values near the solutions guessed at the beginning of Example 9.3.1 leads to iterations that converge relatively rapidly compared with those in the previous examples.   △

Exercises 9.1.

1. Let $g(x) = x^2$ for all x. From a graph similar to that used to trace the fates of fixed-point iterations in Example 9.3.1, determine for what values of x_0 the sequence $\{x_n\}$, defined by $x_{n+1} = g(x_n)$ for all nonnegative integers n, will converge to a fixed point of g and for what values of x_0 it will diverge.

2. Suppose Eq. (9.1) discussed in Examples 9.3.1–9.3.3 is rewritten as $x = (3x^3 + 1)/3$. Putting $g(x) = (3x^3 + 1)/3$ for all x, sketch the graphs of g and g' and use the approach illustrated in Example 9.3.2 to find suitable intervals within which Theorem 9.3.1 can be applied to locate fixed points. Use fixed-point iterations, starting at each end of the intervals you have found, to locate the corresponding fixed points correct to four decimal places. If there are any fixed points that cannot be located in this way, use the graph of g to describe what will happen to iterations that start near these fixed points.

3. Do the same as for the previous exercise but for the function g given by

$$g(x) = \frac{3x - 1}{3x^2}$$

for all $x \neq 0$. Note that in investigating the graph of g' it is not necessary to find exactly the values of x for which $g'(x) = \pm 1$; just locate these values roughly by calculating a few values of $g'(x)$ on either side of the apparent location of these values of x.

4. Show that Newton's method applied to the equation $x^2 = c$, where $c > 0$, leads to the fixed-point problem

$$x = \frac{x^2 + c}{2x},$$

and use graphical methods to show that Theorem 9.3.1 can be applied on the interval $[\frac{2}{3}\sqrt{c}, \frac{4}{3}\sqrt{c}]$ to locate the fixed point $\sqrt{c}$. Find the smallest possible value of L for this interval. In the case $c = 2$, use the fact that $1 < \sqrt{2} < \frac{3}{2}$ to show that 1 lies in the interval $[\frac{2}{3}\sqrt{2}, \frac{4}{3}\sqrt{2}]$ and hence locate $\sqrt{2}$ correct to six decimal places by fixed-point iteration. Calculate, for each step in the iteration, the size of the actual error and the theoretical bound on the error given by Theorem 9.3.1.

5. For the fixed-point problem $x = \cos x$ for all x, use graphical methods to show that Theorem 9.3.1 can be applied on the interval $[0, 1]$. If your calculator can evaluate trigonometric functions, use the initial value $x_0 = 0.5$ and locate the fixed point correct to four decimal places. Also, solve the equation by Newton's method with the same initial value and note the rapid convergence in this case.

6. Suppose

$$f(x) = (x - r_1)(x - r_2) \cdots (x - r_m)$$

for all x, where $m \geq 2$ and $r_1 \leq r_2 \leq \ldots \leq r_m$. Then f is a polynomial of degree m with the coefficient of x^m equal to 1, and the equation $f(x) = 0$ has only real solutions, none of which exceeds the solution r_m. Let $\{x_n\}$ be a sequence produced by Newton's method applied to the equation $f(x) = 0$.

(a) Explain why $f(x) > 0$ for all $x > r_m$, and show by direct differentiation that $f'(x) > 0$ and $f''(x) > 0$ for all $x > r_m$.

(b) Suppose that $x_k > r_m$ for some integer $k \geq 0$. Show that $x_{k+1} < x_k$. By using the mean-value theorem applied to f on the interval $[r_m, x_k]$, show also that $f(x_k) < (x_k - r_m)f'(x_k)$ and deduce that $x_k - x_{k+1} < x_k - r_m$ and hence that $x_{k+1} > r_m$.

(c) From (b) it follows that if $x_0 > r_m$, then the sequence $\{x_n\}$ is a decreasing sequence bounded below by r_m. Hence the sequence $\{x_n\}$ converges to a limit $s \geq r_m$. Prove that in fact $s = r_m$. Illustrate this conclusion for the function given by

$$f(x) = (x + 1)^2(x - 2),$$

for all x, by taking $x_0 = 3$, $x_0 = 10$, and $x_0 = 30$ and carrying out the iterations correct to four decimal places.

(d) In some cases the convergence of the sequence $\{x_n\}$ may be exceedingly slow. For example, if $f(x) = (x - s)^m$ for all x, show that

$$x_{n+1} = \left(1 - \frac{1}{m}\right)x_n + \frac{s}{m}$$

for all $n \in \mathbb{N}$, and deduce that

$$s - x_{n+1} = \left(1 - \frac{1}{m}\right)(s - x_n).$$

[Thus if m is large, so that $1 - 1/m$ is close to 1, then the decrease in the error at each step of the iteration is slight. For example, if $m = 100$, then $e_{n+1} = 99e_n/100$. Note that this is a case where $f'(s) = 0$, since $m \geq 2$, and so we might expect that Newton's method could run into difficulties.]

Chapter 10
Sequences of Functions

10.1 Introduction

The representation of a function as the limit of a sequence or, equivalently, an infinite series is central to many topics in advanced analysis. In this chapter we concentrate on sequences of functions and certain properties of limits that can be gleaned from properties of the sequence terms. We begin with an application that not only motivates the study of sequences of functions, but also highlights some key questions regarding limits.

A first-order ordinary differential equation for a function y is an equation of the form

$$y' = f(x, y), \tag{10.1}$$

where f is a given function of a variable x and y. The equation is usually supplemented with an initial condition

$$y(a) = c, \tag{10.2}$$

where a and c are given numbers. The **initial-value problem** consists of determining a function y that satisfies Eqs. (10.1) and (10.2) for all x in some open interval that contains a.

We gloss over the fundamental questions of the existence and uniqueness of solutions to initial-value problems. It turns out that for most choices of f the problem cannot be solved explicitly. Nonetheless, it can be shown, for example, that if f is differentiable in a neighborhood of (a, c) with respect to x and y, then the initial-value problem has a unique solution. The proof of this result requires

© Springer Science+Business Media New York 2015
C.H.C. Little et al., *Real Analysis via Sequences and Series*, Undergraduate
Texts in Mathematics, DOI 10.1007/978-1-4939-2651-0_10

the use of a sequence of functions, and the solution is the limit of this sequence. Picard proved this result by the method of successive approximations. Specifically, the problem can be recast as an integral equation

$$y(x) = c + \int_a^x f(\xi, y(\xi)) \, d\xi,$$

and this formulation motivates the sequence defined by

$$y_0(x) = c$$

and

$$y_{n+1}(x) = c + \int_a^x f(\xi, y_n(\xi)) \, d\xi$$

for all nonnegative integers n. It is then shown that $\{y_n(x)\}$ converges to a limit $y(x)$ that solves the initial-value problem. The result is local in character: To ensure convergence of the sequence, x is usually restricted to a small open interval containing a.

Convergence questions aside, the claim that the limit is the solution of the differential equation brings to the fore certain questions concerning the properties of the function that is the limit of the sequence. The problem is that more than one limit process is involved, and the order in which these limits are taken must be changed. To prove that the function defined by

$$y(x) = \lim_{n \to \infty} y_n(x)$$

is a solution to the integral equation, we need to justify the following calculation:

$$\lim_{n \to \infty} y_n(x) = c + \lim_{n \to \infty} \int_a^x f(\xi, y_n(\xi)) \, d\xi$$

$$= c + \int_a^x \lim_{n \to \infty} f(\xi, y_n(\xi)) \, d\xi$$

$$= c + \int_a^x f(\xi, \lim_{n \to \infty} y_n(\xi)) \, d\xi$$

$$= c + \int_a^x f(\xi, y(\xi)) \, d\xi.$$

In this calculation, the sequence limit migrates from outside the integral to inside the integrand. There are two limits involved in this manipulation: the limit defining y and the limit defining the integral. The problem in the second line is that we must change the order in which the sequence limit and the integral limit are taken. Intuitive as the calculation seems, it turns out that changing the order of these limits is not always valid.

Indeed, there is a third limit process in the background that stems from the continuity of f and the functions y_n. Given that f is differentiable with respect to x and y in some neighborhood $\mathcal{N}$ of (a, c), a fortiori f is continuous with respect to y (and x) for each $(x, y) \in \mathcal{N}$. Certainly, if $\{w_n\}$ is a sequence of numbers such that $(x, w_n) \in \mathcal{N}$ and $\lim_{n \to \infty} w_n = y$, then the definition of continuity implies that

$$\lim_{n \to \infty} f(x, w_n) = f(x, \lim_{n \to \infty} w_n) = f(x, y).$$

The sequence $\{y_n\}$ consists of functions continuous near a, but is the limit function y continuous near a? For that matter, if y is not continuous, is $f(x, y(x))$ integrable with respect to x in some neighborhood of a?

Suppose that the problems with the calculation above are resolved. The function y thus represents the solution to the integral equation. The original problem, however, involved a differential equation. The integral equation is well defined for every continuous function y; the differential equation requires y to be differentiable near a. It is clear from the definition of the sequence that each y_n is differentiable, but is the limit y differentiable?

The initial-value problem highlights the need to examine conditions under which the order of certain limiting processes can be changed. On a more fundamental level it also raises questions as to whether properties of the sequence terms, such as continuity and differentiability, are preserved in the limit.

Let $I \subseteq \mathbb{R}$ be an interval and suppose the sequence $\{f_n(x)\}$ converges to $f(x)$ for all $x \in I$. We have, in summary, the following questions concerning limits of sequences of functions.

1. If f_n is continuous on I for all n, is f continuous on I?
2. If f_n is integrable on I for all n, is f integrable on I? If so, is

$$\int_a^b \lim_{n \to \infty} f_n(x)\, dx = \lim_{n \to \infty} \int_a^b f_n(x)\, dx,$$

where $I = [a, b]$?
3. If f_n is differentiable on I for all n, is f differentiable on I? If so, is

$$\left(\lim_{n \to \infty} f_n \right)' = \lim_{n \to \infty} f_n'?$$

In this chapter we develop conditions under which the order of limit processes can be changed. We end this section with a few simple examples to illustrate that, in general, the order of these processes cannot be changed.

Example 10.1.1 (Continuity). Let $\{f_n\}$ be the sequence defined by

$$f_n(x) = \frac{1}{1 + nx^2}$$

for all n and all $x \in [-1, 1]$. For every n, f_n is continuous on $[-1, 1]$. If $x \neq 0$, then it is plain that $f_n(x) \to 0$ as $n \to \infty$; however, $f_n(0) = 1$ for all n and consequently $f_n(0) \to 1$ as $n \to \infty$. The limit of the sequence is thus given by

$$f(x) = \begin{cases} 0 \text{ if } x \in [-1, 1] - \{0\} \\ 1 \text{ if } x = 0; \end{cases}$$

hence f is not continuous on $[-1, 1]$. Note that

$$\lim_{x \to 0} \left(\lim_{n \to \infty} f_n(x) \right) = 0 \neq \lim_{n \to \infty} \left(\lim_{x \to 0} f_n(x) \right) = 1.$$

$\triangle$

Example 10.1.2 (Differentiation). Let $\{f_n\}$ be the sequence defined by

$$f_n(x) = \frac{x}{1 + nx^2}$$

for all n and all $x \in [0, 1]$. Then

$$f(x) = \lim_{n \to \infty} f_n(x) = 0$$

for all $x \in [0, 1]$, so that $f'(x) = 0$ for all such x. However,

$$f_n'(x) = \frac{1 - nx^2}{(1 + nx^2)^2},$$

so that

$$\lim_{n \to \infty} f_n'(x) = \begin{cases} 1 \text{ if } x = 0, \\ 0 \text{ otherwise.} \end{cases}$$

Hence

$$\lim_{n \to \infty} f_n'(x) \neq \left(\lim_{n \to \infty} f_n(x) \right)'$$

when $x = 0$.

$\triangle$

Example 10.1.3 (Differentiation). Let $\{f_n\}$ be the sequence defined by

$$f_n(x) = \frac{\sin nx}{n}$$

for all $n \in \mathbb{N}$ and $x \in \mathbb{R}$. Now, $|\sin nx| \leq 1$; consequently $f_n(x) \to 0$ as $n \to \infty$ for all $x \in \mathbb{R}$.

For every $n \in \mathbb{N}$ we see that f_n is differentiable on $\mathbb{R}$. The limit $f = 0$ is also a differentiable function on $\mathbb{R}$. The derivative of f_n, however, is defined by

$$f_n'(x) = \cos nx,$$

so that the sequence $\{f_n'\}$ converges only for special values of x such as 0 and 2π and diverges for most $x \in \mathbb{R}$. $\triangle$

Example 10.1.4 (Integration). Let $\{f_n\}$ be the sequence defined by

$$f_n(x) = 2nxe^{-nx^2}$$

for all n and all $x \in [0, 1]$. For each $x \in [0, 1]$, $f_n(x) \to 0$ as $n \to \infty$; hence

$$\int_0^1 \lim_{n\to\infty} f_n(x)\, dx = \int_0^1 0\, dx = 0.$$

On the other hand,

$$\int_0^1 f_n(x)\, dx = 1 - e^{-n},$$

so that

$$\lim_{n\to\infty} \int_0^1 f_n(x)\, dx = 1 \neq \int_0^1 \lim_{n\to\infty} f_n(x)\, dx = 0.$$

$\triangle$

10.2 Uniform Convergence

The examples in the previous section show that the order in which limits are taken is important. We thus seek sufficient conditions under which this order can be changed. In this section we present the key concept of uniform convergence. In the next section we show that the order of limits for sequences that converge uniformly can be changed.

Let I be a set of real numbers and let $\{f_n\}$ be a sequence of functions defined on I. Suppose that for each $x \in I$ the sequence $\{f_n(x)\}$ converges to $f(x)$. For all $x \in I$, the definition of convergence implies that for every $\varepsilon > 0$ there is an integer N such that

$$|f_n(x) - f(x)| < \varepsilon$$

whenever $n \geq N$. The value of N, in general, depends not only on the choice of ε but also on x. Suppose that we consider two distinct values x_1 and x_2 in I. Since $\{f_n(x_1)\}$ and $\{f_n(x_2)\}$ are convergent sequences, for every $\varepsilon > 0$ there exist integers $N(\varepsilon, x_1)$ and $N(\varepsilon, x_2)$ such that

$$|f_n(x_1) - f(x_1)| < \varepsilon \tag{10.3}$$

whenever $n \geq N(\varepsilon, x_1)$ and

$$|f_n(x_2) - f(x_2)| < \varepsilon \tag{10.4}$$

whenever $n \geq N(\varepsilon, x_2)$. Integers $N(\varepsilon, x_1)$ and $N(\varepsilon, x_2)$ are not necessarily equal, but we could use $N = \max\{N(\varepsilon, x_1), N(\varepsilon, x_2)\}$ to ensure that inequalities (10.3) and (10.4) are satisfied for all $n \geq N$. Evidently, for any *finite* number of points $x_1, x_2, \ldots, x_j$ in I we can always choose N so that if $n \geq N$, then

$$|f_n(x_k) - f(x_k)| < \varepsilon$$

for each $k \in \{1, 2, \ldots, j\}$. It is not clear, however, that we can find an N such that, for *all* $x \in I$,

$$|f_n(x) - f(x)| < \varepsilon$$

whenever $n \geq N$. Generically, it is not possible to find such an N, and this situation leads to the concept of uniform convergence.

Let $\{f_n\}$ be a sequence of functions defined on a set $I \subseteq \mathbb{R}$. The sequence is said to **converge uniformly** to the function f on I if for each $\varepsilon > 0$ there exists an integer N, which may depend on ε but not on any particular $x \in I$, such that if $n \geq N$, then

$$|f_n(x) - f(x)| < \varepsilon \tag{10.5}$$

for all $x \in I$. Clearly, ε in inequality (10.5) may be replaced by $c\varepsilon$ for any constant $c > 0$. In applications we usually take I to be a closed interval. We always assume it to be nonempty.

Note that if there exists a function f such that $f_n(x) = f(x)$ for each $x \in I$ and each n, then the sequence $\{f_n\}$ converges uniformly to f on I.

Suppose that $f_n(x) = a_n$ for each $x \in I$ and each n, and that the sequence $\{a_n\}$ converges to L. Choose $\varepsilon > 0$. There exists N such that

$$|f_n(x) - L| = |a_n - L| < \varepsilon$$

for all $n \geq N$. Therefore the sequence $\{f_n\}$ of functions converges uniformly on I to the constant L.

The geometric interpretation of uniform convergence is straightforward. If the graph of f is drawn for all $x \in I$, then a ribbon of width 2ε can be constructed by curves offset a distance of ε from the graph of f. If $\{f_n\}$ converges uniformly to f on I, then for each $\varepsilon > 0$ there is an N such that the graph of f_n lies in the ribbon for all $n \geq N$. The quantity limiting N is the maximum difference between $f_n(x)$ and $f(x)$ for $x \in I$. This observation motivates the investigation of the sequence $\{M_n\}$ of numbers defined by

$$M_n = \sup_{x \in I} |f_n(x) - f(x)|.$$

Inequality (10.5) shows that M_n necessarily exists for all n if $\{f_n\}$ is uniformly convergent on I.

Theorem 10.2.1. *The sequence $\{f_n\}$ converges uniformly on I to f if and only if $M_n \to 0$ as $n \to \infty$.*

Proof. Suppose that $\{f_n\}$ converges uniformly on I to f. Choose $\varepsilon > 0$. Since $\{f_n\}$ converges uniformly, there is an integer N that is independent of x such that

$$|f_n(x) - f(x)| < \varepsilon$$

for all $n \geq N$; hence M_n exists and

$$|M_n| = M_n = \sup_{x \in I} |f_n(x) - f(x)| \leq \varepsilon$$

for all $n \geq N$. Thus $M_n \to 0$ as $n \to \infty$ from the definition of convergence.

Suppose that $M_n \to 0$ as $n \to \infty$. For each $\varepsilon > 0$ there is an N such that $M_n < \varepsilon$ whenever $n \geq N$; therefore, for all $x \in I$,

$$|f_n(x) - f(x)| \leq M_n < \varepsilon$$

for each $n \geq N$. The choice of N is independent of x and we thus conclude that $\{f_n\}$ converges uniformly to f on I.   □

Corollary 10.2.2. *Suppose that $\{f_n\}$ converges uniformly on I to f. If M is a number such that $|f_n(x)| < M$ for all n and all $x \in I$, then $|f(x)| \leq M$ for all $x \in I$.*

Proof. Suppose that $|f(x)| > M$ for some $x \in I$. Then $|f(x)| = M + \varepsilon$ for some $\varepsilon > 0$. Thus for all n we have

$$|f_n(x) - f(x)| \geq |f(x)| - |f_n(x)| > M + \varepsilon - M = \varepsilon.$$

Therefore

$$\sup_{x \in I} |f_n(x) - f(x)| > \varepsilon$$

for all n, and so we have the contradiction that

$$\lim_{n \to \infty} \sup_{x \in I} |f_n(x) - f(x)| \geq \varepsilon > 0.$$

$\square$

Of course, a corresponding result holds if $|f_n(x)| > M$ for all n and all $x \in I$.

Example 10.2.1. Let $\{f_n\}$ be the sequence defined by

$$f_n(x) = \frac{1 - x^{n+1}}{1 - x}$$

for each $x \in [-1/2, 1/2] = I$. For all $x \in I$, $x^{n+1} \to 0$ as $n \to \infty$, and therefore

$$f(x) = \lim_{n \to \infty} f_n(x) = \frac{1}{1 - x}.$$

Now

$$|f_n(x) - f(x)| = \frac{|x|^{n+1}}{1 - x};$$

consequently

$$M_n = \sup_{x \in I} |f_n(x) - f(x)|$$

$$= \sup_{x \in I} \frac{|x|^{n+1}}{1 - x}$$

$$= \frac{1}{2^n}.$$

Since $M_n \to 0$ as $n \to \infty$, $\{f_n\}$ converges uniformly to f on I by Theorem 10.2.1.

$\triangle$

Example 10.2.2. Let $\{f_n\}$ be the sequence defined by

$$f_n(x) = \frac{1}{1 + nx^2}$$

for all $x \in [-1, 1] = I$. Example 10.1.1 shows that $f_n \to f$ as $n \to \infty$, where

$$f(x) = \begin{cases} 0 \text{ if } x \in I - \{0\}, \\ 1 \text{ if } x = 0; \end{cases}$$

therefore

$$|f_n(x) - f(x)| = \begin{cases} \frac{1}{1+nx^2} \text{ if } x \in I - \{0\}, \\ 0 \qquad \text{if } x = 0. \end{cases}$$

Now

$$M_n = \sup_{x \in I} |f_n(x) - f(x)| = 1,$$

and it is clear that $\lim_{n \to \infty} M_n \neq 0$ as $n \to \infty$. Theorem 10.2.1 thus implies that the sequence does not converge uniformly to f on I.   △

Remark. A sequence $\{f_n\}$ of functions fails to be uniformly convergent to a function f on a set I if and only if there is an $\varepsilon > 0$ such that for all N there exist an integer $k \geq N$ and an $x_k \in I$ for which

$$|f_k(x_k) - f(x_k)| \geq \varepsilon.$$

We illustrate this remark in the next example.

Example 10.2.3. Let $\{f_n\}$ be the sequence defined by

$$f_n(x) = \frac{x}{n}$$

for all $n \in \mathbb{N}$ and all $x \in \mathbb{R}$. Since

$$f(x) = \lim_{n \to \infty} f_n(x) = 0$$

for all $x \in \mathbb{R}$, the sequence converges to a continuous function. However, it is not uniformly convergent: For all $n \in \mathbb{N}$ let $x_n = n$, so that $f_n(x_n) = 1$ and hence

$$|f_n(x_n) - f(x_n)| = 1 > 0.$$

Note also that

$$\lim_{n\to\infty} f_n'(x) = \lim_{n\to\infty} \frac{1}{n} = 0 = \left(\lim_{n\to\infty} f_n(x)\right)'$$

and

$$\lim_{n\to\infty} \int_a^b f_n(x)\, dx = \lim_{n\to\infty} \frac{b^2 - a^2}{2n} = 0 = \int_a^b \lim_{n\to\infty} f_n(x)\, dx.$$

$\triangle$

Our next example shows that a sequence of functions may converge uniformly on every closed subinterval of an open interval yet fail to be uniformly convergent on the open interval.

Example 10.2.4. Let

$$f_n(x) = x^n$$

for each n and each $x \in (0, 1)$. We will show that if $0 < a < b < 1$, then $\{f_n\}$ converges uniformly on $[a, b]$, but it does not do so on $(0, 1)$.

For each $x \in (0, 1)$ we have

$$f(x) = \lim_{n\to\infty} f_n(x) = 0.$$

Therefore

$$\sup_{x\in[a,b]} |f_n(x) - f(x)| = \sup_{x\in[a,b]} |f_n(x)| = \sup_{x\in[a,b]} x^n = b^n \to 0$$

as $n \to \infty$. Hence $\{f_n\}$ converges uniformly on $[a, b]$.

For each $n > 0$ let

$$x_n = 1 - \frac{1}{n}.$$

Then

$$|f_n(x_n) - f(x_n)| = \left(1 - \frac{1}{n}\right)^n \to \frac{1}{e}$$

as $n \to \infty$. Hence

$$\sup_{x\in I} |f_n(x) - f(x)| \geq \frac{1}{e} > 0.$$

We conclude from Theorem 10.2.1 that $\{f_n\}$ does not converge uniformly on $(0, 1)$.

$\triangle$

Theorem 10.2.1 is a useful characterization of uniform convergence, but one must find the supremum of $|f_n(x) - f(x)|$. The problem can sometimes be mitigated by using the sandwich theorem (Sect. 2.5). If there is a sequence $\{K_n\}$ such that

$$|f_n(x) - f(x)| < K_n$$

for each $n \in \mathbb{N}$ and each $x \in I$, then $0 \leq M_n \leq K_n$ for each n. If $K_n \to 0$ as $n \to \infty$, then $M_n \to 0$ as $n \to \infty$. To show that a sequence is not uniformly convergent on I, it suffices to establish a nonzero lower bound for the sequence $\{M_n\}$, valid for all n sufficiently large.

A more serious problem with this characterization of uniform convergence is that it requires a candidate for f. Often there is not an obvious candidate and hence we cannot use Theorem 10.2.1 directly. We thus seek an alternative characterization that does not require a limit candidate. This line of thought leads to a generalization of the Cauchy principle for convergence (Theorem 2.6.9).

Theorem 10.2.3 (Cauchy Principle).   *Let $\{f_n\}$ be a sequence of functions defined on the set I. The sequence $\{f_n\}$ converges uniformly on I if and only if for each $\varepsilon > 0$ there is an integer N, which may depend on ε and I but not on any $x \in I$, such that for all $x \in I$ we have*

$$|f_n(x) - f_m(x)| < \varepsilon$$

whenever $n \geq N$ and $m \geq N$.

Proof. Necessity: Suppose $\{f_n\}$ converges uniformly on I to f. Then for each $\varepsilon > 0$ there is an integer N such that for all $x \in I$ we have

$$|f_n(x) - f(x)| < \varepsilon$$

whenever $n \geq N$. If $n \geq N$ and $m \geq N$, then, for all $x \in I$,

$$|f_n(x) - f_m(x)| = |f_n(x) - f(x)| + |f(x) - f_m(x)|$$

$$< 2\varepsilon.$$

The result follows.

Sufficiency: Suppose that for each $\varepsilon > 0$ there is an N such that for all $x \in I$ we have

$$|f_n(x) - f_m(x)| < \varepsilon$$

whenever $n \geq N$ and $m \geq N$. For every $x \in I$, $\{f_n(x)\}$ is a Cauchy sequence of numbers and therefore convergent. Define

$$f(x) = \lim_{n \to \infty} f_n(x)$$

for all $x \in I$.

Choose any $\varepsilon > 0$ and $x \in I$. There is an N, independent of x, such that

$$|f_n(x) - f(x)| \le |f_n(x) - f_m(x)| + |f_m(x) - f(x)|$$
$$< \varepsilon + |f_m(x) - f(x)|$$

whenever $m \ge N$ and $n \ge N$. Since $f_m(x) \to f(x)$ as $m \to \infty$, we can choose $m \ge N$ (m may depend on ε and x) so that

$$|f_m(x) - f(x)| < \varepsilon.$$

Therefore

$$|f_n(x) - f(x)| < \varepsilon + \varepsilon = 2\varepsilon$$

whenever $n \ge N$. This inequality is valid for all $x \in I$, and N is independent of x. We thus conclude that $\{f_n\}$ converges uniformly to f on I.   □

Note the similarity between the proof of necessity in Theorem 10.2.3 and the proof of Theorem 2.6.1.

In order to prove the uniform convergence of a sequence $\{f_n\}$ of Theorem 10.2.3 on an interval I, it is of course enough to prove the existence of a positive constant c such that

$$|f_n(x) - f_m(x)| < c\varepsilon$$

whenever $n \ge N$ and $m \ge N$.

The following result provides a sequential characterization of uniform convergence of a sequence of functions.

Theorem 10.2.4. *Let $\{f_n\}$ be a sequence of functions on a nonempty set I. Then the sequence converges uniformly to f on I if and only if*

$$\lim_{n \to \infty} (f_n(x_n) - f(x_n)) = 0 \qquad (10.6)$$

for each sequence $\{x_n\}$ in I.

Proof. Suppose that $\{f_n\}$ converges uniformly to f on I. Then for each sequence $\{x_n\}$ in I we have

$$0 \le |f_n(x_n) - f(x_n)| \le \sup_{x \in I} |f_n(x) - f(x)| \to 0$$

as $n \to \infty$. Equation (10.6) follows by the sandwich theorem.

Suppose on the other hand that $\{f_n\}$ does not converge uniformly to f. Then there is an $\varepsilon > 0$ such that for all N there exist an integer $k \geq N$ and an $x_k \in I$ for which

$$|f_k(x_k) - f(x_k)| \geq \varepsilon.$$

In particular, there exist $k_1 \geq 1$ and $x_{k_1} \in I$ for which

$$|f_{k_1}(x_{k_1}) - f(x_{k_1})| \geq \varepsilon.$$

Moreover, suppose that positive integers $k_1, k_2, \ldots, k_n$ have been defined for some $n \in \mathbb{N}$, and that $x_{k_j} \in I$ and

$$|f_{k_j}(x_{k_j}) - f(x_{k_j})| \geq \varepsilon$$

for all j. Suppose also that $k_j < k_{j+1}$ for each $j < n$. Then there exist an integer $k_{n+1} \geq k_n + 1$ and an $x_{k_{n+1}} \in I$ such that

$$|f_{k_{n+1}}(x_{k_{n+1}}) - f(x_{k_{n+1}})| \geq \varepsilon.$$

We have now constructed a subsequence $\{f_{k_n}\}$ of $\{f_n\}$ by induction. Let $\{x_n\}$ be any sequence in I having $\{x_{k_n}\}$ as a subsequence. For instance, since $I \neq \emptyset$ we may choose $a \in I$ and set $x_n = a$ for each positive integer $n \notin \{k_1, k_2, \ldots\}$. Then the subsequence $\{|f_{k_n}(x_{k_n}) - f(x_{k_n})|\}$ of $\{|f_n(x_n) - f(x_n)|\}$ does not converge to 0, and the proof is complete. $\square$

The contrapositive of this theorem is often easier to use.

Corollary 10.2.5. *If there exists a sequence $\{x_n\}$ in I such that*

$$\lim_{n \to \infty} (f_n(x_n) - f(x_n)) \neq 0,$$

then $\{f_n\}$ does not converge uniformly to f.

Exercises 10.1.

1. For each of the following sequences defined on $[0, 1]$, show that the sequence is convergent and determine whether the convergence is uniform:

   (a) $\{x^n\}$.

   (b) $\left\{ \dfrac{\sin(nx+n)}{\sqrt{n}} \right\}$.

   (c) $\left\{ \dfrac{x}{1+nx} \right\}$.

   (d) $\left\{ \dfrac{nx^2}{1+nx} \right\}$.

   (e) $\left\{ \sqrt{x^2 + \dfrac{1}{n}} \right\}$.

2. Show that the sequence

$$\{\sin^{1/n} x\}$$

converges uniformly on every closed proper subinterval of $[0, \pi]$ but not on $[0, \pi]$ itself.

3. Show that the sequence

$$\left\{ \left(\frac{\sin x}{x} \right)^{1/n} \right\}$$

converges on $(0, \pi)$ but not uniformly.

4. Show that the sequence

$$\left\{ \frac{nx}{1 + n^2 x^2} \right\}$$

is uniformly convergent on $[c, 1]$, where $0 < c < 1$. Does the sequence converge uniformly on $(0, 1)$?

5. Use differentiation to find the maximum value of the function

$$f_n(x) = \frac{n^2 x}{1 + n^3 x^2}$$

for all $n \in \mathbb{N}$ and $x \in \mathbb{R}$, and hence show that the sequence $\{f_n\}$ is not uniformly convergent on $[0, 1]$.

6. Let

$$f_n(x) = \frac{n^\alpha x}{1 + n^\beta x^2}$$

for all $n \in \mathbb{N}$ and $x \in \mathbb{R}$, where $\beta > \alpha \geq 0$. Show that the sequence $\{f_n\}$ converges uniformly on $[0, 1]$ if and only if $\beta > 2\alpha$.

7. Let

$$f_n(x) = \frac{1}{x} + \frac{nx}{e^{nx^2}}$$

for all $n \in \mathbb{N}$ and $x \in (0, 1]$. Show that

$$f(x) = \lim_{n \to \infty} f_n(x) = \frac{1}{x}.$$

Use differentiation to find the maximum value of $|f_n(x) - f(x)|$ and hence show that the sequence $\{f_n\}$ is not uniformly convergent on $(0, 1]$.

8. Let $\{f_n\}$ be a sequence of functions defined on an interval I. Show that if $\{f_n\}$ converges uniformly to f, then $\{|f_n|\}$ converges uniformly to $|f|$. Is the converse true?

9. Let $\{a_n\}$ be a convergent sequence and for each n let $f_n: I \to \mathbb{R}$ be a function. Suppose there exists N such that

$$\sup_{x \in I}\{|f_n(x) - f_m(x)|\} \le |a_n - a_m|$$

for all $m \ge N$ and $n \ge N$. Show that $\sum_{j=0}^{\infty} f_j(x)$ converges uniformly on I.

10. Suppose $f: \mathbb{R} \to \mathbb{R}$ is uniformly continuous. Let

$$f_n(x) = f\left(x + \frac{1}{n}\right)$$

for all $n \in \mathbb{N}$ and $x \in \mathbb{R}$. Show that $\{f_n\}$ converges uniformly to f.

11. Let $f: [a, b] \to \mathbb{R}$ be continuous.

(a) Explain why, for each $n \in \mathbb{N}$, there exists $\delta_n > 0$ such that

$$|f(x) - f(y)| < \frac{1}{n}$$

whenever $|x - y| < \delta_n$.

(b) Let $(x_0, x_1, \ldots, x_{k_n})$ be a partition of $[a, b]$ such that $x_{j+1} - x_j < \delta_n$ for each $j < k_n$. For each n and each $x \in [a, b]$ define

$$f_n(x) = \begin{cases} f(x_j) \text{ if } x_j \le x < x_{j+1} \\ f(b) \;\; \text{if } x = b. \end{cases}$$

Show that $\{f_n\}$ converges to f uniformly.

12. Show that the sequence

$$\left\{\cos\frac{x}{n}\right\}$$

converges uniformly on $(-a, a)$ for every a. Is the convergence uniform on $\mathbb{R}$?

13. For all $n \in \mathbb{N}$ let

$$f(x) = \begin{cases} x/n \text{ if } x \text{ is even,} \\ 1/n \;\; \text{if } x \text{ is odd.} \end{cases}$$

Show that $\{f_n\}$ converges on $\mathbb{R}$ but not uniformly.

10.3   Properties of Uniformly Convergent Sequences

The examples in Sect. 10.1 illustrate the need for conditions under which it is valid
to change the order in which limits are taken. In this section we show that uniformly
convergent sequences are well behaved in the sense that the order in which limits
are taken is not important. We begin with continuity.

Theorem 10.3.1. *Let* $I \subseteq \mathbb{R}$ *and let* $\{f_n\}$ *be a sequence of functions that
is uniformly convergent on* I. *Let* c *be a limit point of* I, *and suppose that*
$\lim_{x \to c} f_n(x)$ *exists for all* $n \in \mathbb{N}$. *Then*

$$\lim_{n \to \infty} \lim_{x \to c} f_n(x)$$

exists if and only if

$$\lim_{x \to c} \lim_{n \to \infty} f_n(x)$$

exists, and in this case those limits are equal.

Proof. As $\{f_n\}$ is uniformly convergent on I, we may define

$$f(x) = \lim_{n \to \infty} f_n(x)$$

for each $x \in I$.

Suppose first that

$$\lim_{x \to c} f(x) = L,$$

and choose $\varepsilon > 0$. There exists $\delta > 0$ such that

$$|f(x) - L| < \varepsilon$$

whenever $x \in I$ and $0 < |x - c| < \delta$. The uniform convergence of $\{f_n\}$ on I shows
the existence of an N for which

$$|f_n(x) - f(x)| < \varepsilon$$

whenever $n \geq N$ and $x \in I$. Hence

$$|f_n(x) - L| \leq |f_n(x) - f(x)| + |f(x) - L| < 2\varepsilon$$

whenever $n \geq N$, $x \in I$ and $0 < |x - c| < \delta$. Therefore, since $\lim_{x \to c} f_n(x)$ exists
for all n, we have

$$\left| \lim_{x \to c} f_n(x) - L \right| \leq 2\varepsilon < 3\varepsilon$$

for all $n \geq N$. We conclude that

$$\lim_{n \to \infty} \lim_{x \to c} f_n(x) = L. \qquad (10.7)$$

On the other hand, suppose that Eq. (10.7) holds. For each $n \in \mathbb{N}$, write

$$g_n(c) = \lim_{x \to c} f_n(x).$$

Thus

$$\lim_{n \to \infty} g_n(c) = L.$$

Choose $\varepsilon > 0$. There exists M_1 such that

$$|g_n(c) - L| < \varepsilon$$

for all $n \geq M_1$. Moreover, for each such n there exists $\delta_n > 0$ such that

$$|f_n(x) - g_n(c)| < \varepsilon$$

whenever $x \in I$ and $0 < |x - c| < \delta_n$, and the uniform convergence of $\{f_n\}$ shows the existence of M_2 such that

$$|f_n(x) - f(x)| < \varepsilon$$

whenever $n \geq M_2$ and $x \in I$. Fix $n \geq \max\{M_1, M_2\}$. Then

$$|f(x) - L| \leq |f(x) - f_n(x)| + |f_n(x) - g_n(c)| + |g_n(c) - L| < 3\varepsilon$$

for each $x \in I$ such that $0 < |x - c| < \delta_n$. Thus

$$\lim_{x \to c} f(x) = L,$$

as required.   □

Remark. The hypothesis that c be a limit point of I is needed only to ensure that the limits in question are defined.

Corollary 10.3.2. *Let $\{f_n\}$ be a sequence of functions that is uniformly convergent to a function f on a set $I \subseteq \mathbb{R}$. If f_n is continuous on I for each n, then f is continuous on I.*

Proof. It is immediate from the hypotheses that for each $c \in I$ we have

$$\lim_{x \to c} f(x) = \lim_{x \to c} \lim_{n \to \infty} f_n(x)$$

$$= \lim_{n \to \infty} \lim_{x \to c} f_n(x)$$

$$= \lim_{n\to\infty} f_n(c)$$

$$= f(c).$$

$\square$

The result above gives a sufficient condition for a sequence of continuous functions to converge to a continuous function. The next example shows that it is not a necessary condition.

Example 10.3.1. Let $\{f_n\}$ be the sequence defined by

$$f_n(x) = nx(1-x)^n$$

for all $x \in [0, 1] = I$. It is clear that for all $x \in I$, we have $f_n(x) \to 0$ as $n \to \infty$. The function defined by $f(x) = 0$ for all $x \in I$ is continuous on I and hence the sequence converges pointwise to a continuous function.

This sequence, however, is not uniformly convergent to f on I. Consider the sequence $\{x_n\}$ in I defined by

$$x_n = \frac{1}{1+n}$$

for all $n \in \mathbb{N}$.[1] Since

$$|f_n(x_n) - f(x_n)| = \left| \frac{n}{1+n} \left(1 - \frac{1}{1+n} \right)^n \right| = \frac{1}{\left(1 + \frac{1}{n} \right)^{n+1}} > \frac{1}{2e} > 0$$

for all $n \in \mathbb{N}$, Theorem 10.2.1 shows that the sequence indeed fails to be uniformly convergent on I. $\triangle$

Example 10.1.4 shows that there are sequences $\{f_n\}$ of integrable functions such that $f_n(x) \to f(x)$ for all x in an interval $I = [a, b]$ but

$$\lim_{n\to\infty} \int_a^b f_n(x)\, dx \neq \int_a^b f(x)\, dx.$$

Indeed, it may be that the limit f is not even integrable over I. The next result shows that if $\{f_n\}$ converges uniformly to f on I, then f is integrable and the order in which the limits are taken can be changed.

---

[1]The reader may wonder what prompts the choice of this sequence. In fact, it can be shown using elementary calculus that f_n has a global maximum in I at x_n.

Theorem 10.3.3. *Let* $\{f_n\}$ *be a sequence of functions that are integrable over the interval* $I = [a, b]$ *and suppose that* $\{f_n\}$ *converges uniformly to* f *over* I. *Then* f *is integrable on* I, *and*

$$\lim_{n \to \infty} \int_a^b f_n = \int_a^b f. \tag{10.8}$$

Proof. Choose $\varepsilon > 0$. Since $\{f_n\}$ converges uniformly to f on I, there exists N such that

$$f_n(x) - \varepsilon < f(x) < f_n(x) + \varepsilon$$

for all $n \geq N$ and $x \in I$. Taking the lower integrals over I of both sides of the first inequality, we obtain

$$\int f_n - \varepsilon(b - a) \leq \underline{\int} f,$$

since f_n is integrable over I. Similarly,

$$\underline{\int} f \leq \overline{\int} f \leq \int f_n + \varepsilon(b - a).$$

Consequently,

$$0 \leq \overline{\int} f - \underline{\int} f \leq 2\varepsilon(b - a),$$

and since these inequalities must hold for every $\varepsilon > 0$, it follows that

$$\overline{\int} f = \underline{\int} f,$$

so that f is integrable on I.

We now have

$$\int f_n - \varepsilon(b - a) \leq \int f \leq \int f_n + \varepsilon(b - a).$$

Hence

$$-\varepsilon(b - a) \leq \int f - \int f_n \leq \varepsilon(b - a)$$

for all $\varepsilon > 0$ and $n \geq N$, and so

$$\lim_{n \to \infty} \int f_n = \int f.$$

$\square$

Theorems 7.3.11 and 7.3.14 provide sufficient conditions under which a function is integrable. These results can be used with Theorem 10.3.3 to glean the following corollaries.

Corollary 10.3.4. *Let* $\{f_n\}$ *be a sequence of functions that are continuous on the interval* $[a, b]$, *and suppose that* $\{f_n\}$ *converges uniformly to* f *on* $[a, b]$. *Then* f *is integrable over* $[a, b]$ *and Eq. (10.8) is satisfied.*

Corollary 10.3.5. *Let* $\{f_n\}$ *be a sequence of functions that are bounded and monotonic on the interval* $[a, b]$. *If* $\{f_n\}$ *converges uniformly to* f *on* $[a, b]$, *then* f *is integrable over* $[a, b]$ *and Eq. (10.8) is satisfied.*

Example 10.3.2. Consider the sequence $\{f_n\}$ defined by

$$f_n(x) = \begin{cases} xn^\alpha & \text{if } x \in [0, 1/n) \\ \left(\frac{2}{n} - x\right) n^\alpha & \text{if } x \in [1/n, 2/n) \\ 0 & \text{if } x \in [2/n, 1] \end{cases}$$

for all $n > 0$, where $\alpha > 0$ is a fixed number. The functions f_n are continuous on the interval $[0, 1]$ and hence integrable by Theorem 7.3.11.

We show first that $f_n(x) \to 0$ as $n \to \infty$ for each $x \in [0, 1]$. The result is obvious if $x = 0$. Suppose $x \in (0, 1]$. Let N be any integer such that $N > 2/x$. Then for all $n \geq N$, we have $x > 2/n$; hence $f_n(x) = 0$ and thus $f_n(x) \to 0$ as $n \to \infty$.

Evidently,

$$\int_0^1 f_n(x)\, dx = n^{\alpha-2}, \tag{10.9}$$

so that if $0 < \alpha < 2$,

$$\lim_{n \to \infty} \int_0^1 f_n(x)\, dx = \int_0^1 \lim_{n \to \infty} f_n(x)\, dx = 0; \tag{10.10}$$

however, if $\alpha = 2$,

$$\lim_{n \to \infty} \int_0^1 f_n(x)\, dx = 1 \neq \int_0^1 \lim_{n \to \infty} f_n(x)\, dx,$$

and the sequence $\{n^{\alpha-2}\}$ diverges if $\alpha > 2$. Now,

$$\sup_{x\in[0,1]} |f_n(x)| = n^{\alpha-1},$$

and Theorem 10.2.1 implies that $\{f_n\}$ is uniformly convergent to the constant function 0 on $[0, 1]$ only if $\alpha < 1$. Corollary 10.3.4 can be used to deduce relation (10.10), whenever $0 < \alpha < 1$, without calculating the integral of f_n. None of the results concerning integration of uniformly convergent sequences, however, can be applied when $1 \leq \alpha < 2$. This example thus shows that uniform convergence is not a necessary condition for changing the order of the limits.   △

If $\{f_n\}$ is a convergent sequence of functions that are differentiable on an interval I, Example 10.1.3 shows that the sequence $\{f_n'(x)\}$ may diverge even for all $x \in I$. It is of interest to examine the relationship between these sequences when uniform convergence is imposed. Uniform convergence of $\{f_n\}$ on I does not guarantee the convergence of $\{f_n'\}$ on I, but the next theorem shows that uniform convergence of $\{f_n'\}$ on I guarantees uniform convergence of $\{f_n\}$ on I, provided there is a $c \in I$ such that $\{f_n(c)\}$ converges.

Theorem 10.3.6. *Let $\{f_n\}$ be a sequence of functions that are differentiable on an interval $I = [a, b]$. Suppose that the sequence $\{f_n'\}$ converges uniformly on I and that there exists $c \in I$ such that $\lim_{n\to\infty} f_n(c)$ exists. Then $\{f_n\}$ converges uniformly on I to a differentiable function f, and*

$$f'(x) = \lim_{n\to\infty} f_n'(x) \qquad (10.11)$$

for all $x \in I$.

Proof. Choose $\varepsilon > 0$. As $\{f_n(c)\}$ converges, there exists N_1 such that

$$|f_n(c) - f_m(c)| < \varepsilon$$

whenever $m \geq N_1$ and $n \geq N_1$, and the uniform convergence of the sequence $\{f_n'\}$ on I implies the existence of an N_2 such that

$$|f_n'(t) - f_m'(t)| < \frac{\varepsilon}{b - a}$$

whenever $m \geq N_2, n \geq N_2$ and $t \in I$.

Let $N = \max\{N_1, N_2\}$ and choose $m \geq N$, $n \geq N$ and $x \in I$. For each $t \in I - \{x\}$ we may apply the mean-value theorem to the function $f_n - f_m$ to establish the existence of a ξ between x and t such that

$$|f_n(t) - f_m(t) - f_n(x) + f_m(x)| = |t - x||f_n'(\xi) - f_m'(\xi)|$$

$$< \frac{|t - x|}{b - a}\varepsilon$$

$$\leq \varepsilon. \qquad (10.12)$$

Note that this inequality holds even if $t = x$. Substituting c for x, we therefore obtain

$$|f_n(t) - f_m(t)| \leq |f_n(t) - f_m(t) - f_n(c) + f_m(c)| + |f_n(c) - f_m(c)|$$
$$< 2\varepsilon.$$

Thus $\{f_n\}$ satisfies the Cauchy criterion for uniform convergence. Let this sequence converge to a function f.

Next, for each $n \in \mathbb{N}$, $x \in I$, and $t \in I - \{x\}$ define

$$h_n(t) = \frac{f_n(t) - f_n(x)}{t - x};$$

also let

$$h(t) = \frac{f(t) - f(x)}{t - x}.$$

Fix $x \in I$. Since f_n is differentiable, we have

$$\lim_{t \to x} h_n(t) = f_n'(x).$$

Inequality (10.12) shows that

$$|h_n(t) - h_m(t)| = \frac{|f_n(t) - f_n(x) - f_m(t) + f_m(x)|}{|t - x|}$$
$$< \frac{|t - x|}{b - a}\varepsilon \cdot \frac{1}{|t - x|}$$
$$= \frac{\varepsilon}{b - a}$$

for all $m \geq N$, $n \geq N$, and $t \in I - \{x\}$. Therefore the sequence $\{h_n\}$ converges uniformly to h on $I - \{x\}$. As x is a limit point for $I - \{x\}$ and $\lim_{t \to x} h_n(t)$ exists, we may therefore apply Theorem 10.3.1 to $\{h_n\}$:

$$\lim_{n \to \infty} f_n'(x) = \lim_{n \to \infty} \lim_{t \to x} h_n(t)$$
$$= \lim_{t \to x} \lim_{n \to \infty} h_n(t)$$
$$= \lim_{t \to x} h(t).$$

We conclude that $f'(x)$ exists and

$$f'(x) = \lim_{n \to \infty} f_n'(x).$$

$\square$

In Example 10.1.2 we found a sequence $\{f_n\}$ of functions that converges to the constant function 0 but

$$\lim_{n \to \infty} f_n'(0) \neq 0.$$

In view of Theorem 10.3.6, we conclude that $\{f_n'\}$ cannot be uniformly convergent.

The following example shows that Eq. (10.11) may hold even if $\{f_n'\}$ is not uniformly convergent.

Example 10.3.3. For all $n > 0$ and $x \in [0, 1]$ let

$$f_n(x) = \frac{\log(1 + n^2 x^2)}{2n}.$$

Using l'Hôpital's rule, we find that $\{f_n\}$ converges to 0. Furthermore

$$f_n'(x) = \frac{nx}{1 + n^2 x^2}$$

for all $x \in [0, 1]$, so that $\{f_n'\}$ also converges to 0. Hence

$$\left(\lim_{n \to \infty} f_n(x) \right)' = \lim_{n \to \infty} f_n'(x).$$

However, we can show that $\{f_n'\}$ is not uniformly convergent on $[0, 1]$. Since

$$(1 - nx)^2 \geq 0,$$

we have

$$1 + n^2 x^2 \geq 2nx,$$

so that $f_n'(x) \leq 1/2$ for each relevant n and x. Moreover $f_n'(x)$ attains its maximum value of $1/2$ at $1/n$. Hence

$$\sup_{x \in [0,1]} |f_n'(x) - 0| = \sup_{x \in [0,1]} |f_n'(x)| = \frac{1}{2}$$

for all n. As this result is nonzero, Theorem 10.2.1 shows that $\{f_n'\}$ is not uniformly convergent. $\triangle$

Exercises 10.2.

1. Let

$$f_n(x) = \frac{n^2 x}{1 + n^2 x^2}$$

for all $n \in \mathbb{N}$ and $x \in \mathbb{R}$.

(a) For each x find

$$f(x) = \lim_{n \to \infty} f_n(x).$$

(b) Is f continuous?
(c) Is the convergence uniform?

2. Show that $\{\sin^n x\}$ converges on $[0, \pi]$. Is the convergence uniform?
3. Let $f_n(x) = x^n / n$ for all $n \in \mathbb{N}$ and $x \in [0, 1]$.

(a) Does the sequence $\{f_n\}$ converge uniformly?
(b) Is

$$\lim_{n \to \infty} \int_0^1 f_n = \int_0^1 \lim_{n \to \infty} f_n?$$

(c) Is

$$\lim_{n \to \infty} f_n' = \left(\lim_{n \to \infty} f_n \right)'?$$

4. In view of Corollary 10.3.2 we can say that uniform convergence preserves continuity. Use the following example to show that it does not necessarily preserve discontinuity: For all $n \in \mathbb{N}$, let

$$f_n(x) = \begin{cases} \frac{1}{n} & \text{if } x \text{ is rational,} \\ 0 & \text{if } x \text{ is irrational.} \end{cases}$$

5. For each $n \in \mathbb{N}$, let

$$f_n(x) = \begin{cases} 1 - nx & \text{if } 0 \le x < \frac{1}{n}, \\ 0 & \text{if } x \ge \frac{1}{n}. \end{cases}$$

(a) Does the sequence $\{f_n\}$ converge uniformly?
(b) Show that f_n is continuous for each n but the limit function is not continuous.

6. Let

$$f_n(x) = \frac{2n^2 x}{e^{n^2 x^2}}$$

for each $n \in \mathbb{N}$ and $x \in [0, 1]$.

(a) Show that the sequence $\{f_n\}$ is not uniformly convergent on $[0, 1]$.

(b) Is

$$\lim_{n\to\infty} \int_0^1 f_n = \int_0^1 \lim_{n\to\infty} f_n?$$

7. Let

$$f_n(x) = \frac{2nx}{1 + n^2 x^2}$$

for each $n \in \mathbb{N}$ and $x \in [0, 1]$.

(a) Use differentiation to find the maximum value of each function and hence show that the sequence $\{f_n\}$ is not uniformly convergent on $[0, 1]$.
(b) Is

$$\lim_{n\to\infty} \int_0^1 f_n = \int_0^1 \lim_{n\to\infty} f_n?$$

8. Show that

$$\left\{ x - \frac{x^n}{n} \right\}$$

converges uniformly on $[0, 1]$ but the sequence of derivatives does not.

9. Let $\{f_n\}$ be a sequence of uniformly continuous functions that is uniformly convergent to a function f on an interval I. Show that f is uniformly continuous on I.

10. Let f_n and g_n be continuous on an interval I, and suppose that $\{f_n\}$ and $\{g_n\}$ converge uniformly to f and g, respectively. Show that $\{f_n g_n\}$ converges uniformly to fg and that fg is continuous.

11. Let $\{f_n\}$ be a sequence of functions on an interval I. Suppose that $f_n(c) = 0$ for all n and some $c \in I$, and that $\{f_n'\}$ converges uniformly on I. Show that $\{f_n\}$ converges uniformly and

$$\lim_{n\to\infty} f_n'(x) = f'(x).$$

12. It is crucial that the interval of integration be finite in Theorem 10.3.3. For all $n \in \mathbb{N}$ let

$$f_n(x) = \begin{cases} \frac{1}{n} & \text{if } 0 \le x \le n, \\ 0 & \text{if } x > n. \end{cases}$$

(a) Show that $\{f_n\}$ is uniformly convergent on $[0, \infty)$ and that f_n is integrable on $[0, \infty)$ for each $n \in \mathbb{N}$.

(b) Show that

$$\lim_{n \to \infty} \int_0^\infty f_n = 1$$

but

$$\int_0^\infty \lim_{n \to \infty} f_n = 0.$$

(c) Construct a similar example where

$$\int_0^\infty \lim_{n \to \infty} f_n = 0$$

but

$$\left\{ \int_0^\infty f_n \right\}$$

diverges.

10.4   Infinite Series

The results of Sect. 10.3 can be readily adapted to infinite series of functions. Let $\{f_n\}$ be a sequence of functions defined on a set I and let $\{S_n\}$ be the sequence of partial sums defined by

$$S_n(x) = \sum_{j=0}^n f_j(x)$$

for all $x \in I$. The series $\sum_{j=0}^\infty f_j(x)$ is said to **converge uniformly** on I if $\{S_n\}$ converges uniformly on I. Properties of uniformly convergent sequences can be used to derive analogous results for uniformly convergent series. Moreover the comparison, ratio, and root tests are applicable for uniformly convergent series.

Theorem 10.4.1. *Let $\sum_{j=0}^\infty |f_j(x)|$ be a series that converges uniformly on a set I. For all $\varepsilon > 0$ there is an integer N such that $|f_k(x)| < \varepsilon$ whenever $k \geq N$ and $x \in I$.*

Proof. Noting that the sequence $\{\sum_{j=0}^n |f_j(x)|\}$ converges uniformly on I, we apply the Cauchy principle. Thus for each $\varepsilon > 0$ there exists N_1 such that

$$\sum_{j=N_1+1}^n |f_j(x)| = \sum_{j=0}^n |f_j(x)| - \sum_{j=0}^{N_1} |f_j(x)| < \varepsilon$$

whenever $x \in I$ and $n > N_1$, and the result follows by taking $N = N_1 + 1$. $\square$

Theorem 10.4.2. *Let $\sum_{j=0}^{\infty} |f_j(x)|$ be a series that converges uniformly on a set I. Then $\sum_{j=0}^{\infty} f_j(x)$ converges uniformly on I.*

Proof. By the Cauchy principle, for each $\varepsilon > 0$ there exists N such that

$$\sum_{j=m+1}^{n} |f_j(x)| = \sum_{j=0}^{n} |f_j(x)| - \sum_{j=0}^{m} |f_j(x)| < \varepsilon$$

whenever $x \in I$ and $n > m \geq N$. For each such x, n, m we have

$$\left| \sum_{j=0}^{n} f_j(x) - \sum_{j=0}^{m} f_j(x) \right| = \left| \sum_{j=m+1}^{n} f_j(x) \right| \leq \sum_{j=m+1}^{n} |f_j(x)| < \varepsilon,$$

and the result therefore follows from the Cauchy principle. $\square$

The next result follows immediately from Theorem 10.3.1.

Theorem 10.4.3. *Suppose $\{f_n\}$ is a sequence of functions such that $\sum_{j=0}^{\infty} f_j(x)$ converges uniformly on some set $I \subseteq \mathbb{R}$. Let c be a limit point of I. Then*

$$\sum_{j=0}^{\infty} \lim_{x \to c} f_j(x) = \lim_{x \to c} \sum_{j=0}^{\infty} f_j(x),$$

provided either side of the equation exists.

Corollary 10.4.4. *If $\sum_{j=0}^{\infty} f_j(x)$ converges uniformly on I and f_j is continuous on I for each j, then $\sum_{j=0}^{\infty} f_j(x)$ is continuous on I.*

Proof. The hypotheses and Theorem 10.4.3 show that

$$\lim_{x \to c} \sum_{j=0}^{\infty} f_j(x) = \sum_{j=0}^{\infty} \lim_{x \to c} f_j(x) = \sum_{j=0}^{\infty} f_j(c)$$

for each $c \in I$. $\square$

The function S_n is integrable or differentiable on I if each f_j is integrable or differentiable, respectively, on I. These simple observations coupled with Theorems 10.3.3 and 10.3.6 give the following results immediately.

Theorem 10.4.5 (Term-by-Term Integration). *Let $\{f_n\}$ be a sequence of functions that are integrable on the interval $I = [a, b]$. Suppose that the series $\sum_{j=0}^{\infty} f_j(x)$ is uniformly convergent on I. Then*

$$\sum_{j=0}^{\infty} \int_{a}^{b} f_j(x)\, dx = \int_{a}^{b} \left(\sum_{j=0}^{\infty} f_j(x) \right) dx.$$

Theorem 10.4.6 (Term-by-Term Differentiation). *Let $\{f_n\}$ be a sequence of functions that are differentiable on the interval I. Suppose that*

1. *the series $\sum_{j=0}^{\infty} f_j'(x)$ is uniformly convergent on I and*
2. *there exists $c \in I$ such that the series $\sum_{j=0}^{\infty} f_j(c)$ converges.*

Then the series $\sum_{j=0}^{\infty} f_j(x)$ is uniformly convergent on I to a differentiable function, and

$$\left(\sum_{j=0}^{\infty} f_j(x) \right)' = \sum_{j=0}^{\infty} f_j'(x)$$

for all $x \in I$.

Although the results concerning the uniform convergence of sequences can be readily exported to get analogous results for series, the tests for uniform convergence given in Sect. 10.2 rely on the use of the sequence of partial sums. The sequence $\{S_n\}$ of partial sums can prove to be an elusive quantity to obtain in a form conducive to evaluating limits. Indeed, $\{S_n\}$ can be found in closed form only for a few types of series such as the geometric series. If $\{S_n\}$ is difficult to procure in a useful form, then identifying a candidate for the limit is yet another potentially formidable task. Ideally, one desires a test that avoids these problems and relies directly on the terms of the series. The next result is a comparison test for uniform convergence.

Theorem 10.4.7 (Comparison Test). *Let $\{f_n\}$ and $\{g_n\}$ be sequences of nonnegative functions defined on the set I, and suppose that for all $x \in I$ and $k \in \mathbb{N}$ we have*

$$f_k(x) \leq g_k(x).$$

If the series $\sum_{j=0}^{\infty} g_j(x)$ is uniformly convergent on I, then the series $\sum_{j=0}^{\infty} f_j(x)$ is uniformly convergent on I.

Proof. Choose $\varepsilon > 0$, and let

$$S_n(x) = \sum_{j=0}^{n} f_j(x)$$

and

$$T_n(x) = \sum_{j=0}^{n} g_j(x)$$

for all $n \geq 0$ and $x \in I$. Since $\{T_n\}$ is uniformly convergent on I, the Cauchy criterion implies that there is an integer N such that

$$|T_n(x) - T_m(x)| < \varepsilon$$

for all $x \in I$ whenever $n \geq N$ and $m \geq N$. Thus whenever $x \in I$ and $n > m \geq N$, we have

$$
\begin{aligned}
|S_n(x) - S_m(x)| &= \left| \sum_{j=m+1}^{n} f_j(x) \right| \\
&= \sum_{j=m+1}^{n} f_j(x) \\
&\leq \sum_{j=m+1}^{n} g_j(x) \\
&= T_n(x) - T_m(x) \\
&< \varepsilon.
\end{aligned}
$$

The integer N does not depend on $x \in I$, and therefore $\{S_n\}$ is uniformly convergent on I by Theorem 10.2.3. $\qquad \square$

The next result, although limited in applications as we shall explain later, nonetheless proves to be one of the most useful and convenient tests for uniform convergence of series. It is an immediate consequence of Theorem 10.4.7.

Corollary 10.4.8 (Weierstrass M-Test). *Let $\{f_n\}$ be a sequence of functions defined on a set I. Suppose there exists a sequence $\{M_n\}$ of constants such that $|f_n(x)| \leq M_n$ for all $x \in I$ and all n. If $\sum_{j=0}^{\infty} M_j$ converges, then $\sum_{j=0}^{\infty} f_j(x)$ converges uniformly on I.*

Example 10.4.1. Let $\{f_n\}$ be the sequence of functions defined by

$$
f_n(x) = \frac{\sin nx}{n^2 + |x|},
$$

for all $n \in \mathbb{N}$ and $x \in \mathbb{R}$. We have $|\sin nx| \leq 1$ for all $x \in \mathbb{R}$; consequently,

$$
|f_n(x)| \leq \frac{1}{n^2}
$$

for all $n \in \mathbb{N}$. Now, $\sum_{j=1}^{\infty} 1/j^2$ is convergent, and therefore the series $\sum_{j=1}^{\infty} f_j(x)$ is uniformly convergent on $\mathbb{R}$. $\qquad \triangle$

We know that power series define functions that are differentiable within the interval of convergence of the series and that the derivative can be obtained by differentiating the series term by term. A power series can also be integrated term by term within the interval of convergence. It should thus occasion little surprise that a power series is uniformly convergent within the interval of convergence.

Theorem 10.4.9. *Suppose that the power series*

$$\sum_{j=0}^{\infty} a_j (x-a)^j$$

has a radius of convergence $r > 0$ *and let* α *and* β *be any numbers such that*

$$a - r < \alpha < \beta < a + r.$$

Then $\sum_{j=0}^{\infty} a_j (x-a)^j$ *converges uniformly on* $[\alpha, \beta]$.

Proof. Without loss of generality we can assume that $a = 0$ and $r = 1$ since the general case can be obtained by a translation and scaling of the variable. Let

$$\rho = \max\{|\alpha|, |\beta|\}.$$

The hypotheses show that $\rho < 1$, so that the series $\sum_{j=0}^{\infty} a_j \rho^j$ is absolutely convergent. Moreover, if $x \in I = [\alpha, \beta]$, then $|x| \leq \rho$. Therefore

$$|a_n x^n| \leq |a_n| \rho^n$$

for all $x \in I$ and all n. The uniform convergence of the power series thus follows from the Weierstrass M-test with $M_n = |a_n| \rho^n$.   □

The Weierstrass M-test is a comparison test for uniform convergence. The key feature is that the comparison series consists of terms that are constants. Other series tests that spawn from the comparison test can also be adapted to test for uniform convergence. We give two such results now.

Theorem 10.4.10 (D'Alembert Ratio Test). *Let* $\{f_n\}$ *be a sequence of functions defined on a set* I. *Suppose there exist numbers* $r < 1$ *and* N *such that* f_N *is bounded and for all* $x \in I$ *and* $n \geq N$ *we have* $f_n(x) \neq 0$ *and*

$$\frac{|f_{n+1}(x)|}{|f_n(x)|} \leq r.$$

Then the series $\sum_{j=0}^{\infty} f_j(x)$ *converges uniformly on* I.

Proof. Arguing as in the proof of Theorem 3.7.1, we see by induction that

$$|f_{N+j}(x)| \leq r^j |f_N(x)|$$

for all $x \in I$ and $j \in \mathbb{N}$. The result now follows from the comparison test since f_N is bounded and $r < 1$.   □

The reader should have no trouble proving the next theorem.

Theorem 10.4.11 (Cauchy Root Test). *Let $\{f_n\}$ be a sequence of functions defined on a set I, and suppose that there exist numbers $r < 1$ and $N \in \mathbb{N}$ such that*

$$|f_n(x)|^{1/n} \le r$$

for all $x \in I$ and $n \ge N$. Then the series $\sum_{j=0}^{\infty} f_j(x)$ converges uniformly on I.

The Weierstrass M-test is perhaps the most frequently used test for the uniform convergence of series. Indeed, Bromwich ([4] p. 124) notes that series satisfying the Weierstrass M-test were called "normally convergent" by Baire and that this "terminology has the advantage of emphasizing the fact that the M-test can be applied to nearly all series in ordinary everyday use." A major limitation of the test, however, is that a series must be absolutely convergent at each point in the set I in order to apply the test successfully. More delicate tests are needed to cope with series that are uniformly convergent in a set but conditionally convergent at points in the set. However, we note the following result due to Baire that simply says that every uniformly convergent series of bounded functions on a set I can be made into a "normally convergent" series by a judicious grouping of terms. The proof of this theorem may be found in [4].

Theorem 10.4.12 (Baire). *For every uniformly convergent series of functions bounded on some set, there exists a regrouping of terms such that the resulting series satisfies the Weierstrass M-test.*

Exercises 10.3.

1. Determine whether the following series $\sum_{j=2}^{\infty} f_j(x)$ converge uniformly on the set I:

   (a) $f_n(x) = \frac{\sin nx}{2^n}$, $I = [-\pi, \pi]$.

   (b) $f_n(x) = \frac{x}{n^2 + x^2}$, $I = \mathbb{R}$.

   (c) $f_n(x) = \frac{x^2}{(1+x^2)^n}$, $I = \mathbb{R}$.

   (d) $f_n(x) = \frac{(n+1)x}{e^{nx^2/2}}$, $I = [0, 1]$.

   (e) $f_n(x) = \frac{\log(1+nx)}{nx^n}$, $I = [c, \infty)$ for some $c > 1$ [hint: $\log(1 + h) < h$ when $h > 0$].

   (f) $f_n(x) = \log\left(1 + \frac{x}{n\log^2 n}\right)$, $I = [-c, c]$ for some $c > 0$.

   (g) $f_n(x) = \frac{1}{2^{n-1}x}$, $I = (0, 1]$.

   (h) $f_n(x) = \frac{\sin nx \sin x}{\sqrt{n+x}}$, $I = [0, 2\pi]$.

   (i) $f_n(x) = (-1)^{n+1}\frac{1}{n^x}$, $I = [c, \infty)$ for some $c > 0$ [hint: $n^x = n^{x - \frac{c}{2}} n^{\frac{c}{2}}$].

2. Let

$$f_n(x) = \frac{\sin nx}{n^2}$$

for all $n > 0$ and $x \in \mathbb{R}$. Show that $\sum_{j=1}^{\infty} f_j(x)$ converges uniformly over $\mathbb{R}$ whereas $\sum_{j=1}^{\infty} f_j'(x)$ diverges at 0.

3. For all $n > 0$ let

$$
f_n(x) = \begin{cases} 0 & \text{if } x = 0, \\ \frac{x^n \log x}{n} & \text{if } 0 < x \le 1. \end{cases}
$$

(a) Show that the supremum of $|f_n(x)|$ on $[0, 1]$ occurs at $e^{-1/n}$.
(b) Show that $\sum_{j=1}^{\infty} f_j(x)$ converges uniformly on $[0, 1]$.

4. For all $n > 0$ and $x \in \mathbb{R}$ let

$$
f_n(x) = \frac{n^2 x^2}{e^{n^2 x}}.
$$

(a) Show that

$$
\sup_{x \in \mathbb{R}} |f_n(x)| = \frac{4}{n^2 e^2}.
$$

(b) Show that $\sum_{j=1}^{\infty} f_j(x)$ converges uniformly on $\mathbb{R}$.

5. For all $n > 0$ and $x \in \mathbb{R}$ let

$$
f_n(x) = \begin{cases} \frac{1}{n} & \text{if } \frac{1}{n+1} < x \le \frac{1}{n}, \\ 0 & \text{otherwise.} \end{cases}
$$

Show that $\sum_{j=1}^{\infty} f_j(x)$ converges uniformly but

$$
\sum_{j=1}^{\infty} \sup_{x \in \mathbb{R}} |f_j(x)| \tag{10.13}
$$

diverges. Note that in general if (10.13) converges, then $\sum_{j=1}^{\infty} f_j(x)$ converges uniformly (why?).

6. Show that if $\sum_{j=0}^{\infty} a_j$ is absolutely convergent, then

$$
\sum_{j=1}^{\infty} a_j \sin jx
$$

and

$$\sum_{j=0}^{\infty} a_j \cos jx$$

converge uniformly on $\mathbb{R}$.

7. Suppose that $\sum_{j=0}^{\infty} f_j(x)$ converges uniformly on an interval I to a bounded function $f: I \to \mathbb{R}$. Show that $\sum_{j=0}^{\infty} f(x) f_j(x)$ converges uniformly on I.

8. Let $\{f_n\}$ be a sequence of functions defined on an interval I, and suppose that $f_n(x) \geq 0$ for all n and all $x \in I$. Suppose also that $\{f_n(x)\}$ is a decreasing sequence for all $x \in I$ and that

$$\lim_{n \to \infty} \sup_{x \in I} f_n(x) = 0.$$

Show that

$$\sum_{j=0}^{\infty} (-1)^j f_j(x)$$

converges uniformly on I.

9. Let

$$f_n(x) = (-1)^n (1 - x) x^n$$

for all n and all $x \in [0, 1]$. Show that $\sum_{j=0}^{\infty} f_j(x)$ converges uniformly but $\sum_{j=0}^{\infty} |f_j(x)|$ does not.

10. Show that

$$\sum_{j=0}^{\infty} \frac{1}{j^2 + x^2}$$

is differentiable on $\mathbb{R}$.

Bibliography

1. Ali, S.A.: The mth ratio test: new convergence tests for series. Math. Assoc. Am. **115**, 514–524 (2008)
2. Anderson, G.D., Vamanamurthy, M.K., Vuorinen, M.: Monotonicity of Some Functions in Calculus, Research Report Series, vol. 538, University of Auckland (2005)
3. Bartle, R.G., Sherbert, D.R.: Introduction to Real Analysis. Wiley, New York (1982)
4. Bromwich, T.J.I'A.: An Introduction to the Theory of Infinite Series. Macmillan, London (1926)
5. Carlson, B.C.: Algorithms involving arithmetic and geometric means. Am. Math. Mon. **78**(5), 496–505 (1971)
6. Dunlap, R.A.: The Golden Ratio and Fibonacci Numbers. World Scientific, Singapore (1997)
7. Fulks, W.: Advanced Calculus, 3rd edn. Wiley, New York (1978)
8. Huang, X.-C.: A discrete l'Hopital's rule. Coll. Math. J. **19**(4), 321–329 (1988)
9. Kemeny, J.G.: The exponential function. Am. Math. Mon. **64**(3), 158–160 (1957)
10. Little, C.H.C., Teo, K.L., van Brunt, B.: The Number Systems of Analysis. World Scientific, Singapore (2003)
11. Marsden, J.E., Hoffman, M.J.: Basic Complex Analysis, 2nd edn. Freeman, New York (1987)
12. Moritz, R.E.: On the extended form of Cauchy's condensation test for the convergence of infinite series. Bull. Am. Math. Soc. **44**(6), 441–442 (1938)
13. Piaggio, H.T.H.: An Elementary Treatise on Differential Equations and Their Applications. G. Bell & Sons, London (1920). (Reprinted 1971)
14. Shiu, P.: A generalization of some convergence tests. Math. Gazette **56**, 227–228 (1972)
15. Smith, D., Teo, K.L.: Linear Algebra. New Zealand Mathematical Society, Wellington (1990)
16. Stone, M.H.: Developments in Hermite polynomials. Ann. Math. **29**, 1–13 (1927)
17. Tong, J.-C.: Cauchy's mean value theorem involving n functions. Coll. Math. J. **35**, 50,51 (2004)
18. Zhang, X., Wang, G., Chu, Y.: Extensions and sharpenings of Jordan's and Kober's inequalities. J. Inequal. Pure Appl. Math. **7**(2), Article 63 (2006)

© Springer Science+Business Media New York 2015

C.H.C. Little et al., *Real Analysis via Sequences and Series*, Undergraduate Texts in Mathematics, DOI 10.1007/978-1-4939-2651-0

Index

© Springer Science+Business Media New York 2015
C.H.C. Little et al., *Real Analysis via Sequences and Series*, Undergraduate
Texts in Mathematics, DOI 10.1007/978-1-4939-2651-0

Printed in the United States
By Bookmasters

Printed in the United States
By Bookmasters